Five Acres and Independence

Maurice Kains

Must Have Books
503 Deerfield Place
Victoria, BC
V9B 6G5
Canada
trava2911@gmail.com

ISBN: 9781774641309

Copyright 2021 – Must Have Books

All rights reserved in accordance with international law. No part of this book may be reproduced or transmitted in any form or by any means, electronic or mechanical, including photocopying, recording, or by any information storage or retrieval system, except in the case of excerpts by a reviewer who may quote brief passages in an article or review, without written permission from the publisher.

CONTENTS

CHAPTER	PAGE
1. INTRODUCTION	1

A word about the author, his practical experience, and qualifications suggest reliability of the text.

2. CITY vs. COUNTRY LIFE 4

Advantages and disadvantages; city vicissitudes; dependence upon "income" to supply "outgo"; country stability, productivity; dependence upon "outgo" to supply "income"; self-supporting; occupancy of home in country vs. tenancy of "flat" in city; health, wealth, happiness in country home.

3. TRIED AND TRUE WAYS TO FAIL 7

Too little capital, unfavorable location, uncongenial soil, too large area, inefficient soil preparation and tillage, lack of feeding, big-headedness, inexperience, city hours, laziness, too many pets and guests.

4. WHO IS LIKELY TO SUCCEED? 11

Thinker and worker; owner on the spot; absentee direction; book farming; observation as a teacher; hired help.

5. FIGURES DON'T LIE 15

Striking figures from U. S. Census and Department of Agriculture reports; supply and demand; relation to and contrast with individual owner's problems on productive land.

6. THE FARM TO CHOOSE 18

Soil survey maps; character of soil; nature of plant growth already on the land; depth, drainage, slope, freedom from stones, previous crops and yields, neighborhood crops and yields.

CONTENTS

CHAPTER		PAGE
7.	WHERE TO LOCATE	22

Good roads; their up-keep; snow removal; site with respect to roadside sales; distance from market; schools, churches, electric current, buses, stores, doctors, etc.

| 8. | LAY AND LAY-OUT OF LAND | 27 |

Elevation; aspect; frostiness; impediments such as fences, boulders, stone walls; fields—sizes and shapes; roadways, lanes and paths; arrangement of buildings.

| 9. | WIND-BREAKS, PRO AND CON | 30 |

Importance; types; influence on crops, animals and residence; workability in their shelter; good and bad kinds; saving of fuel; production of fuel.

| 10. | ESSENTIAL FACTORS OF PRODUCTION | 37 |

Good seed; good breed of animal; variety; "strain"; abundant water and available plant food in the soil; rational tillage; ample space between plants and for animals.

| 11. | RENTING *vs.* BUYING | 47 |

Advantages and disadvantages of each; various ways to manage depend upon each; renting with option of buying; buying a small place but working large rented area.

| 12. | CAPITAL | 52 |

Investment and working money; cost of land; rent of property; insurance; equipment; nursery stock and other plants; animals; labor; time needed to get returns.

| 13. | FARM FINANCE | 56 |

Importance of credit; origin of capital; how secured; borrowing for production; usury; fundamental rules for borrowing; character of borrower and business ability; annual inventory and budget; bank cashier as adviser and confidant; safety deposit boxes.

| 14. | FARM ACCOUNTS | 63 |

Planning for production; knowledge of market, and the truth about one's business; records of crops and

CONTENTS

animals individually and of the farm as a whole; account books.

15. **WATER SUPPLY** 66
    Rain water and cisterns; filter cisterns; cistern capacities; cistern cleaning and purification; springs; gravity piping; pneumatic pressure systems; hydraulic rams.

16. **SEWAGE DISPOSAL** 75
    Primitive methods; cess pools; septic tanks; tank construction; personal experience.

17. **FUNCTIONS OF WATER** 82
    Necessity in plant and animal growth; quantity needed by plants; types of water in soil; conservation by tillage and mulching.

18. **DRAINAGE** 88
    Importance; methods; instances to prove their value.

19. **IRRIGATION** 96
    Methods; types of apparatus; assurance of adequate water; success in spite of drouth; use to supply fertilizer and certain kinds of spraying.

20. **FROST DAMAGE PREVENTION** 105
    What frost is; how it affects plants; plant resistance to damage; hardy and tender plants; preventing fall of temperature to or below danger point; forecasting local frosts; methods available.

21. **LIVE STOCK** 112
    Advantages and disadvantages of keeping cow, pig, poultry, rabbits, bees; desirable and undesirable kinds to have.

22. **POULTRY** 116
    Chickens for eggs and meat; ducks, geese, turkeys, pigeons; scrubs *vs.* breeds and strains; housing, feeding, yarding, range, management; hatching *vs.* buying day-old chicks; brooding; sanitation; etc.

23. **BEES** 125
    Honey the principal interest; importance in fruit production; management easy but imperative.

CONTENTS

CHAPTER	PAGE
24. GREENHOUSES	128

Standardized styles preferable to home built; advantages; sizes desirable; avoidance of mistakes; types of houses; ventilation; heating; greenhouse builders' contracts and propositions.

25. COLDFRAMES AND HOTBEDS 136

Invaluable to start seedlings; limitations of each; types of each; how and where to make them; hardening-off plants; electric heating and regulation most desirable.

26. SOILS AND THEIR CARE 146

Nature's soils injured by man; reclamation; types of soils and how to handle them; humus; how to judge soil values; soil erosion and its prevention.

27. MANURES 154

Stable manure best; why; scarcity and cost; fresh *vs.* rotted; dried and pulverized; amounts to apply; functions in the soil; experiences and experiments.

28. COMMERCIAL FERTILIZERS 159

Supplements to manures; organic and inorganic; value of each; cautions in using; composition; most important unmixed ones; functions of each; "mixed goods"; home mixing; determining soil needs; fertilizer distributors; concentrated goods.

29. GREEN MANURES AND COVER CROPS 166

Humus suppliers; importance; savers of soluble plant foods; developers of others; some gather costly nitrogen; prevent soil washing, deep freezing and soil heaving; collect fallen leaves; aid drainage.

30. LIME 171

Functions in soils: neutralizes acidity, makes conditions favorable to growth of bacteria, improves physical properties of soil; experimental data; when and how much to apply; must not be used with manure.

31. COMPOST 175

Importance in gardening, especially under glass; ways to make it; best seasons to start it; handling for best results.

CONTENTS

CHAPTER   PAGE

32. CROPPING SYSTEMS 177
    Rotation—methods dependent upon types of cash crops grown; good and bad sequences; double cropping—companion, succession, marker and partnership crops; examples of each and methods of handling.

33. SOIL SURFACE MANAGEMENT 182
    Effects of tillage; when and how to till; right and wrong implements and ways to use them; deepening soils to increase water capacity and root range; trenching methods; good tools for various purposes.

34. WEEDS 196
    Their significance and control; species and size suggest character of soil; annuals, biennials and perennials; when and how to destroy.

35. TOOLS 200
    Kinds needed dependent upon type of soil and work; essential and non-essential kinds; good and poor styles; storage; cleanliness; oiling, sharpening, etc.

36. RE-MAKING A NEGLECTED ORCHARD 206
    Importance of competent advice before attempting such work; many trees not worth reclamation; how to determine useful ones; tree surgery not desirable from income basis; personal appraisal methods; renovation methods.

37. FRUIT TREE PRUNING 213
    Principles; applications; methods good and bad; times to prune; tree architecture; building strong trees; vine and bush training and pruning; knowledge of flower bud formation and position essential.

38. GRAFTING FRUIT TREES 223
    Simple methods; trees not to graft; best ones and best branches to use; how to get and keep scions; time to graft; grafting waxes; paraffin; repair or bridge grafting to save girdled trees.

39. HOW TO AVOID NURSERY STOCK LOSSES 232
    Buyers, not nurserymen, most often responsible for death of stock; right and wrong handling; loose

CONTENTS

planting; bearing age trees unsatisfactory; young stock best to order; pruning after planting; treatment of Y-crotch trees; staking; label removal.

40. **VEGETABLE CROPS TO AVOID AND TO CHOOSE** . . . 239
Quick and slow maturing kinds, staple and fancy kinds, high and low quality varieties, good *vs.* poor keepers, kinds saleable in several ways.

41. **SEEDS AND SEEDING** 243
Types of seeds; effect of weight on sprouting and the crop; seed testing; age of seed; seedsman's reputation; "special stock" seed; seedsmen's trial grounds; seed growing, selection; sowing times; temperature; depth; etc.

42. **TRANSPLANTING** 253
Stages of development; pre-watering; preparation of soils and flats; lifting, pricking-out, spotting board and dibble; depth, watering, hardening; planting in the open; after-care.

43. **PLANTS FOR SALE** 261
Often highly profitable near town of amateur gardeners; general and special stocks and sales; sales methods; advertising.

44. **SOMETHING TO SELL EVERY DAY** 265
Crops in demand; crops that "work over well"; pickles, jams, jellies, juices, syrups, preserves, "canned goods"; eggs; chickens; ducks; honey; plants; flowers.

45. **STRAWBERRIES** 267
Regular season and everbearing kinds; culture; systems of training; after fruiting, what?; companion and succession crops; quickest fruit to bear; often highly profitable; every farm should have them.

46. **GRAPES** 281
Planting; pruning; training; precocious and annual fruiting; long season of fruiting by successional ripening of varieties and storage.

CONTENTS

CHAPTER		PAGE
47.	BUSH AND CANE FRUITS	291

Raspberry, blackberry, currant, gooseberry, dewberry, blueberry; varieties; culture.

48. SMALL FARM FRUIT GARDENS 299

Does the ordinary farm orchard pay?; investigational experiment; improved methods of cultivation; varieties for home use; sequence of ripening; lay-out of orchard and small fruits.

49. SELECTION OF TREE FRUITS 308

Varieties to choose; type of trade to work for; general market, roadside sales, personal trade; successional ripening to hold trade; filler trees and other fruits; inter-tilled crops to help pay costs of development.

50. STORAGE OF FRUITS AND VEGETABLES 317

Methods, good and bad for various types of crops; root cellars, pits, storage houses, lofts; arrangement; ventilation; cooling; heating; sanitation; fumigation.

51. ESSENTIALS OF SPRAYING AND DUSTING . . . 330

Spraying, dusting and other methods effective when properly used; fruit and vegetable insect enemies.

APPENDICES 341

INDEX 387

THE PURPOSE OF THIS BOOK

When you plan an auto trip, you wisely consult a road map to discover the safest, most direct and pleasantest way to your destination. When you are actually on your way, you follow the signs and obey the signal lights, especially at the cross-roads, the branches and through the cities. Often you may have the choice of several routes and often you may be in a quandary, but by consulting the map and obeying the signs, you ultimately reach your goal.

This book aims to be a "road map" which traces some of the best routes along which you and your family may travel to happy, prosperous and interesting lives. It not only indicates the safest routes but, what is even more important, it particularly warns against blind alleys and side roads that lead to disappointment if not disaster. In this respect it differs from the usual rural life book which depicts only the pleasant features of farming. So for this reason, if for no other, it should be of signal service to you, especially if it prevents your making the serious mistakes commonly made by people who move from the cities and towns to the country.

SPECIAL NOTE

This new edition should be of even greater service to you, reader, as a reference book because many tables and much new data have been added to the Appendix. They have been placed there so as not to interfere with the ease of reading the general text. In most cases their connection with the text have been indicated by reference to the chapters to which they apply. The other instances have no direct textual cross reference. Glance them over now to become familiar with them.

THE AUTHOR

I

INTRODUCTION

MANY a wreck has been the result of taking the family to the country, and afterwards having part or all of it become thoroughly dissatisfied. There are so many rough realities in a life of this kind that it takes the poetry out of the visions of joy, peace, contentment and success that arise in the minds of many.

H. W. WILEY,
In The Lure of the Land.

PEOPLE who think they "would like to have a little farm" naturally fall into two groups; those who are sure to fail and those likely to succeed. This book is written to help both! Its presentation of advantages and disadvantages, essential farming principles and practises should enable you to decide in which class you belong and whether or not you would be foolish or wise to risk making the plunge. In either case it should be worth many times its price because, on the one hand it should prevent fore-doomed failure, and on the other, show you how to avoid delay, disappointment, perhaps disaster, but attain the satisfaction that characterizes personal and well directed efforts in farming.

If your experience in the country so far has been confined to vacations or summer residence and if your reading has been limited to literature that depicts the attractive features of farm life in vivid colors but purposely or thoughtlessly glosses over or fails to emphasize the objectionable ones you will doubtless be shocked at the stress placed in this book upon the drawbacks. My reason for doing this is that I want to present conditions not only as I know them to be but as you are almost certain to find them. "To be forewarned is to be forearmed."

You may already know the country in summer, perhaps in spring or autumn—maybe during all the "growing season"—but do you know what it is to spend the winter in the country? How would you like to be snowed in as my family and I have been so that for *ten weeks* neither you nor your neighbors could use an automobile because of the deeply drifted snow? Can you and your family stand the isolation usually characteristic of farm life? Do you know from experience the meaning of hard, manual work from dawn to dark—and then by lantern-light? Are you prepared to forego salary or income for months at a stretch? I don't seek to frighten you but merely to indicate that though farm life has its joys and satisfactions it also has its drawbacks.

No matter in which of the groups mentioned you place yourself, it is natural that you should ask whether I am a practical man or merely a professor or a writer! Though I must confess to having held professorial and editorial positions, these were because of my familiarity with practical matters. My experience began before my earliest "little red schoolhouse" days and, barring interruptions, has continued until the present.

My boyhood duties included not only the usual chores of the farm and those connected with fruit and vegetable gardening, poultry and bee-keeping, horse and cow care, but canning and pickling, soap and candle manufacture, meat curing and wine making; in fact, practically everything which characterized farm life only a remove or two from pioneer conditions.

As my father, until my young manhood, was a renter of one place after another, I not only learned the disadvantages of this style of husbandry but gained considerable experience by correcting the mistakes of former tenants (and even owners!), especially in making neglected orchards, vineyards and gardens productive, and in learning how to manage a wide variety of soils.

At various times I worked on five farms, on one or another of which the leading features were dairy cattle, sheep, grain, hay, fruit, vegetables and bees. As the owners of these places were good farmers and communicative I learned much from them in addition to how to handle tools and implements effectively. At one time I owned a fruit farm with poultry as a side line, at another I managed the fruit de-

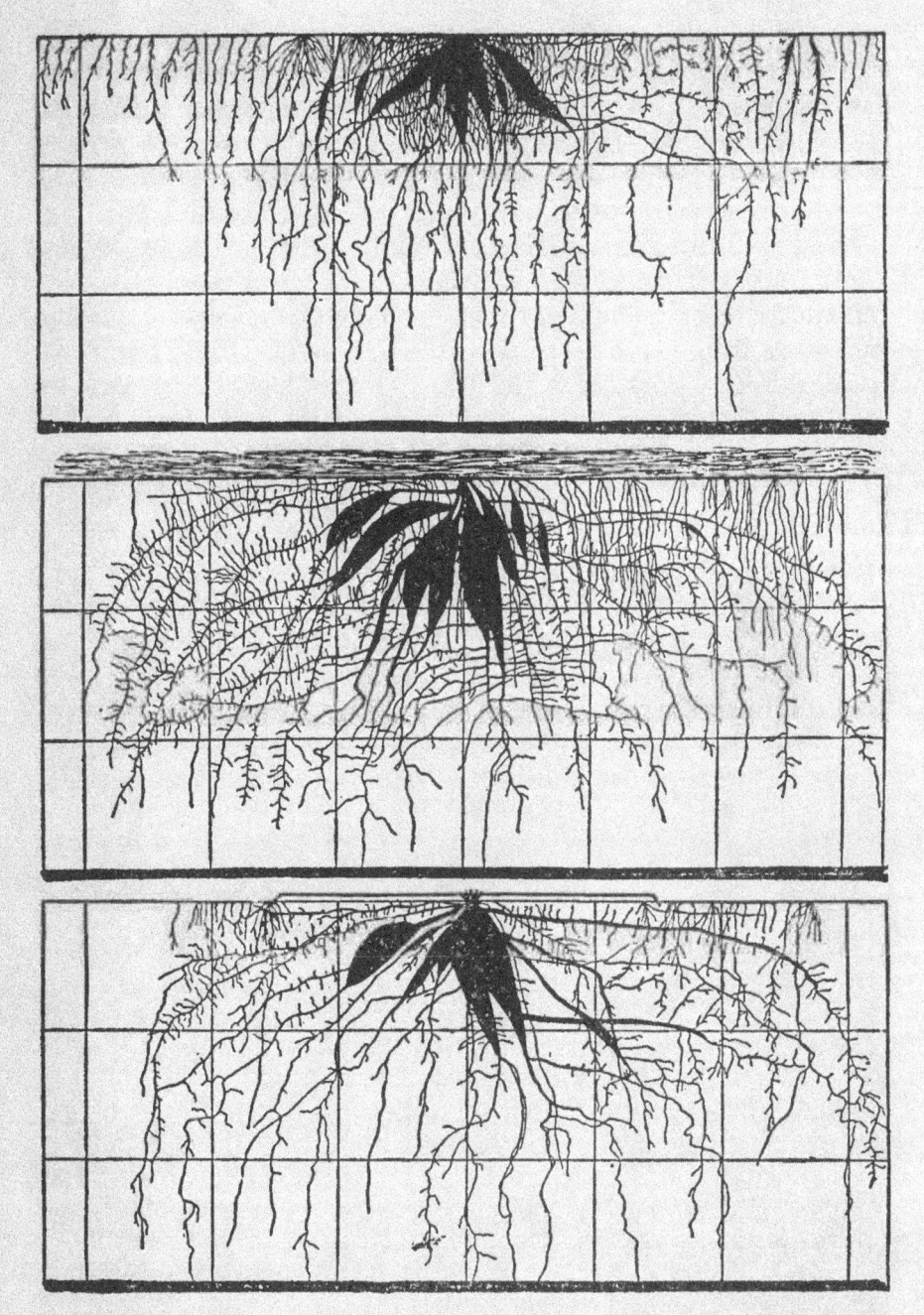

Fig. 1. Root system of sweet potato. Upper, in cultivated ground; middle, under straw mulch; lower, under paper mulch.

partment of a produce-raising concern, at still another planted about fifty acres of orchard and vineyard for a commercial orchardist. As occasion has presented I have also worked in greenhouses and nurseries.

Though, like a politician, I might "point with pride" to some personal successes I would rather present more significant ones made by others. Conversely, as some of my mistakes taught me more than the successes I prefer to hold them up as "horrible examples" (instead of the errors of others!). So you, Reader, may "henceforth take warning by my fall and shun the faults I fell in!"

2

CITY VS. COUNTRY LIFE

FARMING must be a family affair just as much as it has ever been, but the modern way is not to make a drudge of any person, adult or minor. The work of the farm demands system and departments. Each person who is required to perform any of the labor should have it so shaped that it will stimulate energy, sense of responsibility and love for the calling.

C. C. BOWSFIELD,
In *Wealth from the Soil.*

ONE of the most striking characteristics of each "depression period" is the tacit acknowledgment of city dwellers that "the farm is the safest place to live;" for though there is each year a migration from the country to the city and a counter movement to the suburbs and a less pronounced one to more agricultural environment, the movement becomes an exodus when business takes a slump and employees are thrown out of work.

So long as the income continues the employee is prone to quell what desires he may have for rural life and to tolerate the disadvantages of urban surroundings rather than to drop a certainty for an uncertainty; but when hard times arrive and his savings steadily melt away he begins to appreciate the advantages of a home which does not gobble up his hard-earned money but produces much of its up-keep, especially in the way of food for the family.

More than this, however! He realizes at the end of each year in the city that he has only 12 slips of paper to show for his perhaps chief expenditure—rent; that he and his family are "cliff dwellers" who probably do not know or want to know others housed under the same roof; that his children "have no place to go but out and no place to come but in"; in short, that he and they are ekeing out a narrowing, uneducative, imitative, more or less selfish and purposeless existence; and that his and their "expectation of life" is shortened by tainted air, restricted sunshine and lack of exercise, to say nothing of exposure to disease.

Contrasted with all these and other city existence characteristics are the permanence and productivity of land, whether only a small suburban lot or a whole farm; the self-reliance of the man himself and that developed in each member of his family; the responsibility and satisfaction of home ownership as against leasehold; the health and happiness typical not only of the life itself but of the wholesome association with genuine neighbors who reciprocate in kind and degree as few city dwellers know how to do; the probably longer and more enjoyable "expectation of life"; but, best of all, the basis and superstructure of true success—development and revelation of character and citizenship in himself, his wife, sons and daughters.

Which, think you, is the better citizen, the man who pays rent for a hall room, a hotel suite or a "flat," or the one who owns a self-supporting rural home and therein rears a family of sons and daughters by the labors of his head and his hands and their assistance?

In a poignant sense city existence is non-productive; it deals with what has been produced elsewhere. Moreover it is dependent upon "income" to supply "outgo" and in the great majority of cases has nothing to show—not even character—for all the time and effort spent. Country life reverses this order; it not only produces "outgo" to supply "income" but when well ordered it provides "surplus." Nay, further, it develops character in the man and each member of the family. Nothing so well illustrates this fact as *"Who's*

Who in America," a survey of which will show that the majority of the men and women listed in its pages were reared in rural surroundings. Here they learned not only how to work and to concentrate but inculcated that perhaps hardest and ultimate lesson of all education, obedience, succinctly stated in Ecclesiastes: "Whatsoever thy hand findeth to do, do it with thy might."

Of course, decision to renounce city existence for country life is not to be hastily made; however, for the health, the joy, the knowledge, the formation and development of character and the foundation of a liberal education there is no comparison. But what, do you inquire, is a "liberal education?" Let us listen to that great scientist, Thomas Huxley:

"That man has a liberal education who has been so trained in youth that his body is the ready servant of his will and does with ease and pleasure all the work that, as a mechanism, it is capable of; whose intellect is a clear, cold logic engine, with all its parts of equal strength and in smooth working order, ready, like the steam engine, to turn to any kind of work, and spin the gossamers as well as forge the anchors of the mind; whose mind is stored with the great and fundamental truths of nature and of the laws of her operations; one who, no stunted ascetic, is full of life and fire, but whose passions are trained to come to heel by a vigorous will, the servant of a tender conscience; who has learned to love all beauty, whether of nature or of art, to hate all vileness and to respect others as himself."

Where, I ask, can a boy or a girl acquire and develop such qualifications so well as on a farm, well managed by loving parents who are enthusiastic business and domestic heads of the enterprise and who explain and insist upon obedience to the laws of nature as well as those of the land and who live in harmony with their neighbors?

3
TRIED AND TRUE WAYS TO FAIL

ALMOST any farm needs a much larger working capital than the proprietor provides. The more successful the farm is, the more it absorbs or ties up capital. It pays to *hire the extra capital* needed, precisely as one would hire extra teams for ice—or silage—harvest.

DAVID STONE KELSEY,
In *Kelsey's Rural Guide*.

ANYBODY can buy a farm; but that is not enough. The farm to buy is the one that fits the already formulated general plan—and no other! It must be positively favorable to the kind of crop or animal to be raised—berries, eggs, vegetables, or what not. To buy a place simply because it is "a farm" and then to attempt to find out what, if anything, it is good for, or to try to produce crops or animals experimentally until the right ones are discovered is a costly way to gain experience, but lots of people will learn in no other.

Even supposing that the farm discovered is exactly suited to the branch of agriculture decided upon—where is it located? Are there good neighbors, schools, churches, doctors, stores, electric power and bus lines and other features of civilization near by? How are the roads kept, winter and summer? What about taxes? Still more important, how and where can its products be marketed? "Before deciding on a spot for a garden," wrote Peter Henderson, 75 years ago in *Gardening for Profit*, "too much caution cannot be used in selecting the locality. Mistakes in this matter are often the sole cause of want of success, even when other conditions are favorable."

Failure in other instances is due to lack of either "investment" or "working" capital or both; for though one may have sufficient funds to buy and perhaps stock a place, other moneys must be available to carry the venture until the "cash crops" are able to produce them. For instance, though certain vegetable crops and everbearing strawberries may make individual cash returns within a few months of being planted, "regular season" strawberries require 14 or 15 months, bush berries, grapes and asparagus three years; peaches four or five and apples from five to ten or even more! How is one to pay expenses, taxes, insurance; in fact, how is one to live until they pay for themselves and something besides?

This was the fix that an acquaintance got into. As his case is typical a rehearsal of its main features may serve

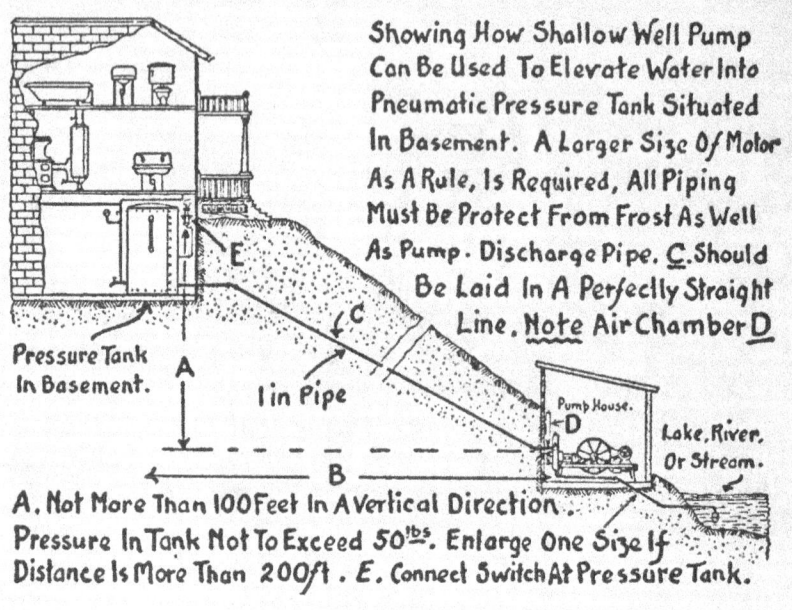

Showing How Shallow Well Pump Can Be Used To Elevate Water Into Pneumatic Pressure Tank Situated In Basement. A Larger Size Of Motor As A Rule, Is Required, All Piping Must Be Protect From Frost As Well As Pump. Discharge Pipe. C. Should Be Laid In A Perfectly Straight Line. Note Air Chamber D

Pressure Tank In Basement. A
1 in Pipe
B
Pump House. D
Lake, River, Or Stream.

A. Not More Than 100 Feet In A Vertical Direction. Pressure In Tank Not To Exceed 50 lbs. Enlarge One Size If Distance Is More Than 200ft. E. Connect Switch At Pressure Tank.

Fig. 2. Electric pumping of water to pressure tank.

as a horrible example and warning to some reader at present headed that same way! He had bought a farm on a good road and good for his purpose but—seven miles from the nearest local market town. There was considerably more land than he needed, especially as nearly a third of it was second growth woodland on which he paid taxes but got no return except a little firewood. The house being an old one and about as well ventilated as a corn crib, was inadequately heated by stoves, so he installed a furnace; it lacked plumbing and electric current so he put these in. These improvements reduced his capital but did not increase his income from the place. With high hopes he planted and cared for 1,000 fruit trees and had plantings of small fruits. For three years he strove to make ends meet but just as the trees were ready to bear their first crop he was obliged to sell—fortunately not at a loss of actual cash but at one of time, effort and hopes.

Another common cause of failure is tumefaction of the cranium, popularly known as "big head!" Though this malady is not limited to people who take up farming it is perhaps most conspicuous and most frequently characteristic of city people who start in this new line, especially the poultry branch. With fine nonchalance they disregard fundamental principles, turn a deaf ear to the voice of experience, adopt crops unsuited to the local conditions or without regard to the market demands, and so on. Usually not until the disease has run its course is there hope for such cases, but after the most virulent ones have been well dosed with ridicule or have paid a heavy fool tax the victims may not only recover and become immune but may in time admit that farmers, like Old Man Noah, "know a thing or two!"

After they have taken up farming, many a city man and his wife—particularly his wife!—have run the gamut of emotions through all the descending scale of delight, gratification, pleasure, surprise, perplexity, annoyance, disgust and exasperation (a full octave!) to discover how popular they have become since moving to the country. Not only do their intimate friends drop in unannounced on fine Sundays but less and less intimate ones even down to people who just happened to live around the block arrive in auto loads and all expect to remain for dinner, perhaps supper also!

This sort of thing is highly unfair, first because the city "friends" never return the courtesy, second because unreasonable amounts of produce—especially chickens, eggs, and butter—are wasted (yes, wasted because there is no *quid pro quo*), and third, because of the work, particularly the wife's.

Sunday after Sunday one wife of my acquaintance made such a slave of herself as cook and hostess that at last her husband laid down the law. In brief he said: "These people come only for your good dinners. They drain your energies and our profits. We must stop both losses." And they did!

As you will probably have to solve the same problem let

me tell you the answer: For Sunday dinners have corned beef and cabbage, beef stew or hash! Good luck to you!!

Among various other ways which help lead to failure are unfavorable soil; undrained land; rocks and stones; wrong crops; improperly prepared and tilled land; too large area devoted to lawns and ornamental planting; excessive time devoted to pets, especially such as occupy areas that should pay profits; inadequate manuring or fertilizing; failure to fight insects and plant diseases and many others.

In farming, as in every other enterprise, success depends primarily upon the man who undertakes it. Not everybody who starts will succeed. On the other hand the man who has the following personal qualifications, no matter what his previous calling or location may have been, stands a good chance of succeeding. Natural liking for the business is the most important asset because it will assure willingness and patience to work and be painstaking, to be open-minded and to be as alert to detect irregularities as to adopt and apply new knowledge.

Farming is a business characterized by abundance of small but essential details which demand close observation and application to prevent loss. In few businesses are cleanliness, orderliness and timeliness of so much importance, for without them weeds, pests and diseases thrive and profits fail to appear. Above all the farmer must be enthusiastic, a condition that will become permanent and characteristic as soon as the business shows a profit.

Of all the many farms I have visited that of a Delaware County, New York, farmer presents fewer natural factors that might suggest success than any other. More than 80% of it is "on edge," rocky, stony, marshy, or otherwise unfit for cultivation. In fact, to quote the owner, "it is just the kind of farm that no sane man would think of buying!" He inherited it from his father who "bought it for its timber and couldn't sell it after logging."

If any farmer ever had excuse to fail this man had, but every year he makes it pay! Moreover he uses less labor than does many another farmer more favorably located. He grows corn and hay on the tillable land, grazes the hillsides and marshes, feeds dairy cattle, makes butter, and gives the skim milk to hens for the production of eggs and meat. These three products—butter, eggs and dressed poultry—which constitute his "money crops"—he sends by express to city customers who pay a premium above market prices quoted on a specified week day. The manure goes back to the land which is thus kept in highly productive condition.

As this man "capitalized his drawbacks"—thought his problem through—and deliberately formulated the plan which has worked well, his case may be taken as an indication that one of the surest ways to succeed is to have a definite plan and a definite goal. To attempt to farm without either is merely a form of gambling. The penalty for one's folly is a loss surer than in a lottery.

4

WHO IS LIKELY TO SUCCEED?

IF A man would enter upon country life in earnest and test thoroughly its aptitudes and royalties, he must not toy with it at a town distance; he must brush the dews away with his own feet. He must bring the front of his head to the business, and not the back of it.

DONALD G. MITCHELL,
In *My Farm of Edgewood.*

IF any one thing is more essential than any other in every branch of farming it is that the owner personally direct all operations. He cannot be an absentee farmer and he cannot entrust his interests entirely to hirelings. However, unless he is experienced he is incompetent to direct any part of the necessary or advisable work; so an even more fundamental essential is that he not only learn to do every kind of work himself but become a keen observer and logical thinker.

To apply these statements to *you:* The important points about making yourself proficient and thinking while performing each operation are that you will thus first teach yourself the *how, why* and *when;* second, by thinking as you work, you can discover quick ways, short cuts and time-savers; and third, you will thus make yourself competent to teach your helpers how to do the work in the ways you have proved to be good, if not best. This will mean both that you will be entitled to their respect as a director and that you will get the work done efficiently and economically.

"Book farming," formerly a term of contempt, is now recognized as a good thing in its place. But without practise no book, not even this one (!) can make you either proficient or competent in any branch of farming. All it can do is to inform you—present ideas, methods, practises, tables, illustrations, etc., for you to digest and utilize as occasion may arise. Until you have tested them by actual practise you cannot *know* how practical they are.

In farming, more than any other business, you must teach yourself, for *every day* during even the longest lifetime will bring its problems and lessons. Reading and listening to lectures, radio talks, etc., though important and often helpful, are poor substitutes for observation and translation of the observations into terms of understanding, decision and action when this last shall be necessary or advisable.

One of the most profitable habits you can form is systematically, every day, to go over at least part of your premises in a leisurely, scrutinizingly thoughtful way, and the whole of it at least once each week throughout the year to reap the harvest of a quiet eye and fill the granary of your mind with knowledge of the habits of helpful and harmful animals, birds and insects; to observe and understand the characteristics of plant growth from the sprouting of the seed through all the stages of stem, leaf, flower, fruit and seed development; to note and interpret the behavior of plants, poultry and animals under varying conditions of heat and cold, sunshine and shade, drouth and wetness, fair weather and foul, rich and poor feeding. Here is not only the best farm school in which to learn the duties you owe your dependants (plants and animals) and yourself for your own best interests, but in which to enjoy the most delightful compensations of farm life; for it gives the thinking observer mastery over his business, brings him *en rapport* with his environment and in tune with The Infinite.

If you are a city man all this at first will be a foreign

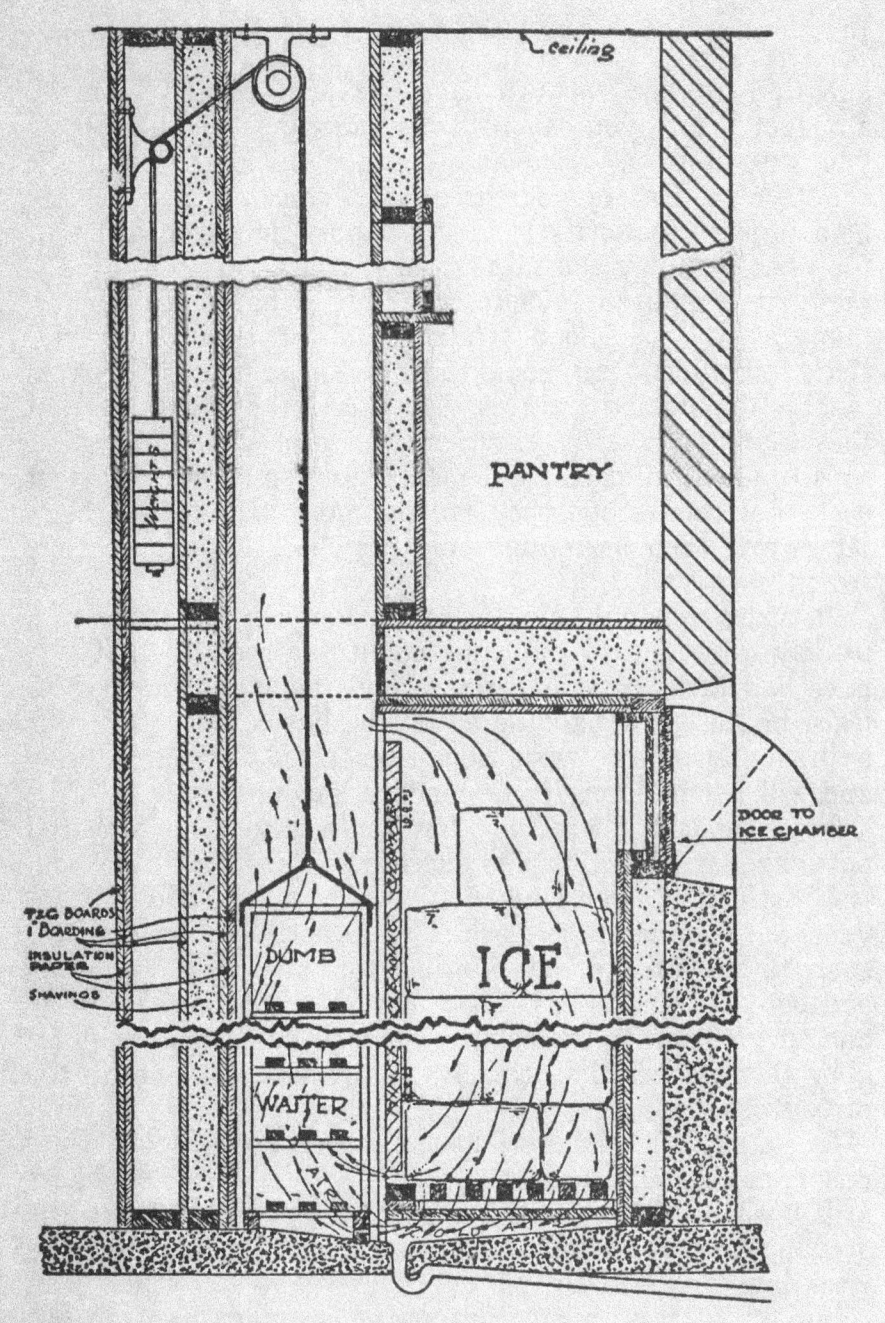

Fig. 3. Household refrigerator and dumb waiter.

language to you; you will have to teach yourself Nature's alphabet before you can read her messages. In this respect you will be at a disadvantage when compared with the countryman. One of your greatest handicaps will probably be assumed superiority to "farm hands" and even farmers; for what you may consider "a good education" too often is really of small use in a country setting. What if farm hands, taken as a whole, do not intellectually compare with city clerks, salesmen and bookkeepers and do not perhaps as quickly understand theories! They have the advantage that they know from daily experience at least something about soils, plants, animals, implements, tools and how to treat each. For this reason they are far more valuable to themselves and their farming employers than are raw recruits from the city.

In order to avoid being jeered at for your ignorance it is usually advisable to tell your country associates that you have had little or no experience and that you are eager to learn because you have taken up the business to earn your living therefrom. You will then be respected for your attitude and will not be likely to be made a laughing-stock.

Even the city laborer who has had a garden for a year if not longer is more likely to succeed than is the "white collar" man of "better education" who has not had the advantage of similar experience. The proportion of successes is surer to be with the laborers than with the clerks, salesmen and bookkeepers. Why? They are trained to *manual work* and they use what brains they have to make their muscles obey them rather than to rely on "someone else to do the dirty work."

Then, too, such laborers have the endurance which white collar men usually lack; they can stand the 12 to 16 hours a day which farming often demands, especially during the growing season. It follows from this that at the start, no weaklings and usually no one "well along in years" can count on success in any branch of farming *for a living;* though, if he have means to tide him over say three or perhaps two years, the outdoor life will "make a new man of him" and fit him for the more arduous but pleasing duties of the business.

5
FIGURES DON'T LIE

He who would keep the innocence of the incipient landed gentry never forgets that figures bear false witness.
RICHARDSON WRIGHT,
In *Truly Rural*.

I AM willing to wager that you are far less interested to learn that the United States produces an average of about 70,000,000 watermelons a year than to remember that one summer you ripened a score or a dozen of higher quality than you can buy! Why? Because the "average" United States watermelon, so far as you are concerned, is a myth! It lacks the reality and especially the personal flavor of the ones you grew yourself! In other words, for you and me statistics are more uninteresting than perhaps anything else we might have thrust at us—as juiceless as dried watermelon!

For this reason I quote them only when I think they are likely to be of value to you. Presumably you are more interested in your own personal achievements or those possible for you to attain than in any maximum, minimum or average of the nation. So I usually cite individual cases because these are likely to be of more practical use to you because within your range of application. If you want national figures you will find them aplenty in the Census Reports and in the *Year Book* of the United States Department of Agriculture.

Now for some figures! These *will* interest you, I am sure, because you can apply and profit by them. According to the Census (!) the "average American hen" lays about 85 eggs a year; yet records made with 17,640 hens (total during 19 years) in the Connecticut Egg-Laying Contests showed an average productivity of nearly double that—164 eggs a year. In the nineteenth year the average was 229 eggs a hen! Still more interesting is the fact that during the present century individual hen egg-laying ability has increased so startingly that several hundred hens have each **a** record of 300 eggs or more laid in a year—some even more than 350 eggs!

Such records are due mainly to breeding but are also greatly aided by sanitation and rational feeding. Though it is probably not advisable for you, Reader, to attempt to match such records at the start, what would you think of yourself if you started with "average American hens," kept

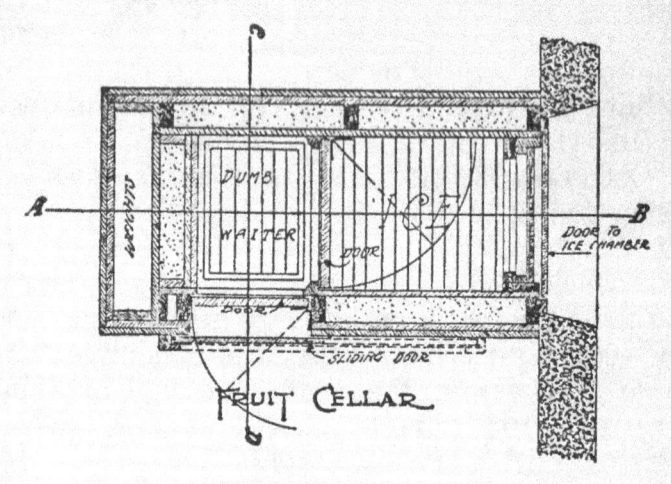

Fig. 4. Plan of ice bunker and dumb waiter.

these in unsanitary quarters and let them "rustle" for their food? Of course *you* are not built that way! The records of those Connecticut contest hens make you covet that kind of stock, especially when you learn that you can buy them as day-old chicks for only a few cents more than "average" chicks would cost. (Chapter 22.)

Take a corn growing example: On a test acre a farmer produced 13½ bushels of corn by using 200 pounds of superphosphate fertilizer. On another acre 85 pounds of nitrate of soda added to the 200 pounds of phosphate increased the yield to 24 bushels. But on an acre where he plowed under a crop of hairy vetch prior to seeding with corn he produced 40 bushels! Many of his neighbors followed his example and increased their yields by an average of 22 bushels an acre at a cost of only $5.—a gain of $4. for $1. spent on seed! Though these cases were in Alabama they may be duplicated in other states by using the right legume as a cover crop.

A third example: In western New York a fruit grower has a Baldwin apple orchard that, when I last spoke with him, had produced 20 profitable crops of fruit in 22 consecutive years. Had this achievement been with Yellow Transparent, Oldenburg, Wealthy or some other variety famous for regular annual bearing it would scarcely attract attention, but because it was with Baldwin, a variety notorious for its "off" and "on" years, it is something not only for the owner to brag about but for other growers to emulate—you especially.

How was it done? Partly by liberal feeding and rational good care but mainly by thinning the partially developed fruit during summer! The effects of this practise are first to get rid of inferior and worm-infested, cull specimens that would have to be discarded anyway at or after harvest; second, to divert the plant food from these inferior fruits to flower-bud formation for the following season's crop and to enhance the size and quality of the remaining specimens; third, to distribute the fruit-bearing area more evenly both over the trees and over the years. Not only did this grower have choice fruit every year (except in the two when frost killed the blossoms) but he paid the cost of thinning by selling the later thinnings for vinegar-making. As further consequences he had choice fruit to sell when his neighbors had little or none and the prices he received were higher than they received for their less attractive, less worthy fruit.

Hundreds, if not thousands, of farmers have increased their yields of potatoes by the method detailed in Chapter 10.

Such instances as these—there are thousands of others—warrant the statement that the productive power of plants and poultry, yes, and of dairy cows, though surmised by the investigative few, tested and proved by their "credulous followers" and adopted by the more progressive farmers, has not even yet been suspected, much less believed by the majority of producers. It also warrants the declaration that, other things being equal, a little farm well tilled will produce more in proportion to the effort expended on it than will one of larger extent under slipshod practise.

6

THE FARM TO CHOOSE

No one should control more arable land than he can maintain in a high state of productivity, the four great factors of which are good seed, suitable moisture, abundant available plant food and rational tillage. In a large majority of cases where failure, or partial failure of an abundant crop is observed the meagre results are due to a partial lack of one of these four fundamentals.

ISAAC PHILLIPS ROBERTS,
In Introduction to *Ten Acres Enough.*

OFTEN too little attention is given to the condition of the soil and the lay of the fields with reference to ease of cultivation.* (Chapter 8.) Crop land in itself is of little value unless it is so situated that it can be made to yield profitable returns through the use of labor and machinery. A farm valued at $100. an acre may be a much better bargain if practically all of the land can be put to profitable use than another farm of equal size purchasable at $50. an acre, of which large areas are practically useless owing to streams, swamps which cannot be drained, or rough stony areas not suitable for pasture.

In choosing a farm, therefore, it is essential that not the total area conveyed by the deed or contract be considered, but the area available for profitable use. Any additional land may be really a liability instead of an asset, since often the returns do not pay the taxes. Many mistakes are made on this point alone, the buyer often thinking that he can crop the land but later finding it more profitable to let it grow brush or woods.

Another factor is ease of cultivation. If the land is steep or broken it is not practical to use improved machinery and it is difficult to harvest crops by older methods. Such fields may possibly be worked but the cost of production is necessarily higher than on fields which permit efficient use of labor and machinery. This difference in cost must be re-

---

* Part of what follows is condensed from *Selecting a Farm,* Farmers' Bulletin, 1088.

flected in lower valuation and hence in lower taxes and interest or rental charges, otherwise the farmer must pay these increased costs out of returns from his own labor.

The physical condition of the soil should receive first attention particularly from the beginner with small capital. Often even on high grade farms soils get into poor condition through a few years of mismanagement. It is easy to cause damage to good land through improper tillage or careless handling. To correct such damage usually takes several years.

One should distinguish between soils which cannot easily be corrected because of naturally poor physical condition and those which through improper management are in poor physical condition but which can be restored by good handling. Putting a soil in prime condition for planting is a fine art that can be fully learned only through long experience in handling the specific type of land involved.

The depth of the soil is of great importance. It should be determined when the land is first examined. Shallow soil is often a liability. Its utility is sharply limited for practically all farming purposes because it is cold and wet in spring, the water table being close to the surface. Later it dries out rapidly and bakes hard. It is quickly affected by drouth. Shallowness is a common defect in many areas. It may not necessarily be due to rock close to the surface but to hardpan.

Depth can best be detected by the soil auger. If it is general or characteristic of the region it may be detected by the tree growth since only shallow rooted trees will be found there such as red and silver maples; whereas, deep rooting trees such as black walnut and hickory will be absent or greatly stunted.

Closely connected with thin soils is drainage. A person choosing a farm should make assurance doubly sure on this point; first as to the natural drainage of the fields and second as to the possibility of draining if artificial drainage is necessary. Most soils need drainage, at least in spots. (Chapter 18.)

In choosing a farm one can rarely find the ideal arrangement of buildings and fields. A farm which may be desirable in many other particulars may be undesirable in this respect. Saving of labor is highly important. Hence the arrangement of fields and buildings may be such that time is lost because of irregularity of fields or because important fields are far from the buildings. The latter fault sometimes cannot be remedied.

In the Eastern States three factors have determined, more or less, the location of the buildings: 1, Water supply; 2, roads; and 3, area of arable land near by. In regions where spring water is available the buildings were generally placed so water could be piped to them. Thus water supply had a greater weight in determining the location than ease of reaching the fields or the highway. Often the best fields are distant from the buildings or the buildings far removed from the highway. Such arrangements greatly depreciate the value of farms.

The arrangement of the buildings themselves, as regards ease of doing chores or other work about them is important. Often planning was for one type of farming which, having been discontinued and another taken up, the buildings are not suited to the specific type of farming now practised. Sometimes conditions may be such that alterations cannot be made economically to improve arrangement. So here is another point to determine when locating a farm.

Social conditions of the neighborhood, though of no moment so far as crop production is concerned, must not be overlooked. If the family includes children good and easily accessible schools are essential. But what are their up-keep taxes? How about churches and other institutions? Stores, lumber and coal yards? These have an important bearing upon the desirability of the farm. So has the character of the permanent residents. Such apparently extraneous conditions may make or mar the home life and comfort of the new farm family and either decide favorably for or against an otherwise desirable property.

No matter whether you decide to buy or to rent be sure that you first thoroughly understand the fundamental principles which govern farming. If you overlook or ignore these you are almost certain to waste both money and time trying to overcome some unsuspected handicap with which you would not have to contend if your farm had been chosen where these principles helped instead of hindered you.

Ask yourself: 1. Is the area large enough to produce a profitable volume of business? 2. Is the soil suited to the crops to be grown or the animals to be reared and to produce a profit? 3. Are the natural resources with respect to crops and animals, and the environment with respect to sales of products favorable to the development of a profitable business?

Mere size of place is not necessarily a factor; for many a farm of several hundred acres pays only a small net income or is run at a loss, whereas many a small one specially adapted, well situated and well managed pays handsomely. For instance, on Long Island scores if not hundreds of market gardeners have grown wealthy on 5 acres or less. Ordinarily and until one has experience, however, when the area is small the gross income is small and so is the net. Usually the deciding factor is the use made of effective implements rather than with hand labor. To be of productive value land, buildings and equipment must contribute their quotas of income.

Factors that contribute to farm income are: 1, Profit on the uses of the land; 2, profit on the working capital; and, 3, profit on the personal and hired labor employed. Capital alone can do nothing; it must be worked; land not used for crops or animals is an expense—for taxes, if for nothing else; area if small or limited in its opportunities to employ labor profitably necessarily curtails income except in such cases as market gardening, but these demand considerable hand labor. In fact, costs of production are proportionately higher for small than for large areas because hand labor instead of power implements must be used.

From all this it is evident that volume of business—size of farm—is perhaps the most important factor to insist upon when choosing a farm. So before you buy make sure, 1, that the property is potentially capable of producing the necessary volume of business. This is suggested mainly by the tillable area and the opportunities for making sales; 2, that costs of production are likely to be economical; 3, that the potential volume of business will yield a profit above all necessary costs and after deducting the value of farm products used by the family. In short, the farm should give opportunity to use capital and labor profitably and to market the products at a profit.

7

WHERE TO LOCATE

REAL property is defined as land and land only; it does not include anything growing out of the land or anything built upon it. . . . The fundamental characteristic of real property is its permanence. Land is used for all the purposes of man, and its value . . . is measured by the degree of such usefulness represented in money. The value is based upon its capacity to produce profit . . . at the prevailing interest rate of money.

EDWARD H. GILBERT,
In *Practical Real Estate Methods.*

WHERE to locate is mainly a personal question. In general, however, you will be wise to choose a property in the climatic region with which you are already familiar; for thus you may avoid costly mistakes due to ignorance of the peculiarities of some other region. The type of agriculture you wish to follow—dairying, fruit growing, poultry raising, market gardening, etc.—should also influence your choice of location. In general you are more likely to succeed where the one you prefer is already well established than elsewhere because such a locality has already been proved favorable and you will have neighbors to help you. An experienced man may risk pioneering new to a locality; for instance, the Western New York grape grower who moved to Delaware and established commercial grape production and thus increased the incomes of himself and the neighbors who followed his example.

Before deciding on a farm one of the most important things to do is to consult the official soil survey map and report on the district in mind. This is issued by the Department of Agriculture and may be seen at public libraries or bought from the Superintendent of Documents at Washington for a small sum. The map shows in color the location and extent of each kind of soil within its area; the report describes the various soils, discusses their adaptability to specific crops and their productive capacity under approved systems of handling. The latter also gives a brief review of local farming and economic conditions at the time of publication and thus supplies data upon which to base a rational system of agriculture.

With the knowledge acquired by this means it will be easier to choose among several parcels of locally available land because the map is large and detailed enough to suggest types of soil in areas as small as even a few square rods! But though easily possible to locate each soil type it is important to consider various other factors which may or may not affect the immediate and subsequent value of each property.

The character of the soil depends first upon the way it was formed; for instance, from weathering of rocks in place as on level and slightly rolling plains; from silt carried by streams and deposited in valley "floors"; from accumulations of vegetable matter in swamps as in muck lands; or from glacial deposits of rocks of many mixed kinds.

Even though the original character of the soil may have been any one of these just mentioned or of others it may have been so altered by man that it is either better or worse than at the start. As a safe rule to follow, it is worse because cropping without adequate return of plant food and humus—the popular short-sighted practise—will have injured its texture as well as reduced its fertility. Such conditions are fairly easy for even a novice to determine by personal examination and asking questions of both owner (or tenant) and the neighbors.

One of the most important things to discover is whether or not the land needs drainage or is already well drained either naturally or artificially. (Chapter 18.) Often this information may be discovered with one's own eyes by noting, 1, whether there are wet spots, often indicated by growths of bullrushes, sedges and other water-loving plants; 2, whether there are brooks or springs, thus suggesting good

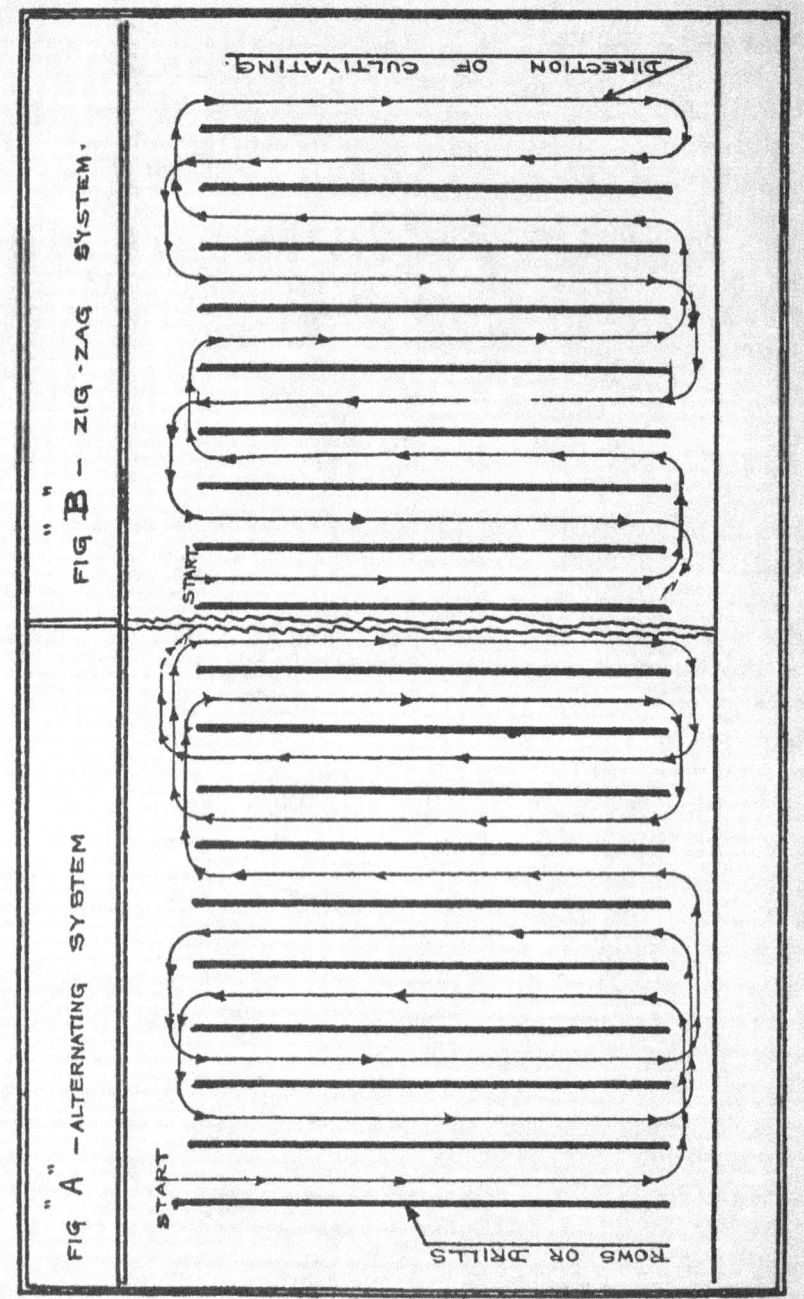

Fig. 5. Time-saving ways of cultivating. (Chapter 8.)

drainage; 3, the elevation of the land itself, either above or below other levels from or to which excess water may flow; 4, the depth of the soil; 5, whether a rock stratum or a "hard-pan" is near the surface, determined by a soil auger or by digging in various places.

In the last case it is necessary to make sure that the impervious layer (rock or hard-pan) can be economically broken so excess water may descend to lower levels and later ascend toward the surface by capillarity so plant roots may get it. Hard-pan, when formed by plowing year after year at a uniform depth while the ground is wet may be broken by subsoiling or by dynamiting. Directions for the latter may be obtained free from manufacturers of dynamite.

One way to estimate the condition of the soil is to inquire what classes of crops have been grown on the land during several years, how fertilized and what the yields have been. If the sequence of crops for years has been hay and grain continuously without adequate manuring or green manuring both the humus and the plant food may be at such low ebb that several years may be needed to reëstablish fertility and therefore profitable crops. On the other hand, if manuring has been liberal—at least 25 tons to the acre annually—or if commercial fertilizers and frequent green manure or cover crops have been the regular orders of management, the land should be in good condition. This will be evidenced by the character of crops growing upon the land.

If there should be no crops to examine the next best thing is to note the character of vegetation already present. Beech, sugar maple, hickory, black walnut and white oak trees of large size and positive thriftiness indicate rich land; white pine, scrub oak, and scrawny trees of most species are typical of poor land; extra thrifty willows, poplars and alder and elder bushes suggest too much water and probable need of drainage.

Weeds, however, are more often telltales than are trees and bushes because they follow cultivation, whereas large trees usually precede it. It is not necessary to identify species, though this is desirable. What does count is the character of growth made by the weeds actually present. Lush, sturdy, very dark green, leafy growths indicate that the plants are well fed, especially with nitrogenous compounds; but if the growth is pale, sickly colored, scrawny and apparently eking out a miserable existence the land is certainly not rich.

If there is abundance of sheep sorrel—small plants with spear-head-shaped leaves with a tart taste—especially if the plants be puny, the land is not only short of plant food but is acid, a condition not favorable to most cultivated crops, but easily corrected with lime. (Chapter 30.) Ox-eye daisy, wild carrot and mullein in abundance and poorly developed indicate lack of humus as well as fertility and prove that the land has been badly mismanaged, for these plants cannot stand either rich soil or rational tillage.

It need hardly be mentioned that if the land is stony, rocky or hilly it is usually less desirable than land not so encumbered. The cost of blasting out rocks and hauling away stones may be prohibitive and the disadvantages and difficulties of cultivating hilly land without having excessive erosion are so great that these areas cannot usually be cultivated at a profit.

If adjacent land as well as that under consideration shows similar characteristics the whole locality may be judged as either highly desirable or undesirable, as the case may be; but when the specific land shows that its apparent bad condition is due to mismanagement as suggested by good crops near-by there will be hope that it may be reclaimed and made as good as the best in the vicinity.

By keeping one's eyes open to all these and other objections and drawbacks even the novice may decide many points for himself. For the most important points to discover at the start are the objections; not the advantages!

8

LAY AND LAY-OUT OF LAND

THE value of real estate rests on two fundamental factors; namely, income capacity, and the certainty that the income will continue. To be able to decide rightly as to whether the income capacity is increasing or decreasing, or is becoming more or less secure; to analyze clearly the causes producing these changes, and to conclude correctly as to the stability of such causes, one must have intelligence supplemented by experience.

IRVING RUTLAND,
In *Practical Real Estate Methods.*

APART from the character and the condition of the soil, the location of a farm and the crops to be grown two of the most important factors that influence profit and loss are the "lay" and the "lay-out" of the land. *Lay* always refers to natural features; *lay-out,* to artificial ones. Items included in "lay" are elevation, aspect (or direction of slope), and level. Each of these has a bearing upon the character of farming that may be conducted.

For instance, other conditions being equal, a farm only say 10' above sea level will be "earlier" and have a much longer season than another in the same latitude but 1,000' above sea level. Similarly a farm 10' above a large body of water such as a lake or a wide river, especially on the lee side, will be less likely to be affected by frosts in late spring and early fall than another on a level, perhaps only 1,000' away. Also a field at the base of a slope or in a "pocket" between two or more slopes is surer to be frosty than the higher surrounding land because cold air flows like water from higher to lower levels.

Aspect generally refers to the tilt of the land. In commercial orcharding, though carefully considered when choosing the site of a large development, it is usually disregarded so far as small areas in the tract are concerned, the planting being continuous regardless of slope. In market gardening, poultry raising and various other branches of farming, how-

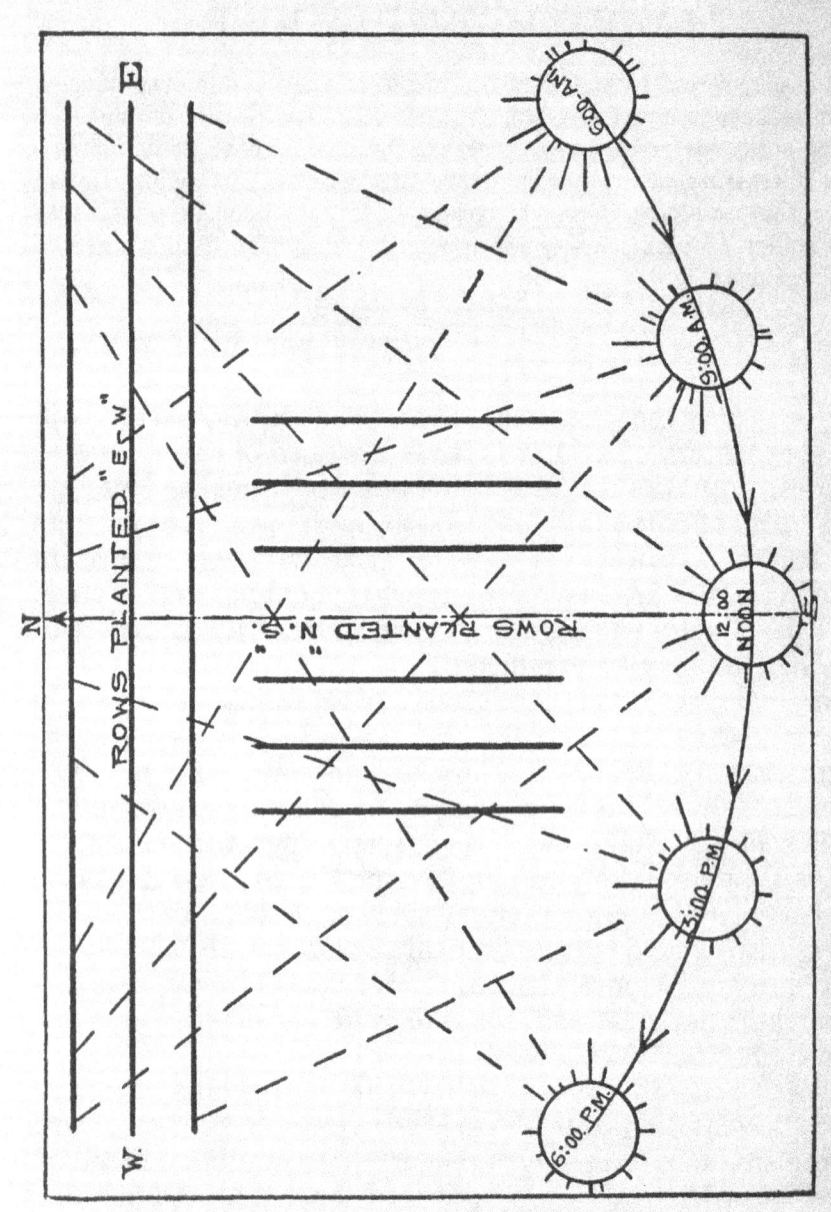

Fig. 6. How sunlight strikes vegetable rows planted from north to south and from east to west.

LAY AND LAY-OUT OF LAND

ever, aspect is highly important. It may even mean the difference between large and small profits or even losses.

A slope toward the south is almost always the warmest so is best suited to greenhouses, hotbeds and coldframes, outdoor early crops and poultry raising. Next come in usual order southeasterly, southwesterly, east, west, northeast, northwest and finally north. The effect of a northerly slope is to retard growth. Hence the flower buds of apricot and peach trees on a northern slope are less likely to be destroyed by frosts which follow warm spells in winter and early spring, than in the same orchard where during the same cold snaps the buds on trees on a south or southerly slope are killed.

The lay of the land modifies production; lay-out varies the cost of operation; both affect profits. When forest lands were first cleared it was necessary to get rid of the wood, so great piles were burned. Other quantities were split into rails to make "snake" or "zigzag" fences. Many fences were also made out of pine and other resinous tree stumps. Rocks and stones had also to be removed before farming could be done well, so these were made into walls from 2' to sometimes more than 10' thick.

As the fields were mostly small and as fences and walls of these kinds require strips of ground 10' to 15' wide they are wasteful of land and highly objectionable. They propagate weeds, shelter insects and animals that attack crops and annually require time, labor and expense to cut the growths of bushes and trees that start in them. The remedy is to get rid of them and thus combine several small fields into larger ones. Often this is easier said than done; for though rail fences may be used for fuel, stump fences and stone walls require considerable work to remove.

Not everybody whose farm is minced into little fields by stone walls is so fortunate as I was when new roads were being built past my property. The stones of the interior and inferior walls were given to the contractor. Thus at no cost to me several small fields were thrown into a few large ones.

Another advantage of getting rid of walls and fences is that the new fields are of far more regular form than the small ones. Irregular fields demand far more time to manage than rectangular ones of similar size. Especially is this the case when the length can be greater than the breadth; for plowing, harrowing and other work may be done from end to end thus reducing the number of turns. Even with the wheelhoe there is an appreciable saving of time in making few rather than many turns. (Fig. 5.)

When full rows of any crop would produce too large a quantity of any one vegetable parts of the row may be shared by other crops that require the same treatment. When it is necessary to have cross paths, breaks of sufficient width may be made in the rows across the area for this purpose. These will therefore not interfere with plowing, harrowing or tillage as would permanent roads, fences or walls: the whole area may be handled from end to end. None but necessary fences, roads or walks should be permanent.

Advantages may be gained by having the rows run north and south when the fields extend that way, because this arrangement favors even distribution of the sunlight. (Fig. 6.)

Other features of lay-out that should be noted when appraising a farm and when laying out the place after buying it are location of the well and the watering equipment in the buildings and elsewhere, the presence and condition (or the absence) of windbreaks, drains, lanes, lawns, shrubbery, orchards, berry patches, vineyards, etc.

9

WIND-BREAKS, PRO AND CON

It is decidedly unwise to plant an orchard and rely for a windbreak on a block of timber owned by a neighbor. One never knows when the neighbor will decide to cut off the timber.

FRED C. SEARS,
In *Productive Orcharding*.

PERHAPS on no horticultural question are the "pros" so obstinately *pro* and the "cons" so stubbornly *con* as on that of wind-breaks. To hear the arguments we might almost suppose that two radically different things were under discussion instead of the same thing—but under different conditions.

Wind-breaks are by no means unmixed blessings or unmitigated curses. In their proper places they will do all their enthusiasts claim; but in wrong situations, the reverse. The whole matter simmers down to the local application and the species of trees or shrubs used.

Properly constructed, a wind-break is a wind-B-R-A-K-E, not a wind-*stop:* it checks the force of wind and thus may gain many advantages, any one of which may be important enough to warrant planting trees or shrubs for that purpose alone. But when a wind-break becomes a wind-S-T-O-P it may become responsible for any or all the damages claimed by its opponents. As the disadvantages are fewer than the advantages let's discuss them first.

When the trees or shrubs in a wind-break include species which harbor orchard or garden insects or diseases they are certain to prove a menace to domestic plantings. For instance, wild cherry encourages peach borer and tent caterpillar, elms are breeding quarters for canker worms; wild roses foster rose chafer; red cedars are alternate hosts for rust disease of apple and quince. But we don't need to plant such trees and bushes nor to let them stand if they are already growing in natural woods near-by.

When wind-breaks are so close to gardens or orchards that their roots rob the cultivated plants or trees of water and plant food, or when their tops shade these plantings unduly they exceed their jurisdiction; but that is our fault, not theirs. We should not have planted them so close, or should have placed our gardens or orchards farther away.

Sometimes wind-breaks may make areas in their lee colder than would be the case without their "protection." Because of this, damage may accompany spring or fall frosts. In such cases the trouble is usually due to the wind-break being a wind-*stop*. This condition may be prevented by spacing the trees far enough apart to allow reduced air movement through them, by reducing the number of trees or shrubs already planted, or by pruning out some of the branches.

The claim that wind-breaks are wasteful of land is by no

means clear cut. Often the land they occupy is worthless for crop production or the advantages to be gained are of far more importance than the quantity of produce that might be gathered from the "waste space."

Now let's look at some of the advantages. If the trees and shrubs are conspicuous when in flower or in autumn colors, especially when evergreens are included among them, they have an ornamental as well as a practical use that makes them doubly valuable. In other cases, if we plant shadblow (or juneberry), mulberry, hackberry, highbush cranberry and other fruit and seed bearing trees and shrubs we may attract birds away from our cultivated berries and yet gain all the advantages of their aid in keeping down insects in our gardening and fruit growing.

We may even insure direct profit from our wind-break by adopting the California practise of planting fruit or nut trees or bushes on the margins of the area we seek to protect. In that state English walnuts, almonds, apricots and figs are popular for this purpose; in the East we might use improved varieties of filberts, black walnuts, northern pecans, highbush blueberries, elderberries, raspberries, blackberries, or sugar maples for maple syrup and sugar making.

When a wind-break has become well established before planting the orchard, the fruit trees will grow straight instead of bent or lop-sided, as they often do in windy sections unless staked. It will also lessen the quantity of windfalls and the breakage of branches when loaded with fruit or ice.

During blossoming time the flowers on protected trees are less likely to be blown off than where unprotected and bees are more certain to do the work of pollenating, thus insuring larger quantities of fruit.

Similarly a wind-break facilitates—even insures—work in the orchard under windy conditions, especially in winter when pruning is done, in spring when spraying might be useless or wasteful or inferior because of unchecked wind, and in autumn when late fruit must be harvested.

Wind-breaks are also important because they reduce evaporation of water from the soil and transpiration from the crop plants, particularly the leaves. Thus they mitigate the effects of drouth and winter injury which often follow a dry summer and a wet autumn. Because of this they also help the plants make better development and enhance the size and quality of thin skinned fruits such as strawberries, raspberries and blackberries.

The harmful effects of winter are lessened by the retention of leaves and snow on the ground, for in the lee of a

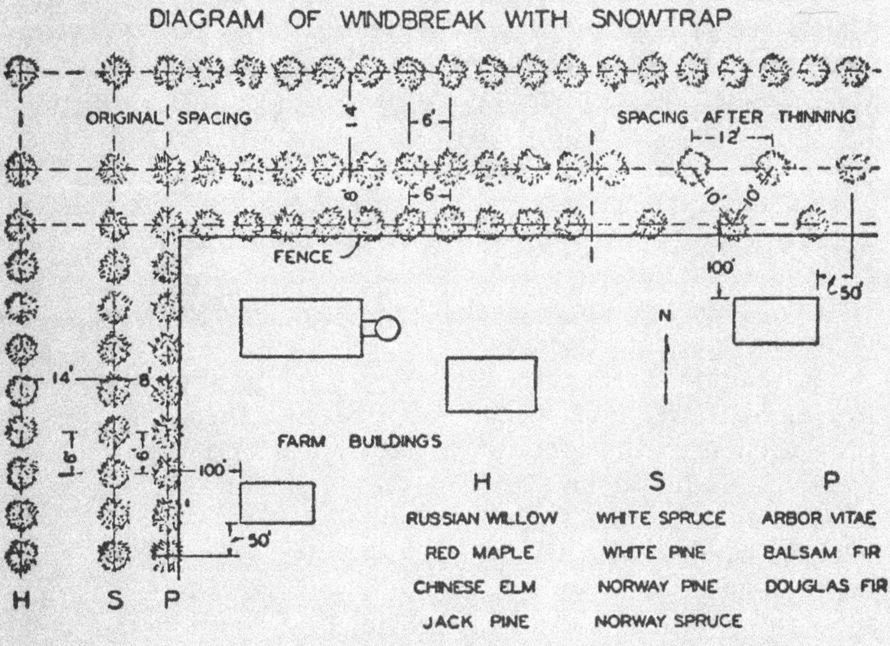
DIAGRAM OF WINDBREAK WITH SNOWTRAP
ORIGINAL SPACING
SPACING AFTER THINNING
12'
6'
6'
14'
8'
FENCE
100'
50'
N
FARM BUILDINGS
100'
50'
H
S
P
RUSSIAN WILLOW
RED MAPLE
CHINESE ELM
JACK PINE
WHITE SPRUCE
WHITE PINE
NORWAY PINE
NORWAY SPRUCE
ARBOR VITAE
BALSAM FIR
DOUGLAS FIR
H
S
P

Fig. 7. Windbreak and snow trap to protect farm buildings.

wind-break where these collect the ground freezes less deeply than where they are blown away, so the roots of fruit trees, bushes and other plants are less likely to be injured because of their presence.

Along the sea and the great lakes shores where sand is often blown for long distances inland, wind-breaks have been found particularly useful to check the destruction of crops and the burying of good soil.

A still more important effect: They greatly enhance the physical comfort and noticeably reduce the cost of maintenance of man and animals whose living quarters they shield from winter winds. Houses so protected require less fuel to maintain comfortable temperatures than do adjacent similar ones not so favorably placed; and animals so sheltered need less food to keep them in condition favorable to the production of work, milk, eggs, and flesh.

As a general thing dense plantings are desirable in inland locations because wind coming over large expanses of ground is likely to be colder in winter and hotter and drier in summer than that which passes over a large body of water. Where sea or lake winds prevail the plantings may be more open so as to allow the air to pass through with less check than through dense plantings. Such plantings also avoid the danger of still air which often occurs on the lee side of a sea or lake shore wind-break.

In general, no wind-break planting should be done before a careful study has been made of the local conditions, especially with respect to air drainage—the flowing of cold air from higher to lower levels. This must be assured to prevent cold spots or "pockets."

For inland plantings the coarser evergreens such as pines, firs, and spruces have special value because they give dense protection near the ground; for seashores and lake shores these trees should be much farther apart but be supplemented by deciduous trees.

Open wind-breaks, where needed at all on low lands, are better than dense ones because cold air naturally settles in such areas—unless means are provided for draining it away to still lower levels.

Planting should be not less than 50', preferably 100', away from the principal area or buildings to be protected. Its influence extends for a distance equal to 20 times its height; that is, trees 30' tall influence the force of wind for 600' on the level. On the protected or lee side of a wind-break of 10' to 30' is a calm zone where snow drifts during wind-driven storms. Hence the necessity of planting far back from buildings.

The length of the planting will depend upon the area and buildings to be protected. It should extend at least 50' beyond the last building, or feed lot area. On rectangular farms where length greatly exceeds width and an L-shaped wind-break would be inconvenient, greater protection may be had

WIND-BREAKS, PRO AND CON

by extending the planting 100′ to 125′ northerly or westerly beyond the buildings.

Wind-breaks need well prepared soil to get a good start. Deep fall plowing and thorough spring discing, especially of heavy soils, are essential. Soils in sod should be plowed early enough in fall to assure rotting of the sod. On light soils the plowed area should be covered with manure both to increase fertility and reduce blowing and drifting of the surface soil in winter and early spring.

Choice and size of trees will depend on type of soil, general region, cost of trees and the primary purpose of protection. The most suitable are conifers. Hardwood trees are not recommended as part of permanent wind-breaks, except under special conditions where a snow-trap or temporary protection rather than a high wind-break may be needed for the young conifers.

Conifers should be at least three years old, once transplanted in the nursery; four-year and even five-year transplants are recommended for white spruce, Douglas and balsam fir. Small seedlings or transplants may be used safely but more cultivation will be necessary until they have grown beyond the danger of weed and grass competition. Hardwood trees one or two years old will usually be large enough. Suitable trees may be grouped according to the kinds of soil to which they are adapted: Norway and Scotch pine for light, sandy soils; Douglas fir, white pine, Norway spruce, Chinese elm and red maple for light loams; white spruce, balsam fir, arborvitæ, cottonwood, ash and sugar maple for heavy loams and clays.

Two or more species of trees in a wind-break provide a more compact growth of foliage than when only one is used, especially where spruce and arborvitæ are used with open-growing white or Norway pines. The possible loss of one species from a future insect or disease epidemic will thus not destroy the wind-break. Russian willow or cottonwood may be used to give early protection while the slower conifers are becoming established in their lee.

Where there is enough space three rows are desirable, otherwise two. For all suitable trees except arborvitæ the rows should be 8′ apart and the trees 6′ asunder. For arborvitæ they should be 6' apart and the trees 4' asunder. On sandy soils where growth will generally be slow the trees should be staggered in the rows; on fertile ones they should be planted in checks because in 12 and 15 years they will crowd at 6' to 8'. Then by removing each alternate tree in each alternate row the remaining trees will be left in staggered positions at wider spacing.

This method will insure compact growth throughout the life of the wind-break. Where arborvitæ is planted, the original spacing should be maintained. Thinning will not be necessary. (Fig. 7.)

Wind-breaks depend for their usefulness largely on the care they receive for the first five or six years after planting. Poultry and livestock must be kept out, perhaps by temporary fencing. A mulch of straw, marsh or salt hay or sawdust 2" deep and a 12" in radius around each tree should be applied within a few weeks of planting. This will hold soil moisture and help to smother weeds. Sod growth around the trees must be prevented.

To avoid heaving by frost, a winter mulch 4" to 6" deep of straw with a low percentage of manure should be applied annually until the lowest branches are at least 2' long. If applied after the ground is frozen and preferably after a light snow is on the ground there will be no danger of mice nesting in it. It should remain on the ground the following summer to add fertility to the soil, prevent evaporation of moisture and smother weeds.

In wind-breaks where spruce and white pine are used some summer shade is desirable during the first two years. Sunflowers, having small root systems, will not compete much with the young trees and they have value for chicken feed and silage. One row should be sown for each row of trees. Three or four plants of a single flower variety are enough between trees.

When the wind-break extends easterly-westerly the sunflowers should be planted 18" south of the trees; in northerly-southerly rows they should be directed in line with the tree rows. Thus the trees will have intermittent sunshine and shade. Conifers in a wind-break should never be pruned: It is desirable to retain those branches that grow near the ground. If trees have enough sunlight they will maintain their foliage throughout most of their lives.

10

ESSENTIAL FACTORS OF PRODUCTION

CONSERVATION in the broadest sense implies neither waste of product nor waste of the forces and conditions which make high production possible. These forces are the environment under which the crop is grown and the inherent hereditary possibilities within the seed or seed-material.

B. M. DUGGAR,
In *Plant Physiology*.

WHOEVER said that "trifles make perfection, but perfection is no trifle" may not have had farming in mind but in perhaps no one walk of life do easily applied principles and rules pay such big dividends as in the production of food and the raw materials of clothing. Strange as it may seem, however, the United States Census and the Department of Agriculture "averages" of production are in countless cases less than 50% those of "record" figures and perhaps not more than 75% of easily obtained "good yields." Suppose we look at a few.

As to milk production, the Department of Agriculture *Year Book* shows an annual average of about 4,200 pounds of milk per cow, yet, many well bred cows yield 20,000 pounds or more a year. It is evident, therefore, that Professor H. H. Wing is well within the truth when he declares that "the average production per animal in the United States is scarcely sufficient to pay the cost of food and labor, to say nothing of interest or profit on the investment."

Why keep such animals when "good grade" cows, the progeny of "record sires," can be raised or bought at reasonable cost with the almost certain assurance that they will produce abundantly and pay good profit?

Much the same argument applies to hogs. "Razor backs" eat ravenously, grow slowly, fatten poorly if at all, and when slaughtered consist largely of hide, bones and offal. Good "grade pigs" whose sires are of almost any breed cost but little more to buy, much less to feed, grow rapidly, fatten well, and when dressed are "almost all meat!"

Figures as to egg laying have been quoted already in Chapter 5.

Similar principles apply to seed. Not only is one variety better than another but individual "strains" are worth more than general stock because more prolific, or disease-resistant or both and therefore more profitable even in the open market and wholly regardless of whether or not they are to be used for seed purposes.

Though many of these strains are developed by farmer-seedsmen (men who grow seed to be used for sowing) there is no reason why you should not develop your own, at least with some of your crops. Many another man has done so, increased his yields and profits, but still more important, has added to his interest and enjoyment of farming.

Doubtless the two crops that farmers have most often improved are corn and potatoes. The methods are so simple and the results so certain that I shall sketch them.

If you can start your corn breeding with ears pick out 50 or 100 of the best you can find in the crib. Be sure they are properly cured specimens of good form and size, well filled and well rounded at both ends and each one pleasing to your eye and your hand. Lay them side by side on a table, critically examine each under a good light in comparison with the others and ruthlessly discard the poorest, the next poorest and so on until only 10 are left. These are to be your nucleus for breeding.

Before you shell the grain from the cobs pick out 10 or 20 individual kernels from various positions on each ear (not including the butts or the tips which should be discarded anyway). Keep each lot by itself so as to make a critical germination test as detailed in Chapter 41, placing the kernels of each ear by themselves. At the end of two weeks you may not only know which ears have given the largest percentage of germination but, if your records are taken daily, you will also know which sprouted quickest and sturdiest. These are the ones to choose for growing in an

ESSENTIAL FACTORS OF PRODUCTION

"ear-to-the-row" test plot, which means that the seed from each ear is to be sown in a row by itself.

During the summer examine the plants to make sure which row or rows produce the sturdiest plants. Especially determine the row which has the most two-ear plants and the smallest proportion of stalks that bear nothing but leaves. Cut out these latter individually and feed them to the cow or make them into compost. At harvest time cut and cure the two-row stalks of this best row by themselves and, for the following year repeat the selection process already described. Also cut and cure the two-ear stalks from the other rows and use the ears for sowing the general field or for sale as seed corn. Each year it will be more valuable than the previous year.

By following this practise combined with critical selection of ears that produce the best shaped and heaviest kernels for a series of years many a farmer has developed a strain of corn noted for its two-ear plants, its small number of barren stalks, its heavy ears and yields—often 10% to to 25% greater production than his original yields.

If the corn you must start with is already shelled, pick out 500 or 1,000 large, heavy kernels of good shape, lay them side by side on a table and discard inferior ones until only the 50% best ones are left. Use these for planting the first year, after which follow the method already described when working with the ears.

With potatoes the method is as simple. Starting with what stock you have, pick out the best shaped, good sized (not necessarily the largest), shallowest eyed tubers in much the same way as described for corn ears; cut each tuber in quarters from end to end but keep each four pieces separate from the others. Plant each piece in a hill by itself then skip each fifth hill so as to keep the four pieces of each tuber in consecutive hills. During the summer treat them all alike, watch for differences of foliage, resistance to disease and other points good and bad, and dig the weaklings for "new potatoes."

At harvest time dig each hill carefully by hand and place the tubers from each four hills together for judgment. Discard the groups of four that produce unsatisfactorily either as to size, number, irregularity or other defect. When at last only the one or two best are left store each lot by itself in a cold but frost-proof cellar or pit and the following year repeat the method. The discarded tubers each year are to be used as general field seed or for sale as seed potatoes. Thus not only will the one or two best lots be improving but the general seed and the field yields will be greater. As with corn, countless farmers have increased their yields at a cost of only a little time and attention to the growing of improved seed.

Whether or not it will pay to carry out similar methods with other crops is largely a personal matter depending mainly upon the amount of each being grown. However, anybody may be the discoverer of a "rogue," a "bud sport," or a "seedling" that by proper handling may be developed into a new "variety" or "strain."

The term "rogue" is applied to those plants which differ from the variety sown in some respect but are certainly not due to accidental mixing. For instance, a yellow podded bean grown from green podded stock, a purple tomato from scarlet tomato stock or vice versa. The term "bud sport" is applied to twigs or branches of trees and bushes whose foliage, flowers or fruits differ from those of other parts of the tree; for instance, a smooth skinned peach (called a nectarine) which appears on a peach tree or a fuzzy skinned nectarine (called a peach!) borne on a nectarine tree. The Golden Delicious, Red Spy, and Red Tompkins King apples are cases of this kind.

If you have a case among bush or cane fruits or strawberries you may benefit from it, first in the production of fruit and second from the sale of plants. Let me cite a case of this kind.

One of my clients who had been growing strawberries by the hill system for years happened to notice that one plant bore nothing but leaves—no berries. This so piqued his curiosity that he examined every hill in his patch and found about 50 plants that bore a liberal quart each. He allowed each of these to produce runners which he planted during early August in a new bed. When these fruited he allowed only the heavy producers to develop runners for the next new bed. At the end of 10 or 12 years he had developed a strawberry strain so locally noted for its prolificacy that he was besieged for plants which he sold at rather "stiff" prices. Thus he received the double profit already mentioned for the time spent in his selection of plants.

By keeping on the watch for such cases you may discover something of greater value than its parent and either yourself profit by the sale of stock or of "scion wood" for propagation or be of service to other growers.

The above cited methods and instances are only a few of the special ways by which you may increase production. Others include high quality seed, germination tests, rational tillage, control of parasites and weeds, thinning of plants in vegetable rows, thinning of fruit on trees, adequate range for livestock and poultry, ample plant food, humus and water in the soil. These and other practises are discussed in other chapters.

In cases where cultivation must be done by horse or tractor and the rows must, therefore, be spaced farther apart than when the wheelhoe is used it is essential, as a time-saving factor, to make the rows long and few rather than short and many so as to reduce the amount of time necessary for turning at the ends. Even so time may be saved by skipping several rows when making each turning because (Fig. 5.) less time is needed to make a long turn than a short one, especially with a horse or a fast moving, heavy or long-radius tractor.

Should a complete row be likely to produce more of any one vegetable than would be needed it should be filled with two or more kinds that require the same general cultural treatment. Figure 8 suggests a way by which the Illinois Experiment Station (Circular 325) suggests that a well-balanced farm garden may be arranged to provide a large assortment and continuous supply of vegetables throughout the growing season, for use fresh, canned and for winter storage.

As will be noticed the plan indicates that sowings are to be made at four different times. This is because of the effects of frost and because seasons vary, some being early, others late in opening. For the latter season the time between the early sowings should be increased in an early spring and shortened in a late one. The first of these sowings should be made about the time that the earliest trees, such as silver and swamp (or red) maple open their buds. See Appendix for lists of vegetables.

By taking advantage of the cool fall weather a second crop of cool-season vegetables may be grown. In the fall garden the vegetables are grouped according to the length of their growing periods and each group planted as late as possible consistent with finishing growth before a killing frost. It must be remembered that fall garden vegetables do not thrive in warm weather and that too early planting will stunt some kinds and cause others to become coarse, woody or pithy and unfit for use.

Complaints of poor vegetable yields by growers who practise clean tillage, and reports of satisfactory ones by others who used other methods have led to experiments whereby such practises could be tested side by side for comparison. One of the most comprehensive of these series was conducted by H. O. Werner and reported in a 40-page bulletin (No. 278) of the Nebraska Experiment Station. Some of the conclusions are condensed as follows:

Most vegetable crops can be increased and improved by irrigation. Straw or paper mulches are also useful. Irrigation will be found desirable at some time in practically every season and often in many seasons. Except for hastening seed germination in a dry spring, irrigation is seldom needed before July and not after August. It has been found profitable with all vegetables but especially with those growing during these two months.

Needless or excessive irrigation early in the life of the plants might cause the development of shallow root systems. However, vegetables should be kept growing steadily. Knobby, growth-cracked, hollow, rough-shaped, double and otherwise undesirable vegetables are produced when growth is uneven, especially when a period of abundant moisture follows one of prolonged drouth. When vegetables are approaching edible maturity the quality, especially firmness, flavor, sugar content and keeping quality, will be improved by irrigation water used sparingly.

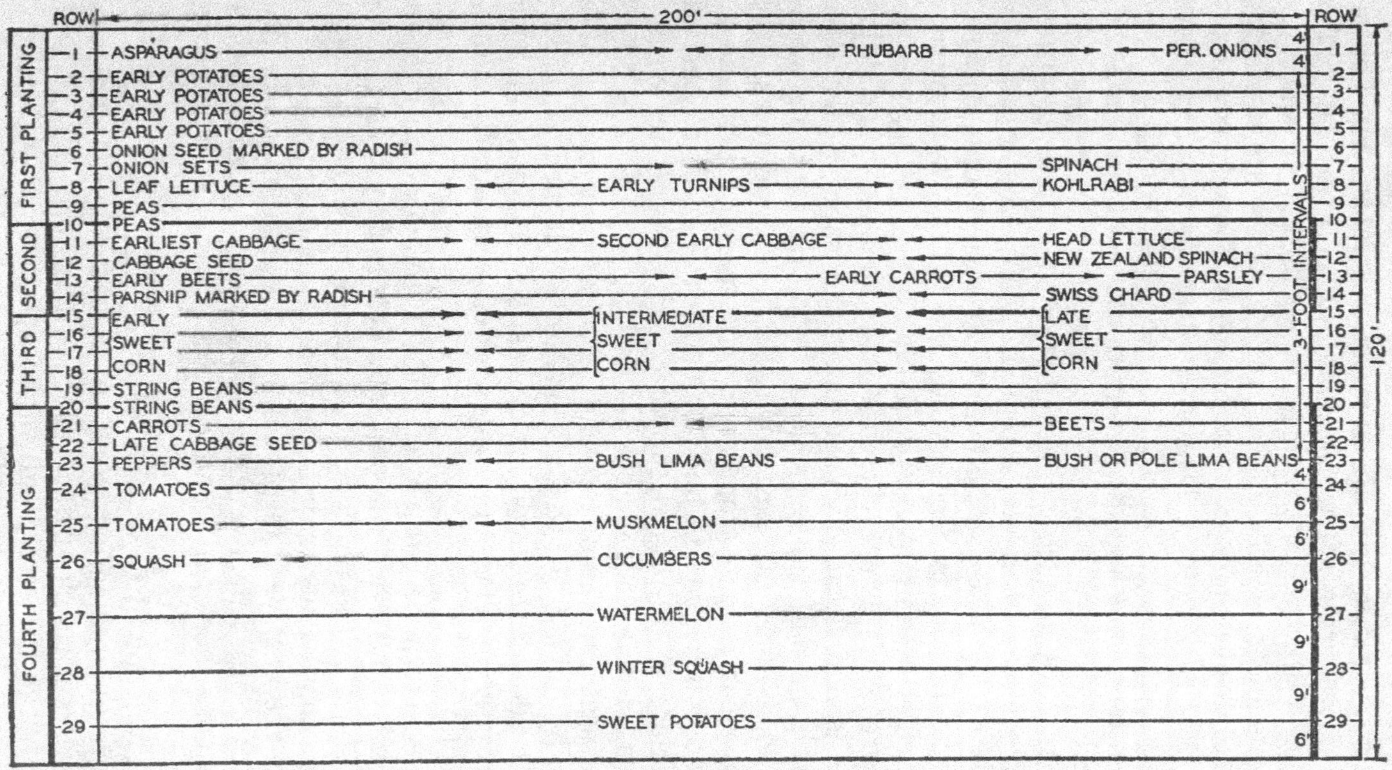

Fig. 8. Well-balanced long row farm garden for family of six. (See Appendix for varieties.)

One inch of water, in one rain or from irrigation, should maintain vigorous growth of most vegetables for five to seven days during hot weather and 10 to 15 days in cooler weather. With smaller applications much more frequent irrigations and a greater total amount of water are necessary to maintain the same growth, and the final results are likely to be less satisfactory. Heavy applications of 2" or more are also less desirable, because the soil will be poorly aërated for a time and the loss from rotting, blight, etc., will be increased.

When applied during the night by the overhead system with a pressure of about 35 pounds, 1" of water can be applied for 25' to 30' on both sides of the irrigation line in about six hours. On a hot, windy day when evaporation is great, eight hours might be required.

Where water under pressure is available the sprinkling system is best because it distributes the water evenly, does not pack the soil, can be used on sloping land without washing, and requires the least labor. With one line of pipe the gardener can irrigate from six to ten times as much land as the line will irrigate in one position, the line being carried from place to place and supported on posts or temporary supports, wherever it is to be used.

The problem of moisture can be solved to a considerable degree by the use of two plats, one to be fallowed each summer. In addition to supplying a moist soil this plan provides many of the benefits ordinarily derived from a well planned crop rotation.

The soil may be fall plowed and left rough over winter to catch snow and avoid run-off. Snow fences may be set up to catch snow in the portion to be used for the late or long-season crops. They may be placed in another part of the garden the next year to provide for the rotation of the position of crops. Effective snow barriers may be made with a row of corn shocks or even by several rows of standing corn plants.

The function of cultivation with vegetables is to conserve moisture by eliminating weeds, to close up cracks and provide a loose, rough surface which will absorb rainfall and prevent "run-off." Deep cultivation destroys many roots, reduces the yield of most vegetables and is unnecessary. Shallow surface cultivation is recommended for all vegetables, especially in unirrigated soils and in dry seasons.

Mulching gardens with straw or other litter such as hay or manure is a practical way to increase yields and produce vegetables of the best quality. The benefits are greatest with long-season crops and in dry years. Though straw mulches have increased the yields of nearly all vegetables their use is not recommended with the early short-season crops such as leaf lettuce, peas, spinach, seeded onions, cauliflower and early cabbage. With root crops such as carrots, beets and parsnips their use does not appear advantageous and with transplanted onions is of doubtful value. The difficulties of applying the straw more than offset the advantage which most of these crops might gain.

Straw mulching has been found desirable with all long-season crops except sweet corn. As a means of increasing yields and quality it is often almost as good as irrigation. Because of the lowering of the temperature and the increased frost hazard, straw should not be applied until the plants are well established.

A mulch of 2" to 4" is adequate; deeper is unnecessary and undesirable. Between 10 and 15 tons of straw are needed for mulching an acre, or about 500 pounds for 2,000 square feet. The labor of applying straw is largely balanced by the reduced labor of weed control. At the end of the season straw mulches should be removed or burned because of the unfavorable effect upon the soil when such a large amount of dry organic matter is plowed under. This is most serious with unirrigated or sandy soils. [Instead of burning, it is probably better to use this straw to make artificial manure. M. G. K.]

With potatoes the straw mulch should be applied before the plants come through the soil. With other crops, such as tomatoes, eggplants, and other transplanted vegetables, before transplanting or after the plants are well established, preferably at the latter time.

If the spring plantings have been properly grouped as indicated in the plan, the green onions, spinach, leaf lettuce, turnips, kohlrabi and peas will occupy adjacent rows and will all be harvested in ample time to allow the ground to be cleaned up and prepared for the planting of fall vegetables. It will also prevent the development of a weedy, unsightly patch.

The second and third plantings for the fall garden may occupy the area of the early potatoes and onions, the early cabbage or the early sweet corn.

In order to take advantage of crop rotation the arrangement of rows should be reversed each second year.

The first planting of fall vegetables should consist of kinds which require a growing period of 95 to 105 days: Cabbage, cauliflower, Brussels sprouts, Italian broccoli, endive, kale, grown from seed sown six or seven weeks previously for transplanting. Also pe-tsai (Chinese cabbage), rutabaga, round beets and short carrots, sown where the plants are to remain.

The second sowing should include kinds that require 65 to 80 days to mature: Turnips, winter radishes, kohlrabi, head and cos lettuce, round-seeded peas (not wrinkled seeded). These last rarely do as well in the fall as in spring but are usually worth while.

The third sowings should consist of only those kinds that mature in 45 to 55 days: Spinach, mustard, peppergrass, forcing radishes, leaf lettuce, fetticus.

Where coldframes are available these last may be had until Christmas or even later with only slight protection on coldest nights. They will all stand slight frost if not allowed to thaw too rapidly or to be exposed to the direct rays of the sun while frozen. Watering with cold water while frozen will "draw the frost" without injuring them.

The successful garden is well planned; includes a large variety of vegetables; by succession sowings supplies adequate quantities in the best eating and canning stage from early spring until winter and by storage and canning until spring; yields at least two vegetables, besides potatoes, daily throughout the year; pays better for the labor it requires than does any other equal area; is kept free from weeds by surface tillage while the seedlings are small; and being heavily fertilized, fall plowed or dug, left rough over winter and worked down by spring care to make a first class seed bed.

II

RENTING VS. BUYING

THE life of the husbandman of all others is the most delectable. It is honorable, it is amusing, and, with judicious management, it is profitable. To see plants rise from the earth and flourish by the superior skill and bounty of the laborer fills a contemplative mind with ideas which are more easy to be conceived than expressed.

GEORGE WASHINGTON.

NO ONE reared in the country need be told whether it is better to own or to rent farm property; he knows! But the city dweller who is planning to take up any branch of farming may well consider the pros and cons of both leasehold and ownership before he decides upon either. So far as the beginner is concerned, renting farm property has several outstanding advantages, among which the perhaps most important are:

1. No capital need be invested in permanent features such as land, buildings, fences, orchards, vineyards, repairs or improvements; these are for the landlord to supply because they are permanent fixtures on the land. If the tenant makes repairs or improvements, builds houses, or plants orchards he cannot remove them or charge their cost to the landlord unless a specific agreement in writing has been signed beforehand by both parties to that effect.

2. Such capital as the tenant does invest may all be for machines, tools, animals, portable houses and other "loose" property which he may take with him when he moves.

3. All crops he plants may be annuals which, because they mature in one season or less, may all be sold the same season.

4. Day-old chicks may be bought in spring and sold in the fall, thus, if desired, leaving no stock to winter over.

5. Even though the first season may not make ends meet —as is more than likely!—the beginner may often be able to decide from this one season's experience whether or not he and his family are likely to adapt themselves congenially to farm life. However, one *growing* season (say from April

to October inclusive) is not a fair test. The winter is the trying time—December to March—when, except from eggs and milk, there is usually no income and when the rigors of cold weather must be met with fortitude.

To sum up its advantages: Renting is chiefly valuable for the opportunities it affords to gain experience before making a permanent investment. Until the beginner has passed his self-imposed novitiate he will be wise not to indulge in ownership or to spend his probably hard earned money in buildings, permanent crops or improvements for the following outstanding reasons:

1. Because of some feature, attractive or conspicuous at the time of preliminary visits, decision to buy may have been too hasty, the result being that the property is perhaps found to be less well adapted to the desired purposes than other properties in the near neighborhood.

2. Though buying a farm is easily and often too quickly done, selling it is usually slow, difficult and often done at a loss, even when the property has been improved and thus made more valuable by improvements in the buildings, new orchards, vineyards, berry patches, and perennial crops such as asparagus and alfalfa.

From the above it is evident that the prospective farmer should never sign a contract or agreement offering to rent, buy, place an option, or rent with option of buying until, 1, he has had disinterested advice as to the value and adaptability of the land for specific types of farming and, 2, making sure that every legal item is thoroughly understood and complied with, particularly *by the seller*.

"Disinterested advice" is generally paid for by retainer or per diem fees, plus expenses; for instance, traveling, meals and lodging when these are necessary. Though such fees may seem large the money is generally well spent, particularly when the counselor or consultant for good and sufficient reasons advises his client not to buy certain properties; for instance, inadequate water supply.

In my own consultant practise I have "turned down" at least twice as many farms which clients had asked me to report upon than I have recommended purchasing for reasons which in every case my clients had not seen but which, when pointed out, they recognized as well founded in every case. Some reasons that I recall and that may be of service to my readers are the following; Inaccessibility, inadequate water supply, land unsuited to the purposes of the client, isolation (not a house in sight), impracticability of hauling spraying apparatus up a steep grade in an orchard already planted (the client and his wife had admired the view of the hills and the distant Hudson River!), too large a proportion of untillable (swampy or stony) land to tillable, and too great cost to remove brush and stumps from neglected land. After receiving several adverse reports, one client decided not to risk his money in farming!

I can recall only one case where *everything* was favorable —good buildings, modern conveniences, excellent orchards on favorable soil, both well managed and bearing profitable crops, shipping facilities first class, good neighbors—everything! Why, then, did the former owner wish to sell? I asked him after making my inspection. Family tragedy. The buyer got a bargain and has made good.

Among the most important items which a prospective farm purchaser should see to are clear title, deed, survey of the property, mortgage (if any) and its conditions, transfer of insurance and premiums paid by the seller to date of sale, and taxes. (Chapter 12.)

An instance in my own experience illustrates the importance of making a thorough search of title records. A farm that I bought some years ago had been owned by father and son since about 1850. For this reason I accepted a warranty deed. But when I wanted to sell it the prospective purchaser found that there was no record of the payment of a $500. mortgage recorded in the 1840's—before even the father of the man who sold to me was born! So the sale was held up until the legal requirements had been met; namely, 30 days' advertising in the county papers to discover if possible the heirs of the former mortgagee. None appearing, the court assumed that none were still living, that in all probability the mortgage had been "satisfied," and that, as in "the old days" was often the case, the papers had been destroyed without record. Though the man from whom I bought the farm had never heard of the mortagee or the mortgage he paid the court and advertising costs because of the warranty deed, his agreement being to give me clear title to the property.

But that is not the best way to handle a property transfer! Every buyer of real estate should insist upon an "unclouded" title, proved by a thorough search of the county records.

Having found a property which seems desirable and which the counselor or consultant reports upon favorably it is advisable to visit the cashier of the local bank to discuss plans, ask advice as to further procedure and the recommendation of a good attorney to search title and draw up the necessary papers (Chapter 13).

Three courses are open: 1, out-and-out buying; 2, buying a smaller place than actually needed and renting supplemental land; 3, renting, either for cash or on shares. Which one is best for *you* cannot be offhandedly decided by anyone, but suggestions as to the use of your capital and the desirable size of a farm should be helpful in any case, for available capital is generally the deciding factor.

If your funds are limited and if you must derive your whole income from the land it would almost surely be safer to start as a renter. However, if you know that any specific farm is just what you want and where you wish to live and if you are not wholly dependent for a livelihood upon what the farm will produce the first few years it may be advisable to buy, even though your capital is small.

Before you decide on the second plan suggested make sure that land suited to your purposes and located near-by can be rented on reasonable terms. When such can be had you may be able to utilize your small capital to better advantage and to operate a larger area than under the first plan and may enjoy all the satisfactions of owning your home place which you may improve as your means and time will permit.

Generally a farm is rented more as a business venture than as a home; hence special attention should be given to importance of good soil, sureness of market, and size as related to efficiency of operation. You must also remember that when renting a farm the income must be enough to pay the landlord a return on his investment as well as your own income for your labor and the capital you have invested in equipment, livestock, etc. Hence the farm must have the basis of good income so as to make a profitable return to both parties concerned.

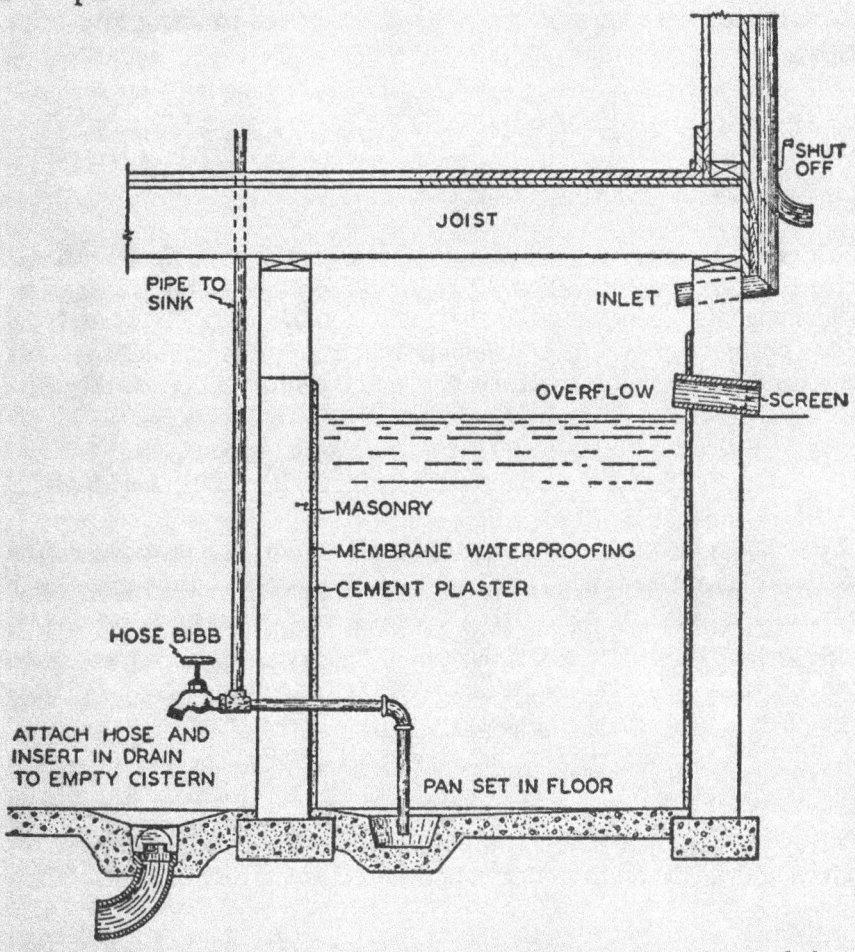

Fig. 9. Concrete cistern. Note how bottom slopes to sump whence sludge may be siphoned to drain.

Farm management investigations show that tenant profits are usually in direct proportion to the size of the business. The tenant generally supplies the animals and equipment so a good sized business allows him to use his capital efficiently.

Buying a small area and increasing the size of the business by renting adjoining land permits operating a good sized business with limited capital and having greater security than if you start with no property of your own but rent the whole area. As the plan might lead to your buying the adjacent land, you should study the whole problem in advance of making any move so as to avoid making any mistake.

12

CAPITAL

THE small amount of capital required to begin farming operations, creates great misconception of what is necessary for commercial gardening; for, judging from the small number of acres wanted for commencing a garden, many suppose that a few hundred dollars is all sufficient for a market gardener. For want of information on this subject, hundreds have failed, after years of toil and privation.
PETER HENDERSON,
In *Gardening for Profit*.

PERHAPS no question is so difficult to answer satisfactorily as: How much capital should I have to start my little farm? Instead of attempting to set a figure (which probably would not apply in one case out of scores) suppose we consider a few specific crops and items that have more or less direct bearing on the general subject of capital.

In apple orcharding a paying crop can rarely be expected under seven years, more often it is 10 to 15. As many varieties bear satisfactorily only in alternate years they will rarely yield more than 15 crops in 37 to 40 or 45 years from planting. When such yields do occur other orchards will probably also have heavy crops so prices may be too low to make much profit. With pears and most nuts the times to reach bearing are about the same.

Peaches generally begin to be profitable in the fourth or fifth year. Some commercial growers who take special care keep their trees in profitable production for more than 20 years; but the majority count on only 10 or 12. Though the peach would naturally bear every year, an annual yield cannot be counted upon because cold winters and spring frosts often destroy the buds, flowers or newly formed fruits; so if a crop is gathered oftener than twice in three years it is "just so much velvet."

Cherries and plums, which rarely become profitable before five years, are more regular annual bearers than apples and pears and are naturally longer lived than peaches—15 to 20 years for well managed sour cherries and plums and 30 or more for sweet cherries. The main objections to them are the cost of picking and their proneness to brown rot of the fruit and foliage.

Oranges, lemons and grape-fruit require five years to bear profitable crops but, unless injured by frost, they usually bear well annually and for many years.

Currants and gooseberries begin to yield, usually, during the fourth or fifth year. As they are perfectly hardy they should bear well annually, provided they are properly managed, and continue for 10 to 12 or 15 years.

Raspberries, blackberries and dewberries generally start to pay during the third year and bear annually for six to ten years or more.

Strawberries, when 14 or 15 months old bear their most profitable crop. When a second crop is borne by the same bed it is not only lighter and inferior to the first but, because of necessary hand weeding, it costs relatively more to produce and often brings lower prices.

Among vegetables, asparagus cannot be expected to yield a paying crop until the third or fourth year, but when well fertilized it should yield annually for at least 10. The bed at my boyhood home yielded abundantly for more than 40 years to my certain knowledge, but it was fed lavishly. Other perennial vegetables should yield well the second or third season but they are much shorter lived.

Because of the times necessary to bring these plants into bearing the grower must be patient, sanguine and assiduous in his attention. They indicate also that he must have either ample reserve capital, have ways of hiring it during these periods of development or must pay the necessary yearly expenses by income from the place itself. Hence the reason for growing annual vegetable crops between the trees, bushes and vines until these woody plants need all the space.

Parenthetically it may be said that many a man, by growing vegetables and strawberries has developed his orchard, vineyard or bush berry patch through these periods with a balanced account and even a small profit before the fruit crops began to pay; but, conversely, probably far more have had a deficit; in fact, many have failed and lost their money, their property and, what is worse still, their reliance on themselves. It takes cash, confidence, courage, cultivation and patience to develop a fruit plantation of any kind. So the small farmer with limited means may prudently hesitate to go in for tree fruit growing, except on the basis of home supply or at most for a strictly local sale of his products. He will more likely succeed if his money crops are small fruits and vegetables and his other sources of income are dressed poultry, eggs, honey and perhaps pork and milk.

From all this it is evident that the amount of capital necessary to equip a place for crop production of any kind varies widely. The most important factor to reckon with when estimating is *the man himself*. He must consider his experience, his ability, his teachableness, his capacity to work himself and to direct others, and so on. Other important factors are locality, cost of land, character and adaptability of soil, size of place, use of greenhouses, hotbeds and coldframes, crops to be grown, ways of marketing, equipment necessary, labor to be hired or supplied by the family, style of living to be maintained, etc.

In the growing of annual vegetables which naturally requires smaller capital per acre than fruits and perennial vegetables, the estimates of experienced men vary from $25. to $500. an acre! The first is obviously too low for the majority of crops; the second too high except where considerable glass is employed. Small places require proportionately far more capital per acre than do large ones. However, the higher of these figures need not frighten even the man of small means and limited experience—provided that he will make haste slowly. Of course, the more limited the capital and the less the experience the smaller the area that can or should be handled, even when the man owns more land than he attempts to work. One is far more likely to succeed when 50, 25 or even a smaller percent of the available area is thoroughly well worked than when the seed, fertilizer cultivation and especially the labor are spread out so thin that the crops get inadequate food and attention and weeds gain and maintain supremacy over them. The experience gained in producing a small crop of high quality that commands a good price is sure to foster one's confidence and enthusiasm; whereas to fail because of too large-scale operation is sure to quash both.

In grain, hay and other field crop farming the rent or interest on the land cost may be 50% of the total expense of growing and harvesting; but in the production of inter-tilled crops such as vegetables the land rent or interest may be only 10%; that is, the total cost of bringing an inter-tilled crop from seedage to sale may be five to ten times as great, depending largely upon the amount of tillage, fertilizing, spraying and handling necessary.

Properly to work five acres of closely planted vegetables for a market within three to five miles and where greenhouses, hotbeds and coldframes are used (as they should be) not less than a man to the acre can be kept busy throughout the year with more or less extra hands, depending upon the class of crop during the harvesting periods. To attempt to get along with fewer is almost sure to invite disaster.

From all this it is evident that when the available capital is limited to say $1,500 or less with which to pay rent or interest on land cost, buy equipment, fertilizers, manures, seed, labor and other essential items, it is far safer to plant only two or three acres in inter-tilled crops than to sow the whole area to them. The balance of the land need by no means be left idle: it may be sown to green manures as an investment in its own improvement and, later, the profit of the owner. Every square foot should be made to give a good account of itself.

In view of all this it is important to consider what returns to expect from the various crops. The usual way is to calculate on the acre basis; but this is objectionable because it gives a false impression. A better way is the labor-hour basis. For instance, an apple orchard may show a gross return of $100. an acre and a profit of $50. Looks good, eh? A crop of oats may bring in $25. gross an acre and a profit of $7. "Not so good," you say! Hay, considered a "cheap crop," may yield *only* $15. gross and $6. profit an acre. "Humph! I'll have none of that!" you may ejaculate. But let's look at the labor-hour calculation. The apples show a labor-hour profit of perhaps 25¢, the oats 30¢ and the hay 60¢! Thus the labor-hour method shows which crops pay best, which poorest and which are not paying, even when the acre-basis seems to show a profit! All this has a bearing upon the capital necessary to operate a place, the costs, yields, profits and other items necessary for production being estimated beforehand.

Vegetables and day-old chicks may pay well for themselves during the first season and strawberries the second but on a newly started place they cannot be expected to carry the whole load. So if the first year's balance sheet shows that expenses have been met by production income the owner may congratulate himself as a rare exception to the general rule. In such cases the second and subsequent years are likely to be more encouraging, especially when some of the plantings are of bush and cane berries, grapes and asparagus.

13

FARM FINANCE

It is a fundamental error to suppose that farming is neither a business nor a profession. It is a business which requires the highest business talent, it is a profession which requires the best technical skill. ... No other profession requires such a variety of learning, such an insight into Nature, such skill of a technical kind in order to be successful, as the profession of farming.

Harvey W. Wiley,
In *The Lure of the Land.*

CREDIT is a powerful agency for good for whoever knows how to use it; dangerous for those who do not. Like land, capital is a factor which requires use—investment. Hence its extensive employment in farming has developed because machinery demands it. Though in the past little money was needed to start farming, today the cost of good land and machinery require far greater sums.

Capital, as money, originates in only one way—consumption of less than produced. You may spend a dollar for amusement or for a tool. In the first case you may satisfy your desires—"get the worth of your money;" in the second you become an investor, a captialist, to the extent that you use the tool to create more dollars. If you do not have a dollar your only chance to become such an investor or capitalist is to borrow the tool or the money with which to buy it. No matter which of these plans you adopt the lender of either must place confidence in you—give you credit.

Capital for farm equipment may be secured in only two ways; by consuming less than one produces so as to accumulate it, or borrowing it. The disadvantage of the former is that one is handicapped until the necessary equipment is accumulated; the advantage of the latter is that this delay can be eliminated and the tools used at once to pay for themselves. This is the only advantage credit has in any business. For those who know how to use it, it is a powerful aid.

However, credit is not to be invoked without due calculation. Machines wear out, seasons are unfavorable, prices fall, unforeseen events occur, but payment time is inexorable. It comes with fatal certainty. Debts, unless paid, may bankrupt the borrower just as surely.

Borrowing money for production is no more dishonorable than borrowing tools for the same purpose. What has stigmatized it as such in the past is borrowing to pay the cost of living, still more of "living high." To borrow for production shows business enterprise as honorable in its degree as borrowing to build a railway, found a factory, or buy goods to stock a department store.

In the past usury or interest was condemned because the money borrowed was too often used for non-productive purposes. Modern thinkers and teachers favor interest obligations when the borrowing is for production; the only insistence is that the legal rate be fixed by the state. Except where the rate is exorbitant interest gives little trouble; it is principal that does. When a man borrows $100. at 6% for a year, the debt to pay when due is $106.; when he can borrow from another source at 5% the debt is $105., a difference of $1. This is not merely $1., but just that much capital to be either saved for later use or ignored and wasted. Good business saves just such and even smaller items. However, the point here is not the difference of $1. but the $100. original capital which is identical in both cases.

The only safe way to use the borrowed money is to establish means of payment beforehand. For instance, if the $100. at 6% is invested in fertilizer which increases crop yields of $125., not only the interest but the principal could be paid out of earnings and the difference ($19.) credited as so much increased capital. But should the returns on the borrowed money be only $75. the question will arise as to how to pay the principal, to say nothing of the interest. Hence the investment of the $100. so as to return $125. instead of $75. is more important than to be able to borrow money at a lower interest rate or without any interest at all! No unproductive enterprise is a safe one on which to borrow.

This discussion leads up to the fundamental rules to be observed when borrowing on credit. Rule Number 1 insists that before seeking farm credit *the borrower must be sure that the project for which the money is sought will produce more money than will be needed to pay both principal and interest;* for except in rare instances, it is bad policy to borrow for anything that will not pay for itself.

Should money be borrowed for, say, fertilizer under promise to pay before the crops can be harvested and sold, difficulty may arise to pay it. In such cases three events may occur to satisfy the debt: 1, Money may come from some other source; 2, the loan or the note may be extended; 3, the creditor may sell out the borrower. The first violates the good business precept that each part of a business should pay its own expenses and a profit; the second asks a favor of the creditor and thus more or less disrupts business arrangements; the third almost invariably results in a more or less heavy loss for the borrower.

Rule 2. *Have the repayment date of principal fall when*

FARM FINANCE

most convenient to the borrower; i. e., when the borrower is most likely to have cash to meet the obligation; as, for instance, when a crop will be marketed.

Closely associated with this rule is Rule 3, duration of the loan, how long it is to run. Should the loan be for a year's fertilizer it should not run for longer, for should the first crop not pay for it money must come from some other source, otherwise the borrower will never be able to pay for

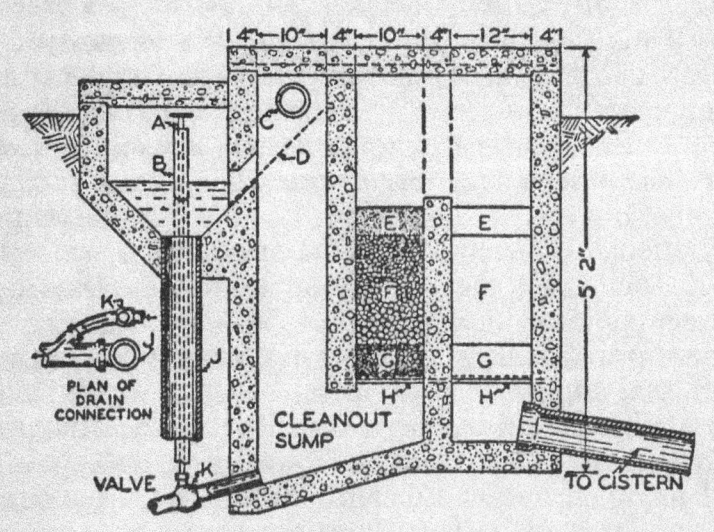

Fig. 10. Upward-flow filter for cistern. Dimensions 40" x 48" x 62". A, Sump drainage handle. B, Valve sleeve. C, Inlet. D, Wire screen. E, Sand. F, Charcoal. G, Coarse gravel. H, Grate with wire cover. J, Waste drain. K, Sump drain.

it, thus violating Rule 1. The period should not be shorter than the growing season of the crop because this would violate Rule 2.

Should a machine that will last, say, 10 years be bought on borrowed money each crop upon which it is to be used should pay not only the annual interest charge but a part of the principal—in this case 10%. Small loans of this type should give no trouble and should not require special arrangements for repayment, but large loans, for instance to buy land, build barns or make costly improvements, may entail financial stress, so any method whereby this may be relieved should be considered. One of the best is the long time loan—but not so long as to outlive the improvement!

Should an improvement be estimated to last 10 years it is good judgment to repay the loan in fewer years; for if the improvement will not last that long it will never pay for itself. Conversely, seldom can an improvement pay for itself in 1 or 2 years; so unless the borrower have other means he would invite strain to agree to too early payment. *The period a debt is to run should be closely related to the productive life of the improvement for which money is borrowed.* Thus the borrower may avoid the necessity of renewing the loan, also subjection to an unscrupulous lender who might refuse to renew a short-time loan, but foreclose.

Rule 4. In long time loans *provision should be made to reduce the principal in installments*. This may be done in either of two ways: 1. The note may state that the borrower may (optionally) repay part of the principal on any interest date. In such cases the amount of interest is also reduced with each payment until the whole debt is canceled. 2. The note may provide for a stated rate of amortization by annual or semi-annual fixed payments, each of which includes interest to its date of payment and also a part of the principal until all is paid. Every borrower of a long-time loan should insist on one or other of these plans. Amortization tables for loans of various durations and rates of interest are available at banks and other financial institutions.

Though Rule 5 is obvious; namely, that as low as possible an interest rate should be secured, its application is not always evident. Interest rates depend upon the law of supply and demand. When the number of lenders is large and loanable money is abundant in a community the rate will be low because borrowers can insist upon lowered rates; under reverse conditions the loaners can raise the rates. Hence the advisability of increasing the number of loaners or the loanable capital, not by denouncing either, but by attracting loaners to the neighborhood. As numbers and available capital increase borrowers may get more favorable terms.

Farmers must disabuse their minds of the idea that land is the best security. It is not! The character and business ability of the borrower in farming, just as in every other business exceed all else. Such men meet their obligations promptly and without legal proceedings, so credit conditions in the community improve as the numbers of such men increase because the right kind of lenders flock to such communities and the "sharks" decamp. The right kind dislike to foreclose mortgages or resort to legal technique. They merely want their principal and interest according to agreement. However, one farmer working alone can rarely do much to attract lenders, but coöperation can. Even so small a group as half a dozen men of sterling integrity, each with confidence in and respect for the other, may work out the problem of local farm credit to the advantage of the whole community.

Suppose we review some ways the business farmer and the local bank may serve each other. Probably "working capital" is the greatest need of the greatest number of farmers. The fact of its usual under-supply, particularly at the start, is the most restricting of all factors to success. Even when it is available it is often badly managed. Properly handled, money makes money; improperly, it loses.

To determine ways to make money in farming the annual budget and the annual inventory are of prime importance. When the farmer starts business and at the beginning of each of his business years (which may be calendar or his own fiscal year, say March 1) he makes his inventory, then estimates his probable gross expenses and income for each month and for each crop or department so as to determine in advance at what time he is likely to be pinched for money, when he will have surplus, when he must borrow and when he can repay. Knowledge of business methods teaches him that hiring money is the same as hiring labor. So he shows both his budget and his inventory to the cashier of his local bank and arranges for loans perhaps months before he will actually need to borrow.

In seeking such working capital he is not asking, but conferring a favor because the banks make money work for them by earning interest and discount on their loans. By loaning money on business enterprises, especially the production business of farming, the bank enables its borrowers to pay their petty bills such as labor and repairs and thus keep their own and the community affairs running smoothly. Still further, as such moneys when borrowed on notes rarely cost as much as 7% at discount rates, the user may make money by paying cash for his bills at 2% discount instead of net at 30 days. This means a 24% gain on the money so spent. Upon deducting the cost of the note (7% or less) from the 24% earned on the discounted bills the remainder is 17% or perhaps 18%. This not only means that amount profit on the borrowed money so used but the enhancement of the farmer's reputation for prompt payment—always an asset.

But the cashier of the bank is able and glad to serve the farmer in many other important ways. He can replace the attorney in countless cases—and *at no cost*. In fact, so far as business transactions are concerned he is far better able to advise wisely in financial affairs than are most attorneys! He has such business at his tongue tip all the time. Not only so but he serves his bank when he advises his patrons in their business matters. All he requires is full frankness—knowledge of all conditions that may bear directly or indirectly upon each specific case.

Some of the ways in which he may serve and benefit the farmer are in selling or buying in distant markets by bills of lading and sight drafts, advising as to investments of all kinds, arranging partnership agreements, buying or selling property, joint ownership in specific equipment or breeding stock and countless other business projects.

Finally, the bank deposit vault is the safest place in which to store valuable papers such as promissory notes, deeds, mortgages, insurance policies, contracts, stocks, bonds, and inventories of all buildings on the farm property, these last each brought up to date annually. Always these papers should be kept in a safety deposit box in a fireproof bank. Though the value of negotiable paper is evident and though copies of deeds, mortgages and some other papers may be made, the importance of keeping other irreplaceable documents in the bank deposit box cannot be over-stated; for insurance policies and inventories are convincing in case of fire. Never should they be kept where they may be destroyed, stolen, misplaced or lost.

14

FARM ACCOUNTS

FARMING is largely commercial, not self-sufficing. The farmer is dependent upon the sale of commodities to pay his debts and his taxes and buy many necessities. . . . It is true that the farmer generally has food. . . . But he also has difficulty in meeting debts and taxes. Often the spectre of possible loss of his farm through foreclosure hangs over him. He sees himself unable to provide the reasonable comforts he would like his family to have.

DOWELL and JESNESS,
In <i>The Farmer and the Export Trade.</i></div>

EVERY farmer seeks to make his farm pay, to gain more than a mere living, to enjoy the comforts of life and to provide for old age. To do all this demands careful planning months and often years in advance of marketing. At the opening of each season the farmer who plans a production program may feel that he has planned well but at the close of the year may find that he has erred, or that growing conditions were unfavorable. High prices at planting and breeding time usually prove a poor guide when planning the operations for the following season.

Crops to grow, area to plant, live stock and how much to keep are problems which demand knowledge of market requirements and conditions, when the produce is to be sold, the advantages and disadvantages of various competing regions, knowledge of price trends and the potential production of the individual farm. Most farm products are supplied by many farmers working independently and competing with one another in the market. Usually the keenest competition does not come from producers in other districts but from the neighbors.

Each farmer, therefore, should know what these competitors are planning to do. He should learn to utilize the facilities created by agricultural experiment stations and the Department of Agriculture as to the combination of the various crop and livestock enterprises and economic conditions. These unbiased agencies strive to give true information concerning the leading farm enterprises to enable farmers to plan their work intelligently with respect to market demands, changing economic conditions and probable price trends at the time when sales are planned to be made. Information is available through diverse publications and state, county and community channels. The county agricultural agent will also supply information.

More important still, every farmer should know the truth about his own farm business, be able to plan production programs and adjust his program to meet changing economic conditions. He who discovers his less economical and productive methods and either corrects or discards them usually makes his farm pay better than before. Keeping records is the most effective way by which to determine the truth about himself and his business, his less economical and less productive methods, practises, crops and live stock.

Of primary importance in these respects is an annual farm inventory, for the following reasons:

1. It shows the net worth or total value of all property above liabilities and tells whether or not the enterprise is being run at a profit or a loss, and how much. 2. It shows how the total investment is apportioned among the diverse branches of the business. 3. As cash on hand, taken alone, is not a safe guide as to earnings, an inventory prevents drawing false conclusions as to prosperity at the close of the year. Often a comparatively small amount of cash discourages a farmer who has done well but whose earnings are tied up in some kind of property. Conversely, a large amount of cash on hand may have come from a decrease in inventory of other property. 4. An inventory kept up to date enhances credit relations with banks and other loaning concerns when money must be borrowed to carry on the business. 5. Should adjustment be necessary after a fire, it is highly valuable. 6. It is fundamental to the keeping of any accounting system.

Farm accounts reveal the less productive, less economical methods and practises and where income may be augmented; but they are of little or no value unless analyzed and the results studied. Among important items they should show the following facts: 1. Net earnings the farmer pays himself

FARM ACCOUNTS

for his labor and management. 2. Gross income or total amount received on sales of crops, live stock and live stock products and increases in inventories each year. 3. Volume and increase in business as a whole and from each department—live stock and individual crops. 4. Total operating expenses. 5. Cost per $100. of income from each department as a check upon expense control. 6. Total live stock. 7. Productive animal units per area or the proportion of stocking to the land stocked. 8. Acres pastured per animal unit

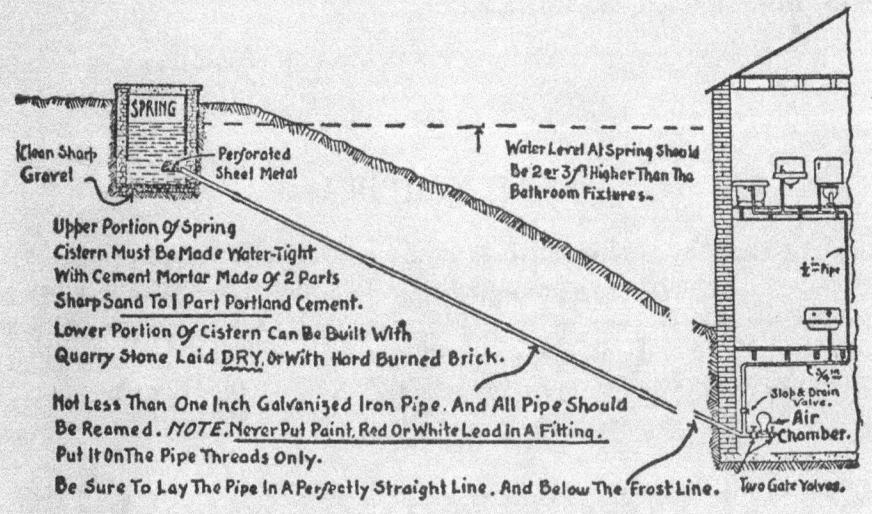

Fig. 11. Gravity system of water supply. (See also Fig. 12.)

or the economy of pasturage. 9. Receipts from stock departments. 10. Receipts per unit whether animal or crop, as a check upon the quality of product. Low returns per unit indicate that quality is below par.

11. Records of how well horse, tractor and man labor are being used. 12. Productive labor, or the average number of hours—horse, tractor and man—necessary to manage each crop and class of live stock annually. 13. Total number of available days' labor annually. 14. Use of man labor compared with available days' labor to care for each crop and class of live stock. 15. Number of men needed annually to run the farm. 16. Crop-acres per man or number of acres grown per man. 17. Productive animal units per man, or number of head per man. 18. Crop-acres per horse. 19. Days

of productive horse-labor to care for each crop and department of live stock. 20. Days of work, per horse or tractor each year. 21. After the record has been analyzed the less economical and less productive methods and practises may be studied, amended, or replaced by better ones, and what can and can not be done to improve conditions may then be considered.

Every farmer should have his account book, take inventories, keep records of receipts and expenses as suggested and thus increase his income.

15

WATER SUPPLY

For the Lord thy God bringeth thee into a good land, a land of brooks of water, of fountains and depths that spring out of valleys and hills; a land of wheat, and barley, and vines, and fig trees, and pomegranates; a land of oil olive, and honey; a land wherein thou shalt eat bread without scarceness, thou shalt not lack any thing in it.
Deuteronomy, VIII, 7–9.

So far as family water supply is concerned I may fairly claim to have graduated! At various times I have lived in houses where the primitive rain barrel furnished family needs and reared mosquitoes; where the shallow cistern provoked profanity every winter because holes had to be chopped in the ice and from which the water had to be lifted by a "sweep," "the old oaken bucket," or hauled, hand over hand, by rope and pail; a "chain-pump"; where a deep, unprotected cistern was built without provision for drainage and had to be cleaned of noisome sludge, dead toads, mice and other gruesome ingredients every summer; where there was a "filter cistern" which could not be cleaned(!) because of inaccessibility; where an attic tank filled direct from the roof collected leaves, soot, dirt and bird droppings; and where, in several houses, the water had to be pumped by hand either to a tank in the garret or a pressure tank in the cellar. Such experiences prompt me to include a

WATER SUPPLY

chapter on family water supply and to urge the installation of modern apparatus in every farm home.*

Rain water is so superior to most well and spring water and is such a money-saver that every farm home should have a supply, if for no other purpose that for the family washing. It requires no "softener," uses less soap and is pleasanter to work with than even the best water that has come in contact with the ground. Objections to the latter include the brownish or whitish scale that forms in kettles

Fig. 12. Enlarged view of "two-gate valves" shown in Figure 11.

because of its content of lime, magnesia and iron. This scale is a poor conductor of heat thus making necessary the use of excessive fuel. It also clogs "waterbacks" in kitchen stoves and "heating coils" in furnaces, often causes leaks and sometimes explosions.

Lime and magnesia form "curd" or "sludge" with soap and washing powders and spot or stain clothes being washed, and water that contains iron turns brown upon standing, stains pails and makes clothes yellow or stained when laundered. Bleaching such stains shortens the life of the cloth.

To obtain an adequate supply of rain water for household use provision must be made to collect it, to eliminate contaminating materials by screening and filtering, to provide storage and to make the water readily usable. The roofs of buildings are the most common sources of supply, but they are subject to various types of contamination and since water is a good solvent it is quickly polluted. Hence, in collecting a rain-water supply the roof must be thoroughly cleaned by rain before any water is allowed to collect in the cistern. Also the down-spouts always should be disconnected from the cistern by cut-off valves except when water

* Part of what follows concerning cisterns was published by A. M. Buswell and E. W. Lehman in Circular 393 of the Illinois Experiment Station.

is to be stored. At regular intervals they should be inspected to make sure they are uncontaminated.

The filter cistern is used chiefly to remove foreign matter carried in suspension in the rain water. Coarse material such as leaves is first removed by a screen before the water reaches the filter proper. The filter should have surface area enough for sedimentation before filtration; be so constructed that sediment and the filtering material may be easily removed.

Though the commonest type is the downward flow or gravity style a far better type is the upward-flow style with drains shown in Figure 10. This is a concrete box filled with alternate layers of coarse gravel, charcoal and sand from which a drain discharges the water when not flowing into the cistern and which drains the filter when not in use. A screen removes all coarse material and a hose or a few pails of water will cleanse the inlet side without disturbing the cistern supply. The easily removable water-tight top is for removal and renewal of the filtering material.

Storage may be either above or below ground. Where there is little danger of freezing the former is satisfactory, and where the roof is high the tank may be placed high enough to give pressure without pumping. Such tanks may be of wood, galvanized iron or sheet lead lined and placed inside or close to the house whose roof collects the water.

The commonest storage for rain water is the masonry underground cistern. The walls must be made water-tight to prevent leakage and to avoid the entrance of seepage water. Never should it be located near trees because the roots might crack the walls. Trouble of this kind is most likely to occur with plaster-on-earth masonry.

Cylindrical and rectangular cisterns are both satisfactory. By providing a pipe at the lowest point of the bottom practically all the sediment may be siphoned out when there is a drain to a lower level (Fig. 9) or pumped out otherwise.

Cistern sizes to serve all practical household purposes depend on three factors: 1, Requirements of the family. 2, amount and frequency of rains; 3, area of roof to collect water. The quantity of water depends on the kind of house

WATER SUPPLY

equipment and the number of persons in the family. When water is pumped by hand less is used than when a gasolene or an electric pump does the work. Five gallons daily per person is a fair estimate where ample hard water under pressure is also available for other purposes. Far more should be provided where there is enough roof area to collect it.

Study of local annual rainfall will help in calculating the size of a cistern when the roof area is adequate. At least 33% of the rainfall should be deducted for leakage, evaporation and to wash the roof. Enough storage capacity should be provided to store water that falls during the rainy period to meet needs during the dry time, better still to have sufficient storage for six months.

Water in a new or a repaired cistern is likely to be hard because of dissolved lime from the plaster or the cement. Thorough curing and setting of the plaster will reduce this difficulty. Curing may be hastened by sponging, spraying or washing the surface of the new plaster or cement with a strong solution of baking soda before allowing the cistern to fill with water, or by filling it with well water and allowing it to stand a few days before emptying it, repeating the treatment if necessary.

Where this condition has not been anticipated and the cistern has been allowed to fill with rain water, the quality of the water may be improved by dissolving one to two pounds of baking soda in a gallon of water, then thoroughly mixing it with the cistern water. It is not advisable to drink water treated in this way because it is highly alkaline, but it may be used for washing.

Even with a cistern carefully constructed and well cared for, it is not always possible to keep the water free from undesirable substances. Soot from coal does not wash easily from roofs so even though the first part of each rain is drained away, the water collected in late fall and early spring is likely to be dark; it gets very dark during winter. This color may be removed by adding carefully prepared solutions of alum and soda to the water to form a sludge which when it settles will carry the color with it. Wrong proportions will give poor results. Make two solutions as follows:

1. Dissolve ¾ pound of baking soda in 1 gallon of water.
2. Dissolve 1 pound of alum (potassium aluminum sulphate crystals) in ½ gallon of water. ("Filter alum"—aluminum sulphate—is cheaper than alum and may be used if available. Use ½ pound to ½ gallon of water. Do not use "burnt alum.")

Determine the amount of water in the cistern by multiplying the area by the actual depth of water (1 cubic foot contains 7½ gallons). For each 30 gallons of water add ½ pint of solution No. 1 and stir. Next add ¼ pint of solution No. 2 for each 30 gallons and stir again. Allow 24 hours for the precipitate to settle to the bottom, after which the water above the sludge will be clear.

The relatively small amount of sediment precipitated will not cause inconvenience. Mud and dirt that always accumulate in a cistern should be removed, together with the precipitate, once each year when the supply of water is lowest.

In spite of precautions leaves, mice, toads or insects sometimes get into cisterns and produce bad odors. The proper thing then is to clean the cistern thoroughly. In dry years, when this waste of water may be serious the water may be deodorized and made satisfactory for all uses except drinking, by treatment with chloride of lime which burns up the odor. Mix about a tablespoonful in a porcelain, glass or crockery dish with two or three tablespoonfuls of water; rub the lumps with the spoon, then add about a quart of water. Stir thoroughly and pour the solution into the tank, mixing it intimately with the cistern water by stirring with a long board or a paddle. If this treatment is not adequate, use a second or a third dose if necessary. This chemical is harmless unless used in excessive amounts.

A spring may often be the source of a water supply. When higher than the point of use pipes placed below the frost line may be laid so the water will flow by gravity. Though the system usually works well, trouble is sometimes met by the gradual reduction and finally stoppage of the flow. Generally this is because air has collected, little by little from bubbles in the water and lodged in some irregularity of the pipe. The less the grade, the smaller the

WATER SUPPLY

pipe and the slower the flow the greater the likelihood of such trouble because the bubbles will not be carried down by the current but gradually work back to a high spot.

To avoid difficulty the pipe should not be smaller than 1". Stand pipes may be tapped in at high spots to allow the air to escape. The entrance to the pipe should be 6" or more above the bottom of the spring well and be protected

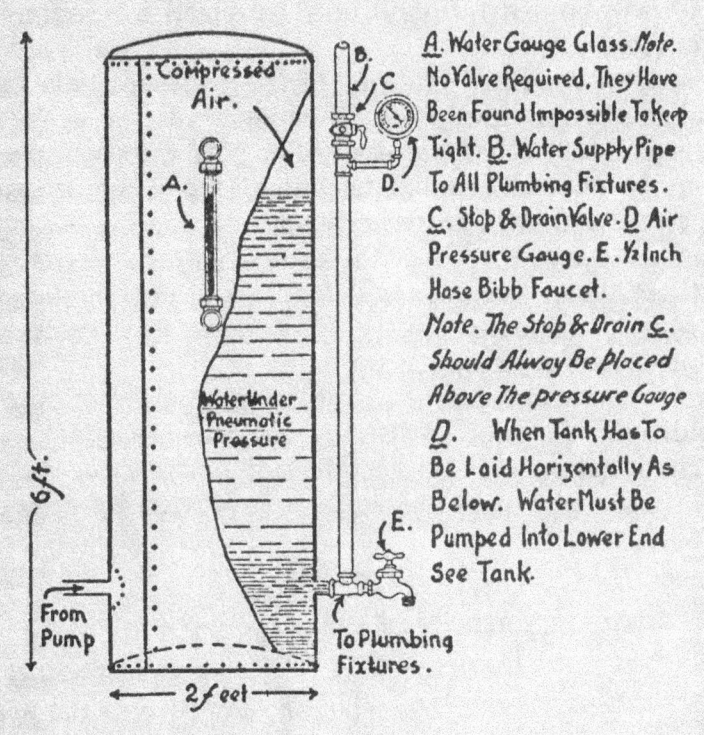

Fig. 13. Erect pneumatic pressure water tank.

by a screen to prevent anything but water getting in. The pipe itself should be laid as straight as possible, with neither dips, rises, nor avoidable angles. All joints should be well leaded and screwed tight so as to cover all the threaded parts.

That this kind of system works well where conditions are favorable I can fully attest because one place owned by a cousin and another rented by my father when I was a boy were supplied by springs considerably higher than the house. The water was conveyed in "pump-logs" and in the latter case the pressure was great enough to throw a stream over the house. It had its source more than half a mile distant.

The pressure or pneumatic water system is doubtless most popular of all used in farm homes because it is efficient, simple to install, easy to operate, reasonable in cost and adaptable to a wide range of conditions. The one I installed on my farm to supply both house and barn (150 feet away) was pumped by hand until an electric company supplied current. (Fig. 13.)

As water is pumped into one of these closed tanks the air is condensed to a fraction of its original volume, so its pressure is increased. When thus half filled with water the pressure is about 14 pounds to the square inch above the pressure of the open air and will so be indicated on the pressure gauge attached to the tank. When the tank is two-thirds its capacity full the gauge pressure will read about 28 pounds and at three-fourths capacity about 42 pounds, the usual maximum pressure for ordinary service.

Where conditions are favorable the hydraulic ram * or the hydraulic "engine" is usually the most satisfactory and least costly apparatus to install and operate for pumping water. (Fig. 14.) The machine is a device by which the

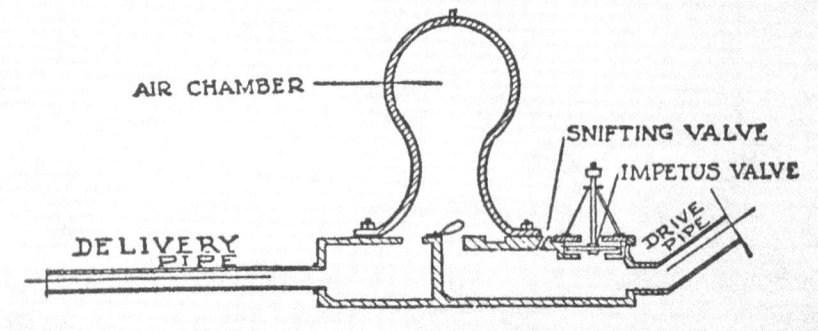

Fig. 14. Diagrammatic section of hydraulic ram.

momentum of a column of water flowing in a pipe forces a small part of itself to a higher level than its source. It may be used wherever water in sufficient volume can be piped from its source to a lower level of several feet. Some styles

* Much of what follows is condensed from Kentucky Experiment Station Bulletins.

WATER SUPPLY

are built to use impure water for operation but pure water for delivery.

Water flows through the drive pipe to the ram and out through an impetus valve. It gains velocity until its momentum closes the valve, whose sudden closing forces the moving column of water to open a valve in the bottom of an air

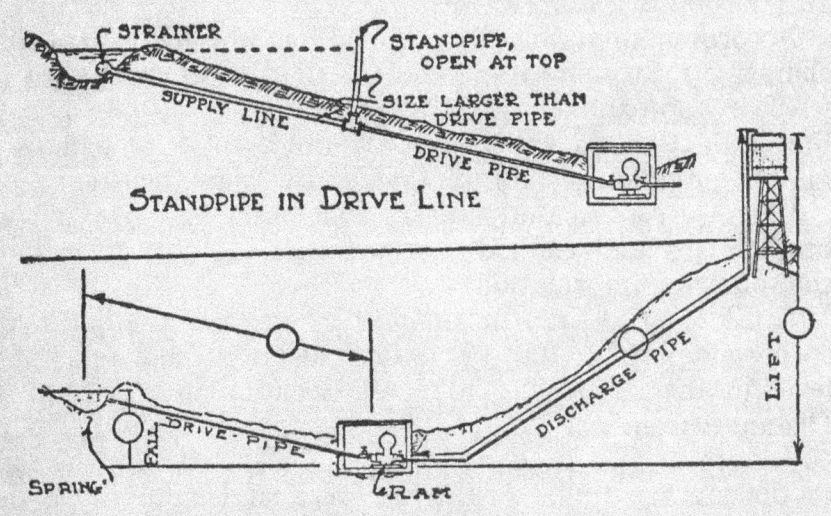

Fig. 15. Relation of hydraulic ram and pipe lines.

chamber, drive water in and compress the air until the water column has spent its energy. Then the air-chamber valve closes and the compressed air forces water through the delivery pipe to the reservoir. When the air-chamber valve closes the water column in the drive pipe rebounds slightly, removes pressure from the impetus valve which then opens and re-starts the cycle. This repetition continues as long as the drive pipe is full of water. In order to work the following conditions must be met:

The fall from the supply to the ram must be at least 2' and the minimum flow 1 gallon a minute; but the fall necessary depends on the volume of water needed and the height it is to be pumped.

The fall may be measured with a carpenter's or a surveyor's level or with a garden hose used as a siphon from the source to the point where the ram is to be located. In the last case the hose must be lifted at its lower end until the water no longer runs out but stands at the opening which is pointed upward. The distance from this point to the ground is the height of the fall.

The lift or height to which the water is to be raised should not be more than 10 or 12 times that of the fall. This height is the vertical distance between the level of the ram and the top of the reservoir.

According to the ratio of the fall to the lift a ram discharges approximately 4% to 15% of the water supplied to it. The approximate quantity discharged a minute equals the product of the fall in feet by the number of gallons a minute supplied to the ram, divided by twice the lift.

When writing manufacturers concerning the size of ram needed, cost and method of installation, be sure to include the following information:

1. Flow of spring, or number of gallons a minute the spring supplies. 2. Fall or vertical distance from the water level in the supply basin to the location of the ram. 3. Distance from the source of water to the ram location. 4. Length of pipe needed to deliver the water from the ram to the storage tank. 5. Lift, or vertical height above the level of the ram to the level of discharge at the reservoir. 6. Maximum number of gallons of water needed daily to meet all demands.

Should a ram work imperfectly or stop any one of several causes may be responsible. If the flow is insufficient in the drive pipe, close the gate valve on that pipe at the ram until enough water has accumulated; then open it and start the ram by pressing down the impetus valve, allowing the water to escape, then permitting the valve to rise again. After several such repetitions the ram should operate automatically. By shortening the stroke of the impetus valve the ram may be made to use less water. This may be done with a simple adjustment. Sand or twigs or leaves sometimes get into and clog the impetus valve. Thorough flushing and cleaning both valve seats is indicated. When the valves become worn they must be replaced. Usually they last for several to many years if the water is free from sand or soil.

Perhaps the commonest cause of stopping is insufficient air in the air-chamber. Pressure in this compartment makes the water absorb air, so unless new supplies are taken in the apparatus will stop. Rams generally have snifting valves to take in air with each stroke but sometimes these become clogged. They are easily cleaned with a small wire. With each stroke they suck in a little air which rises in the air-chamber and thus maintains pressure.

16

SEWAGE DISPOSAL

IT IS now a well-established fact that all excreta contain myriads of seeds which in growing will destroy all animal and vegetable matter, converting them into liquids and gases. These seeds ... which are known by various names such as germs, microbes, bacteria, etc., war upon each other relentlessly. ... If we give the friendly ones a chance they will kill off the unfriendly or disease-producing kind. ... Therefore, we should help our friends by giving them such surroundings that they can rapidly multiply and destroy not only sewage but the disease-producing enemies.

DR. ELLIS M. SANTEE,
In Farm Sewage.

ON the first farm that I bought there were no "conveniences." So as we had long been accustomed to city conditions we installed a pressure water system, plumbing and a septic tank. Though neither my son nor I had ever worked with concrete or dynamite or seen a septic tank we dug the pit, blasted out some boulders, built the tank and laid the drains ourselves according to specifications in Dr. Santee's book.

As the only place we could locate it was in the vegetable garden we reinforced the cover with such quantities of scrap iron that heavy horses plowed across it in safety. The installation is in its twentieth year and has always worked perfectly. It has never needed to be opened. In fact, to quote Dr. Santee, such a tank, properly built and not abused should never need to be cleaned out during the lifetime of its builder. Unlike the cesspool (which fills up and must be replaced or cleaned periodically) the first cost is the only cost.

As the details of construction of this installation are no longer clear in memory, as some of them might not be of service to the reader and as improvements have been made in methods since mine was built I therefore synopsize the following paragraphs and reproduce the illustrations from experiment station publications, mainly from New Hampshire Extension Circular 98.

Where no municipal sewage system is available the septic tank is generally the most satisfactory way to dispose of household wastes. Anyone who will give thought and attention to the important details may build one. The standard type may be built as follows:

A plain, rectangular, tight box, preferably of concrete, 5' or 6' deep, 1½ times as long as the breadth, receiving only household sewage and liquid waste from the kitchen, which enters by an inlet pipe submerged as shown in Figure 16, which has a slow and regular flow to the outlet pipe 2" lower than the inlet pipe. The liquid (called the "effluent") which flows from the tank is but little discolored and carries no visible solid matter. The best method of its disposal is through ordinary drain tile laid as subirrigation. The cover, which must be tight and at least a foot above the scum which forms on top of the liquid in the tank, should be removable, at least in part, so as to allow access to the inside when necessary.

Figure 16 shows the action of a septic tank. Sewage enters from the house through a 4" glazed, tight jointed sewer pipe which should be as short as convenient, with a slope of at least 2½" and not more than 10" in 10'. It is undesirable to have the incoming stream enter the tank violently and thus disturb the contents. The fitting used at the entrance should be an elbow, a Tee or a Y-branch, as shown, so its opening will always be submerged, and so the incoming stream will be deflected downward. This direction of flow may be aided by plank baffle boards, as shown. A similar pipe fitting, placed at the discharge end 2" lower than the inlet pipe, should be vented as shown in order to allow the gas to escape.

SEWAGE DISPOSAL

As the sewage flows slowly through the tank the solid matter settles to the bottom and leaves an almost clear effluent which is disposed of by subirrigation. Bacterial action within the tank converts most of the solid matter (called sludge) into liquid or gas, thus reducing its volume until only a trifling amount is finally left at the bottom.

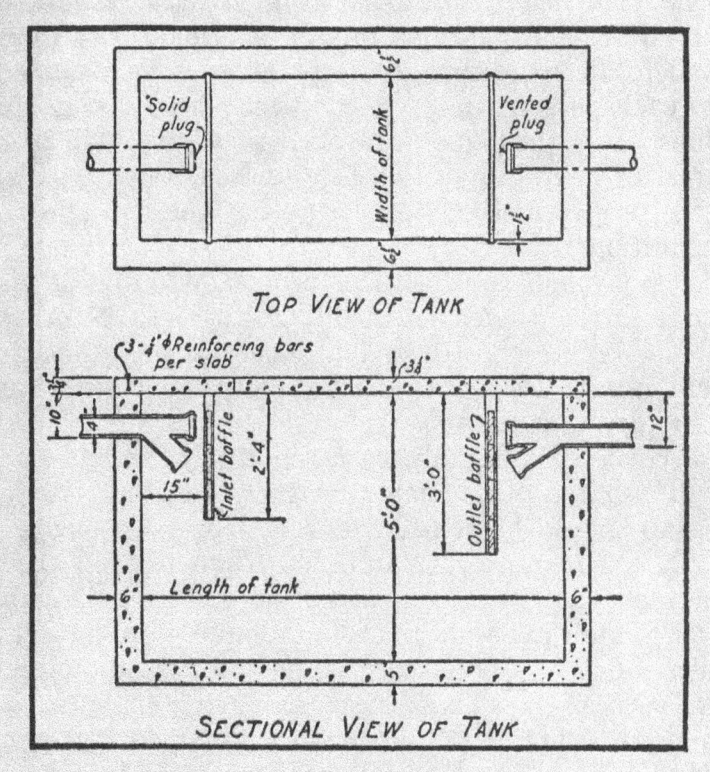

Fig. 16. Dimensions and forms for single chamber septic tank.

Though the effluent is generally clear, it must be disposed of carefully or it may give offense or even cause disease if it comes in contact with a water supply. The best method of disposal is by means of land tile laid as follows:

Tile not less than 3" in diameter is laid in a trench 18" to 24" deep, with open joints, protected over the top with tarred paper to prevent the entrance of soil or sand which would tend to clog the pipe. The slope of the tile in the disposal bed should not exceed ½" to the foot. This requires careful laying to secure best results. A 1" x 4" grade board nailed to stakes driven in the bottom of the ditch will prove helpful in obtaining a uniform slope.

By using boards 12' long and setting the down stream end 3" lower than the end toward the tank the correct grade may be easily made. The board is to be left permanently in position. The length of drain (which may be laid out in several branches) depends on the character and ability of the soil to absorb drainage. Light, loose soils require about 30' to each person who uses the tank; heavy, clay soils, as much as 75' to the person. Careful installation will be repaid in care-free operation of the disposal bed. Figures 18 and 19 indicate proper methods of laying out the tile under varying conditions.

In heavy soils it is sometimes necessary to dig the disposal trench a little deeper, back-filling with gravel or cinders about 6", before the tile is laid, thereby providing a more efficient absorption area. In such cases dry wells may be used to advantage.

The cover of the tank, cast separately, should be made carefully so the slabs may fit tightly together. Preferably each slab should be reinforced with iron rods, heavy mesh wire fencing or similar materials. When placing the cover on the tank (except the one or two to be used as a manhole) the top of the tank wall must be smeared with cement paste about as thick as rich cream; so should the adjoining edges of the successive slabs, except the manhole ones. A blue print of construction details may be obtained free from the the Portland Cement Association of Boston, Mass.

Both an inner and an outer form must be constructed, except where the ground is so firm that it will not cave in when excavated vertically. Where this is possible the outer form may be dispensed with. Brush the concrete face of the forms with crank-case drainings to make removal easy and prevent excessive warping and destruction of the boards. (Fig. 17.)

The concrete mixture must be water-tight, and capable of resisting the attack of sewage. Since the strength of concrete depends on the amount of water used, it follows that the kind and amount of sand and pebbles used is of secondary importance, so long as the mixture is plastic and work-

SEWAGE DISPOSAL

able. The two extremes of "soup" and "crackerjack" must be carefully avoided.

A trial mixture is recommended; namely, Portland cement, 1 bag; sand, 2 cubic feet; pebbles, 3 cubic feet; water 5 gallons when the sand is moist, 6 when dry. If this trial mixture is too sloppy, add a little more sand and a few

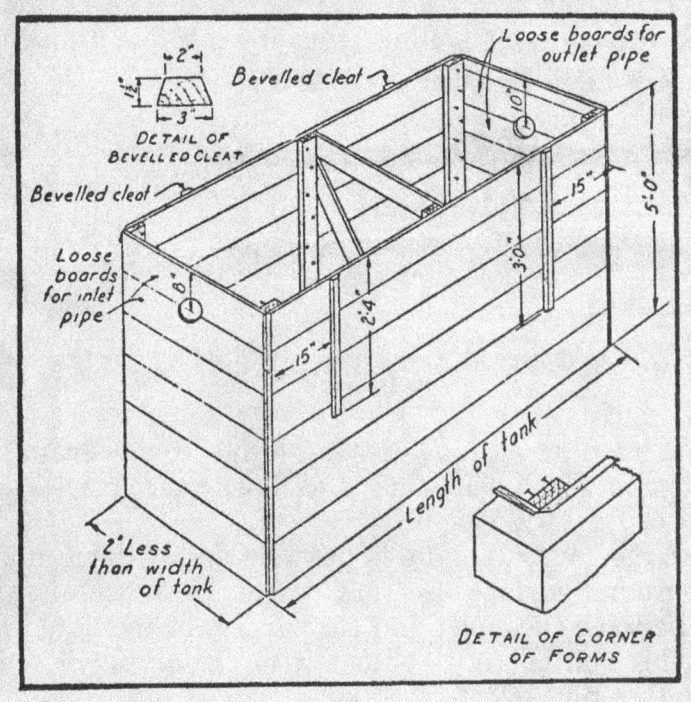

Fig. 17. Form for septic tank construction.

more pebbles until it becomes plastic and jelly-like; if too stiff to be workable, add a little cement and water. To make subsequent batches uniform, vary the amounts slightly until the proper proportions and consistency are attained.

Gravel as dug from the pit, called "bank-run," usually contains too much sand, so is not economical to use as found. It should be passed through a ¼"-mesh screen and the resulting sand and gravel re-combined in proper proportions as suggested.

Care must be exercised to use sand free from loam or decaying vegetable matter. When such is used it reacts against the cement and prevents the making of good, water-tight concrete. To test sand for organic matter dissolve three rounded teaspoonfuls of lye in one quart of clean water in a Mason jar, thus making a 3% solution of caustic soda. In another clean glass bottle place 2" of sand, add enough soda solution to cover the sand 1½" deep, shake well and allow to stand 24 hours. If the sand is clean the solution should be nearly clear, but if dirty it will be dark, even coffee color or darker. Such sand is unfit to use for concrete, though if no better is available it may be washed.

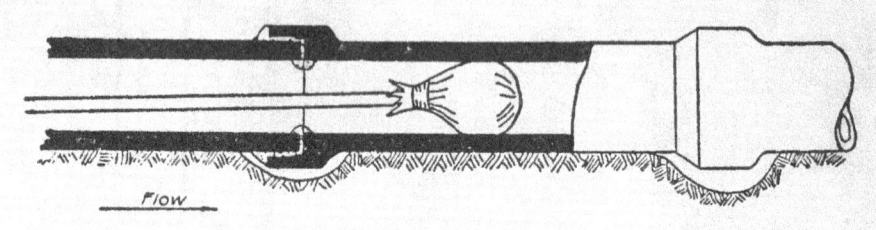

Fig. 18. Swab to remove excess concrete in vitrified tile line from house to septic tank.

Some further cautions: Never locate a tile drain, especially one in which "effluent" is to flow, where tree roots can reach it or they will clog it.

Make the cover as tight as possible and without any vent to the ground surface. Be sure that a gas space of at least 12" is between the cover and the surface of the liquid in the tank. This and the non-contact with the air favor uniform temperature and effective action.

If possible have the tank wholly below the ground surface with at least 12" of earth on top of it. If this cannot be arranged otherwise, heap earth over it to even more than that depth to assure uniform temperature.

Never connect a septic tank to eaves or "leaders" or other excessive flow of water; all such water must be excluded.

Never connect factory wastes to the tank as they might interfere with operation. Consult a sanitary engineer in such cases before making connection.

Locate the tank as near the house as possible so as to avoid chilling the water unnecessarily during cold weather. The exit drain may be extended as far as necessary or be branched.

SEWAGE DISPOSAL 81

When branched the branches should start from a distributor box or tank made preferably of concrete.

Always make the drain pipe from the house to the tank with vitrified tile with joints tightly packed in concrete or melted tar to prevent leakage. If the sewage is to contain

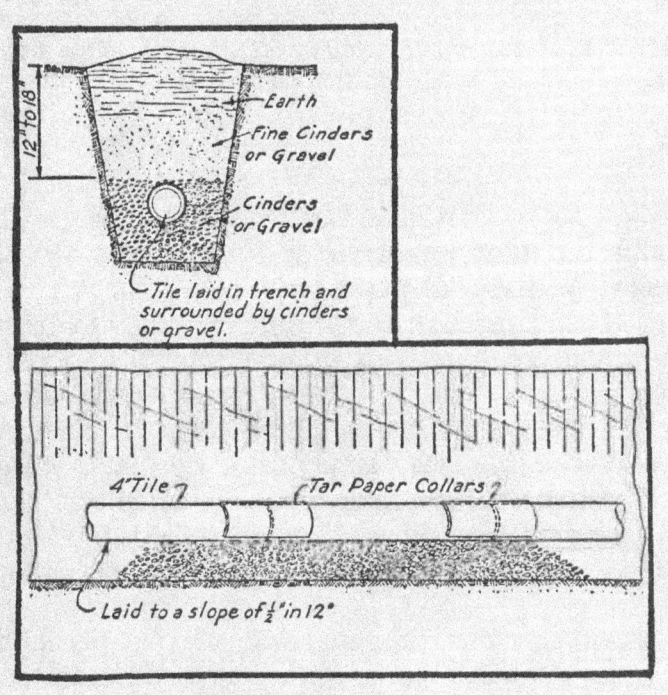

Fig. 19. Discharge drain from septic tank.

grease the tile from the house should lead first to a "grease trap" in which, rather than in the tank, this material will collect so it may be removed.

Various state boards of health have issued regulations which cover the following typical points: Effluents of septic or other settling tanks, or similar sewages, shall not be used to water growing vegetables, garden truck, berries or low-growing fruits, or to water vineyards or orchard crops when windfalls or fruit lie on the ground; and no effluents, sludge or screenings shall be permitted in ditches or pipes which may be used to irrigate such crops. Nursery stock, cotton and field crops may be watered with effluent provided that no milch cows are pastured on the land while moist with sewage, or have access to ditches carrying such.

17

FUNCTIONS OF WATER *

Any soil can supply plants with all the water they need, and as fast as they need it, so long as the moisture within the soil is not reduced below one-third of the whole amount that it can hold.

F. H. Storer,
In *Agriculture*.

Though many factors are essential to plant growth, perhaps the most important is water. No other plays so many roles. It dissolves plant food in the soil; carries these solutions to and through the plants; supplies hydrogen and oxygen which combine with other elements to form sugar, starch, oil, plant tissues, and many other compounds; keeps plant cells distended, thus enabling them to perform their functions; regulates the temperature of plants (and incidentally of the air) by transpiration from the foliage; and carries food constituents and soluble plant products from part to part inside the plants for storage, assimilation or growth.

The amount of unassimilated water living plants contain at any time is much greater than that of all other constituents combined—from 60% to 95% by weight. This indicates the importance of an abundant supply of water because these percentages must be maintained in order that plants may live.

Growth demands still more than the percentages mentioned. In addition to the amounts assimilated, a constant *current* of water must be maintained from the soil through the plant, then into the air as vapor. This amount is enormously greater than that assimilated or than that in the plant at any one time. In hot, dry, windy weather some plants transpire in 24 hours amounts of water equal to or even greater than their own weight! The amounts are also influenced by such factors as character of foliage (large or

---

* This chapter consists mainly of excerpts from the author's *Modern Guide to Successful Gardening*.

small, unprotected or protected by hairs, thick skin, etc.), by character of climate, size and development of plant and amount of available water supply in the soil.

Studies of plant growth have proved that each pound of "dry matter" in a mature crop has required from 200 to 900 pounds of water to develop, the usual range for common crops in the Northeastern States being from 300 to 500 pounds! During the growing season an average crop requires from 1 to 2½ pounds for each square foot! If this were present on the surface all at one time it would mean a depth of 3½" to 9"! Hence the necessity of an adequate supply of water for plant development.

Water exists in the soil in three conditions; hygroscopic, free and capillary. *Hygroscopic water* is of no importance in crop growing because apparently plants cannot utilize it. It is neither moved by gravity nor by capillarity since it is not in liquid form.

Free water, that recognized by sight and touch, fills hollows during rains and sinks into the ground to the level of "standing water" (the "water-table"). In swamps this is at or above the ground surface; in shallow and undrained soils perhaps only a few inches below; in deep and porous soils often many feet beneath. This "free" water supplies springs and wells but is of use to plants only when near enough for capillarity to draw it up to the region of plant roots. When too near the surface it may injure cultivated crops by saturating the ground, for then it excludes air, makes the soil cold, prevents instead of favors nitrate formation, encourages formation of compounds poisonous to plants, retards decomposition of organic matter, and so on. It is corrected by drainage (Chapter 18).

Capillary water, the thin film of water that surrounds soil particles, is the main source of water in plant growth (bog and water plants excepted) and is the most effective form for dissolving and holding soluble plant nutrients. As its movement is from moist to less moist parts its general direction is upward from the water table to the air. The distance it can move depends largely upon the character and arrangement of the soil particles and the air spaces.

Soils of close structure (clays, adobes and heavy loams)

when cultivated properly usually have increased air spaces, greater capillary water movement and water-holding power than when mismanaged. In loose and coarse soils (sandy and peaty) the air spaces are already great, so stirring decreases capillary movement and water-holding power.

You may control the capillary water supply by decreasing water losses from the soil, by increasing water-holding power (draining off superfluous free water), and by direct addition of water (irrigation, Chapter 19).

One of the most important ways to prevent loss of water is by tillage (Chapter 33). The other is mulching. As originally understood, a mulch (formerly spelled mulsh) is an application of manure or any other loose material such as leaves spread upon the soil surface to protect the roots of newly planted trees, shrubs, tender plants, etc. Today it is extended to include earth kept loose by surface tillage to check evaporation.

Though this last has become almost universal, some scientists doubt that a soil mulch alone lessens evaporation because, they declare, stirring the surface brings moist soil from below into contact with the air and wastes the water contained in it. Thus, they claim, evaporation is increased. Doubtless this is true when tillage is both deep and frequent but where only the surface inch or less is stirred by weekly raking, better crops are produced than in adjacent ones in which the practise is not followed.

When such loose materials as buckwheat hulls, shredded corn stover, chopped straw or granulated peat moss are available and cheap they are more effective than loose soil in checking moisture losses from tilled areas and when dug into the soil they add humus. For mulching trees and berry plants coarser materials may be used—marsh hay, straw, leaves, corn stalks, etc.

The most recent type of mulch is specially made, impervious black paper spread upon the ground and between or through which young plants are set. It is sold direct by the manufacturers and through garden supply stores.

It has been so successfully used by Hawaiian sugar and pineapple growers that experiments in the United States and Canada have been conducted, mainly with vegetables, because it is claimed that this material will shed water into the soil, conserve water in the soil by checking evaporation, increase germination, greatly reduce or even eliminate weeding and cultivation, increase soil temperature, hasten maturity, increase yields and produce larger, higher quality, cleaner crops. But results have varied so widely that the question is still open. Hence, to obtain additional comprehensive information, A. E. Hutchins conducted investigations during three years to determine effects of mulch paper in comparison with those of clean culture and to secure incidental information. His conclusions (slightly condensed) are quoted from Bulletin 298 of the Minnesota Experiment Station as follows:

1. A beneficial effect appears to be exerted by the paper. In the experiments the increases obtained do not, in most cases, appear to pay for the additional cost. 2. Mulch paper seems most beneficial with warm-season crops. 3. It seems to hasten maturity of certain vegetables and may be profitable with crops that have a relatively high market value.

4. It also appears to be most beneficial under conditions unfavorable to the optimum development of the crop such as poor soil, deficient precipitation and low temperature. As there is no precise way by which climatic conditions can be predicted in a given locality, its value from this standpoint can be determined only after the growing season is past. Under favorable growing conditions, often little beneficial and sometimes a detrimental effect is produced.

5. Its effect varies with local climatic conditions, with each crop grown and, to some extent, with different varieties of the same crop. Therefore each grower must determine the value of paper for his particular crops and for his local conditions. 6. Warm-season crops of high acre value and yield grown intensively are most likely to give the best results. 7. Paper should not be used on low value crops. 8. It largely eliminates weeds in the covered area and thus conserves the moisture and fertility they would use. It also cuts down the cost of cultivation but this item is offset to a large extent by the added cost of laying and caring for the paper and by the additional labor involved in planting and transplanting when it is used.

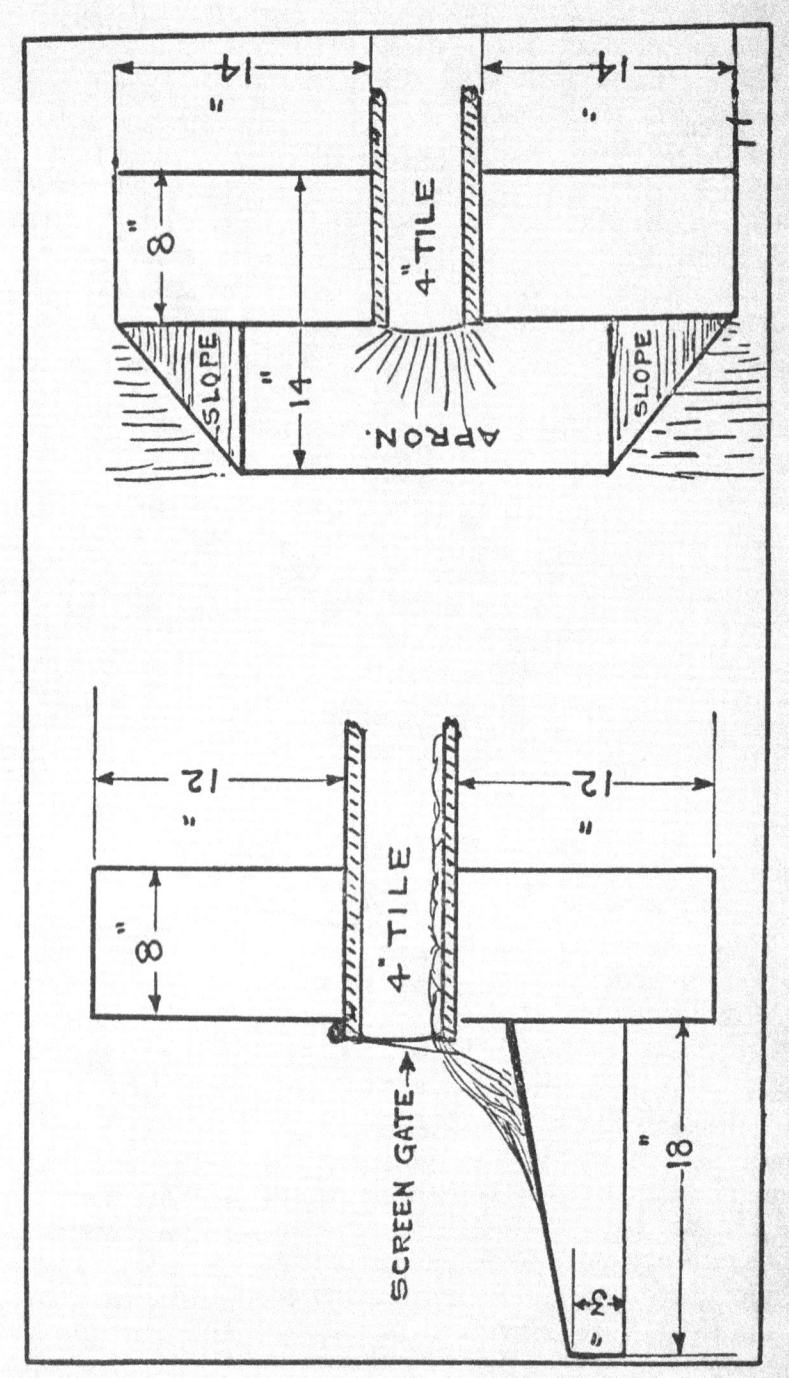

Fig. 20. Head wall, screened drain outlet and apron to prevent erosion.

FUNCTIONS OF WATER 87

You may increase the supply of capillary water in the soil by drainage, fertilizing, adding humus and by watering or irrigation.

Soluble salts in fertilizers in the surface foot of soil tend to draw water from lower levels to dissolve them and to dilute the solutions. As evaporation of the water in these solutions is constant at the soil surface water flows upward to maintain a balance. Hence a fertilized surface soil is normally moister than an unfertilized one.

Doubtless the most feasible way to increase water supply in the majority of soils is to maintain the crumb-like structure that favors the ascent of capillary water from below. The most important factor in attaining this end is not mechanical operations such as plowing, digging or cultivating but humus in adequate supply. Tillage operations permit particles of humus to wedge themselves between the mineral particles where they act as sponges to soak up and hold water raised by capillarity from standing water below. Thus humus becomes the most important reservoir of water for plants (Chapter 26).

In spite of all cultural precautions so far discussed to assure and conserve water supply artificial methods of watering are often necessary or at least advisable, especially in localities where long dry spells are common. (Fig. 28.) Some of these methods (watering pot and hose) are capable of exceedingly limited serviceable application; others are infeasible under various specific conditions (furrow systems on sloping ground); and still others are applicable anywhere that water is needed (overhead irrigation).

In view of the figures already quoted, the limited use of watering pot and hose is evident. Moreover, each entails work that may be avoided by more adequate methods. As ordinarily used—sprinkling only the surface—these two garden accessories do more harm than good. Mere sprinkling encourages the roots to develop near the surface where they are almost certain to be injured by summer heat, especially if watering is neglected for a few days in dry weather.

If the same amount of water were applied once a week instead of a little every day the soil would be soaked (theoretically) many times as deeply, the roots would develop deeply and the plants benefit in proportion. Still better than hose is irrigation. (Chapter 19.)

18

DRAINAGE

THERE is a false notion . . . that where water does not lodge on the surface of a soil, it is "dry enough." . . . *Stagnant moisture,* either in the surface or subsoil, is highly injurious—ruinous to fruit trees. . . . All soils, then, not perfectly free from stagnant moisture, both above and below, should be *drained.*

PATRICK BARRY,
In *Barry's Fruit Garden.*

UNLESS soil is well drained there is no use trying to produce profitable crops. Yet it is astounding how many farmers and fruit growers and even truck farmers try to make a living off such soils. Undrained lands are not merely wet but cold and often acid. Because of the wetness they cannot be worked nearly as early in spring as well drained lands; when seed is sown it germinates poorly, unevenly or not at all; such plants as start develop roots near the surface and when summer comes they suffer because they cannot then reach water which by that time is at a much lower level. Crops are therefore late, poor and unprofitable.

Where land is not naturally well drained, artificial drainage is a necessity. It is often the one factor that makes all the difference between bankruptcy and profit.* From a financial standpoint, underdrainage can be considered as a long-time investment. Tile drainage of agricultural lands is a comparatively expensive improvement and the capital expended in drainage work cannot be recalled or transferred, but owing to its permanent nature a properly installed drainage system should continue to return dividends for many years. Just how many is problematical, but in a general way

* From this point forward parts of what follows in this chapter are condensed from Bulletin 360, by F. L. Ferguson, Director of Drainage, of the Ontario Department of Agricultural Engineering.

DRAINAGE 89

tile drains may be said to last a lifetime. Underdrainage, like most other long-time investments, will often pay dividends sufficient to return the original capital in two or three years.

All soils not naturally drained require drainage. These usually are cultivated areas with fair surface drainage but with heavy subsoil; heavy clay soils with little or no surface

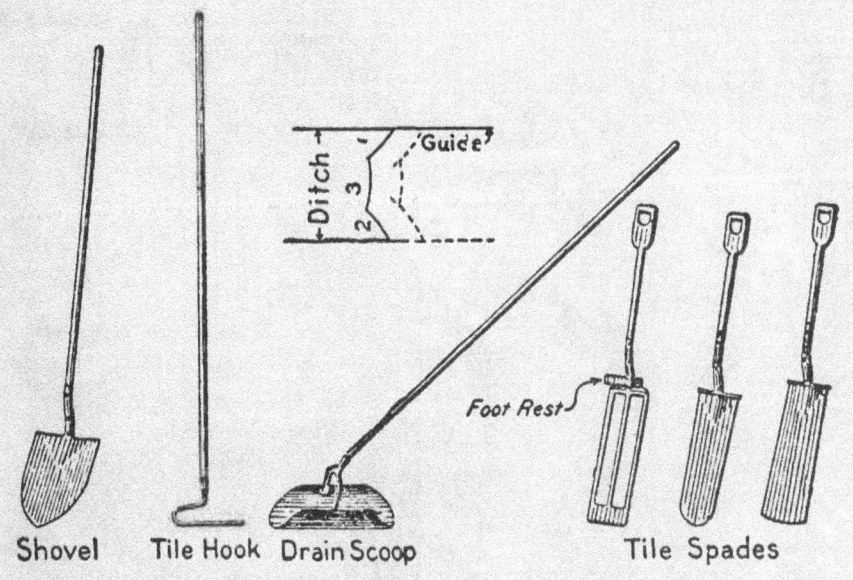

Fig. 21. Popular tools for ditch digging. Above, correct way to dig ditch.

drainage; rolling areas with impervious subsoil; areas, large and small, saturated long enough each year to destroy the physical condition of the soil and to interfere with spring seeding and harvesting operations; pot-holes and swamp areas.

Drainage benefits crops by permitting earlier spring seeding; making the land more easily worked; encouraging higher percentage of seed germination; making the soil more favorable to root development, hence providing more moisture and plant food (Fig. 25); reducing winter killing; controlling weeds more easily; and by increasing crop yields.

Whether the area to be drained is large or small, whether one drain or many may be needed, it is generally advisable to have the area surveyed and a plan drawn to scale. Such

90   FIVE ACRES

a plan will show the proper locations for all mains, submains, laterals and silt basins; the number and size of tile for each drain; the possible grades and depths; the best possible outlet or outlets and the locations of obstructions, buildings, trees, fences, etc.

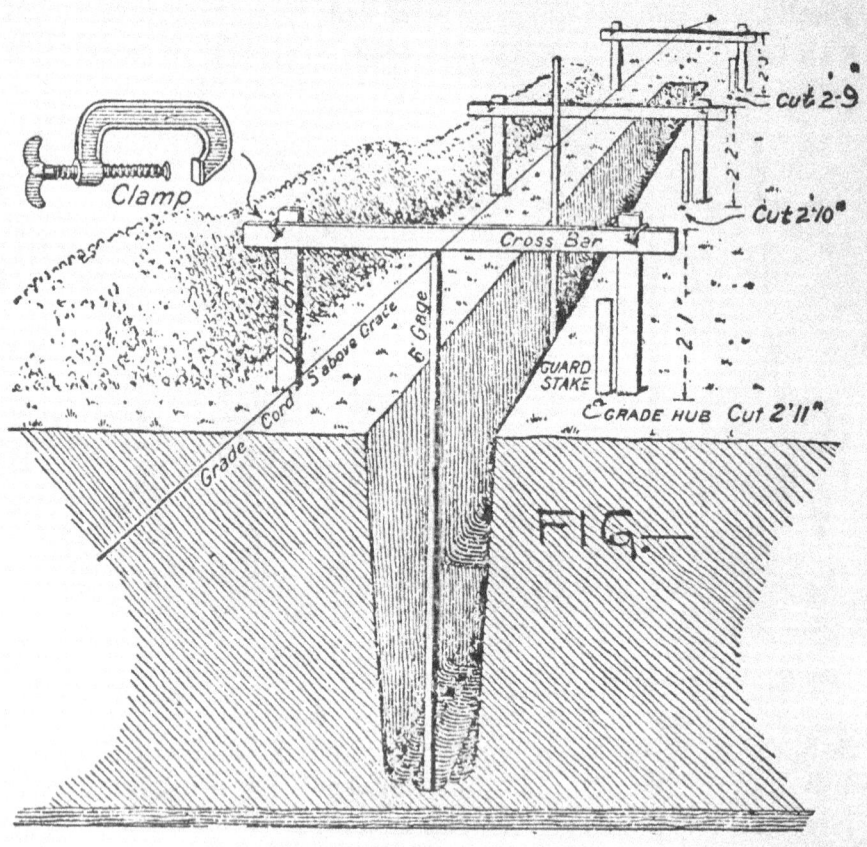

Fig. 22. Drain digging. Survey stakes (right) with gage frame posts beside them. Cross-bars made level. Gage rod and cord used to check depth.

With this information the farmer will be in a position to put in as much as his time, help or financial conditions will permit. The main drains should be installed first and the others added at will. Further, he may call for bids on a contract and have a basis upon which to calculate the cost of the system. Since the same area often lends itself to several different drainage schemes, the most economical may be chosen.

DRAINAGE 91

Figure 24 shows three methods of draining the same 10-acre field in which it is assumed that the slope of the land is equally favorable to each system and under each will be equally well drained. However, as figured out Plan A (herring bone system) costs $45.98 an acre; Plan B $42.59 and Plan C $41.65. Moreover, Plan B is more difficult for the machine operator to install because of the number of short

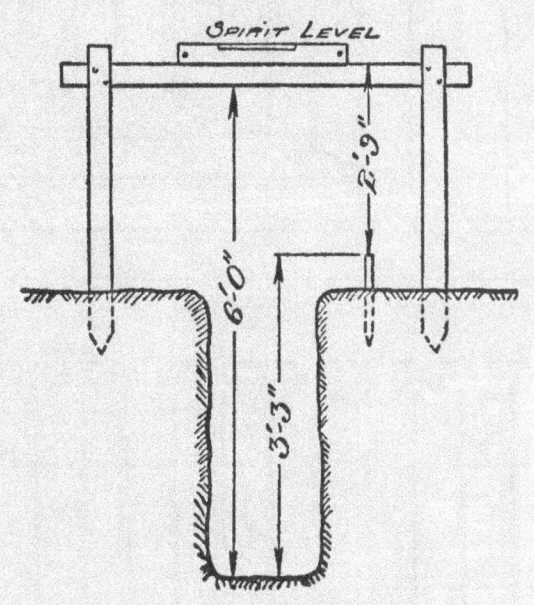

Fig. 23. Gage and line method of ditching.
Use carpenter's level to level cross-bars.

drains. A study of these plans shows the advantage of having a drainage survey made by a competent engineer, whose duty it is to install the most efficient and economical system possible.

The cost of drainage work varies, first, with the system required—in some cases a "system" is necessary, in others merely a few drains; second, the nature of the crops to be grown; third, the kind of digging; fourth, the location of the tile plant; and fifth, whether or not power machinery is available.

Tile may be installed either by hand or by a ditching machine. The latter, when properly operated, is quicker, more efficient and usually more economical. When hand work is

92   FIVE ACRES

necessary, one can make use of an ordinary plow, a ditcher, a ditching machine, or other digging machinery now on the market. In any case the following points should be kept in mind:

The grade must be uniform so there will be no depressions

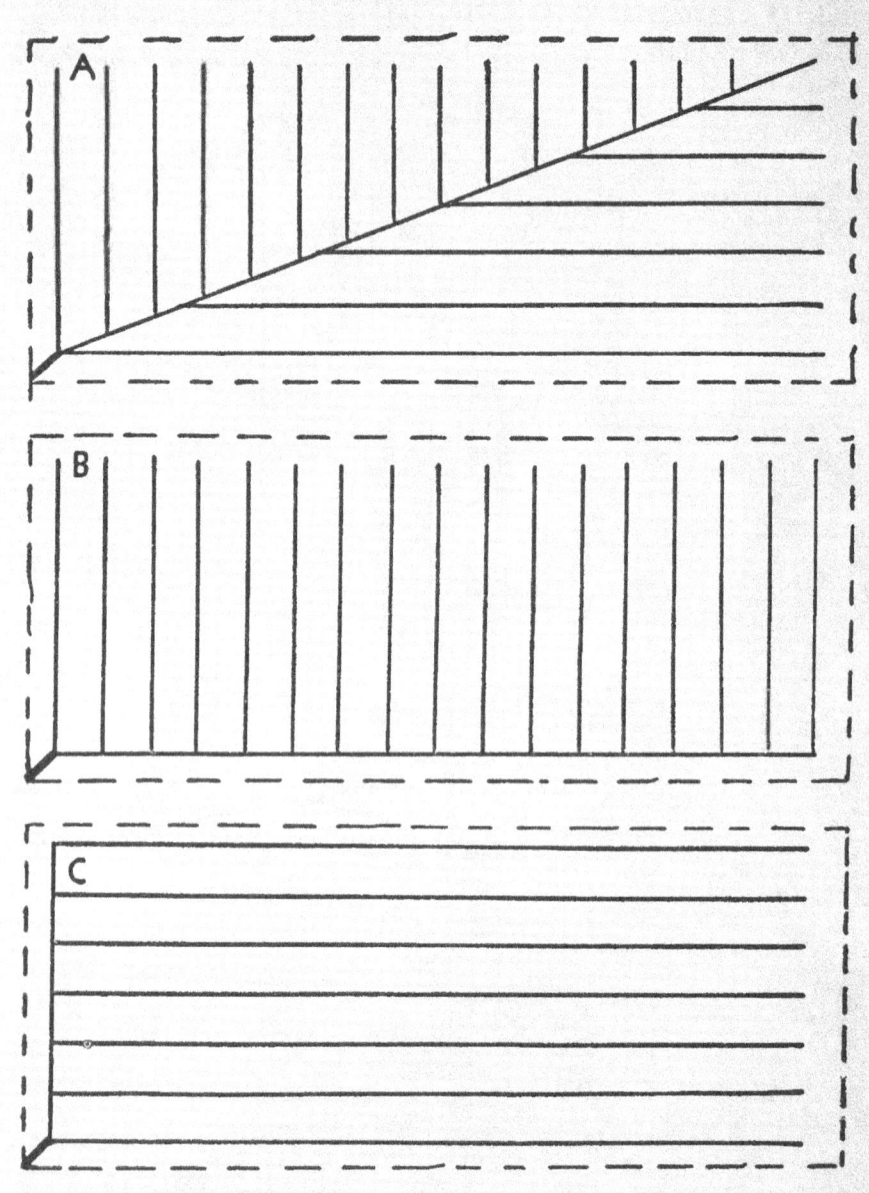

Fig. 24. Ways to lay out drainage system. (See text.)

to collect sediment; a solid bottom is essential—tile laid on muck or other soft material are likely to shift out of alignment and to obstruct the flow of water; stones and other obstructions encountered in the trench must be removed, the holes being carefully filled and tamped to give a solid bottom; only the best tile should be used; blinding and backfilling must be carefully done to prevent the breakage of tile

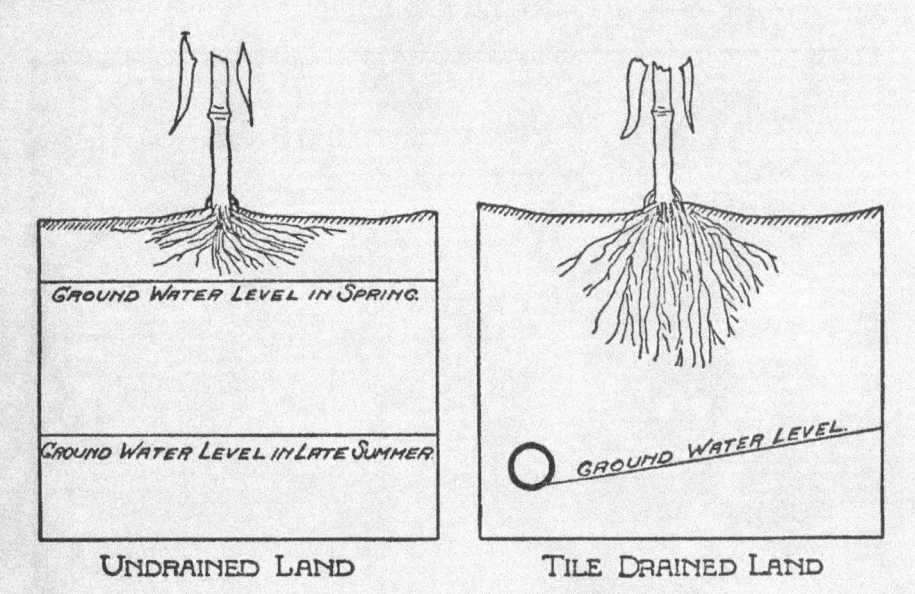

Fig. 25. Effects of drainage on root development.

by rolling in stones or by horses stepping in the trench; junctions and outlets must be carefully constructed; and trees likely to block the tile with their roots must be removed.

Ordinarily, tile should be laid as closely together as the cut ends will permit. In heavy clay soils a little opening is not objectionable, but in sands or sandy loams it is necessary to fit them closely together. In the latter it is often desirable to cover the upper half of the tile with tar paper to prevent the sand from entering. Whether tile are laid by hand or with a hook is not important so long as they are kept in proper alignment and well graded. Junctions should be carefully constructed in order that no obstruction may offer resistance to the flow of water. A few inches of soil, preferably the surface soil, should be placed over the tile as soon as they are laid to make sure they will not shift by accident. Backfilling should be done as early as possible because soil often becomes baked after rain or in other ways is hard to move.

Depths and distances apart of drains depend almost entirely upon the nature of the soil. The lighter the soil the deeper and farther apart; in heavy soil they must be placed closer together and somewhat shallower.

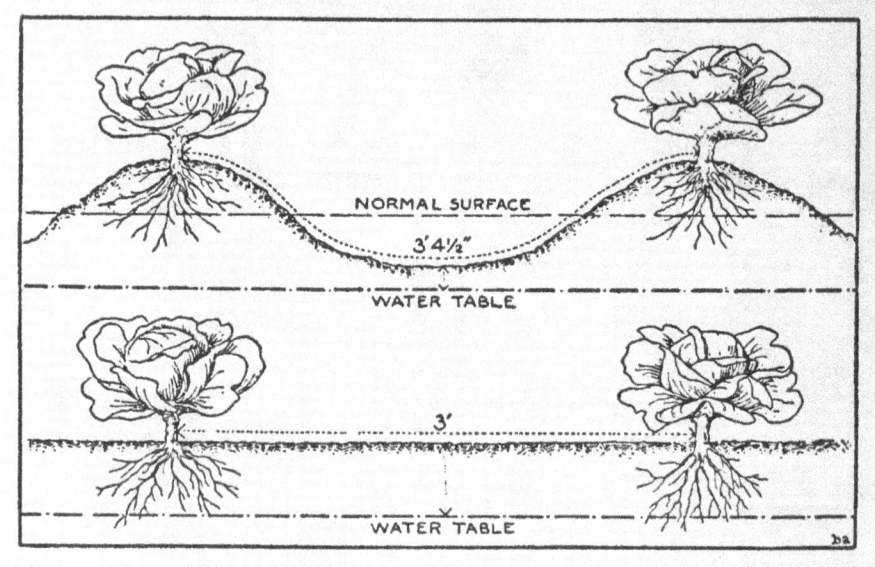

Fig. 26. As the upper diagram shows, plants in raised beds and "hills" usually suffer for lack of water except in poorly drained soils. The lower diagram shows roots nearer water supply.

Of all problems, quicksand is perhaps the worst. The following suggestions may be helpful: Undertake quicksand drainage during the driest part of the season; if possible, after opening the drain into quicksand leave it until the water drains out so the sand may solidify and the drain be more easily completed; in some cases it is better to remove the last foot or so by hand, as action of the digging wheel seems to provoke the trouble; sod thrown in the trench and around the joints is practical for short distances; straw, sawdust, shavings, gravel and cinders are often used to good advantage; cemented sewer pipe may be found practical in some cases; a silt basin (Fig. 27) should be constructed on

DRAINAGE

the line of tile as soon as possible after passing through the quicksand area and the deposits of silt removed when necessary.

Tree roots seldom interfere with a drain unless it carries seepage or spring water during the dry season. All trees such

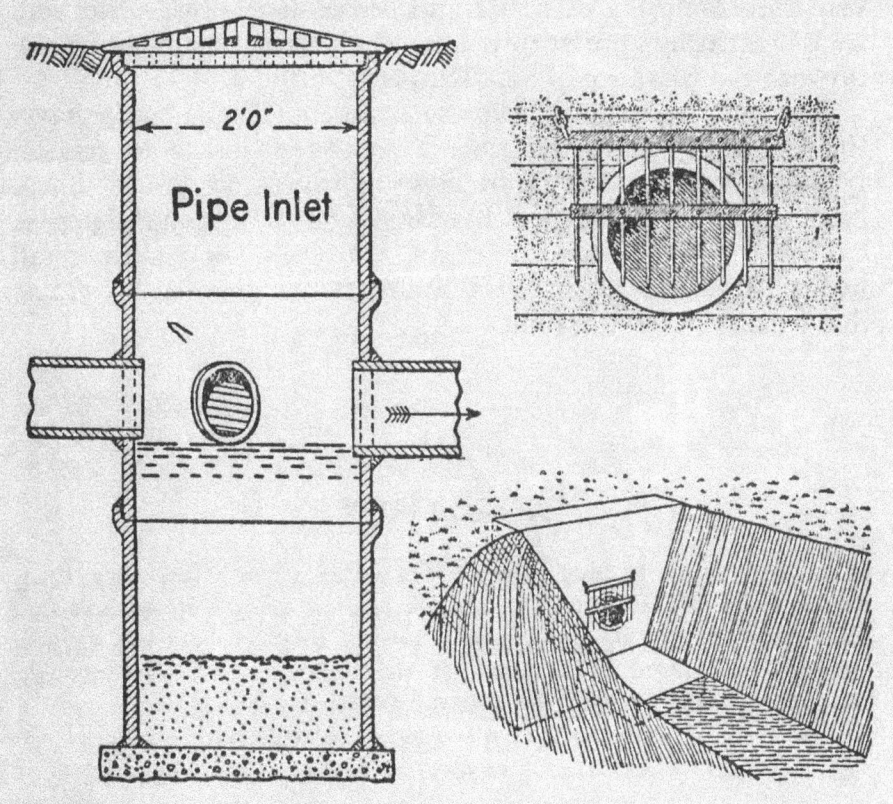

Fig. 27. (Left) Silt well and storm water catch basin. (Right, below) Head wall and outlet. (Right, upper) Screen gate, ¼" rods spaced 1½" between centers.

as willow, poplar, soft maple, elm and elder bushes should be removed from the location of a drain. They are likely to clog the tile with roots. Where it is desirable to leave a shade tree, cemented sewer-pipe should be used for at least 50' on each side of the tree position. In orchards and permanent crops, cut-off drains should be installed to remove all seepage water as this has a tendency to feed the tile drain during summer and thus give trouble by root development in the tile.

96   FIVE ACRES

Always the outlet of a drain should be well protected. The last 8' or 10' should consist of a piece of iron pipe or sewerpipe. A retaining wall may be a necessity; so may be a concrete or stone block to spread the water and prevent it from eroding the earth at the exit. Protection of the mouth to prevent the trampling of cattle and horses is also essential and so is a grating, preferably hinged or hung from above, to prevent the entrance of small animals. (Fig. 27.)

Tile may be made of clay or concrete. They should have the following characteristics: Smoothness inside to reduce friction and promote rapid flow; hardness to assure durability, good shipping and handling qualities; good shape—cylindrical—not warped; clean cut ends to assure good fitting; freedom from burnt limestone to prevent breakage due to slaking of the lime.

19

IRRIGATION

IT IS much wiser to give plenty of water once every few days, than a little each day. The latter method keeps the surface moist, and the roots naturally rise to the moisture, so that they are near the surface and will be injured by the heat of the following day. Give enough water to go deep or else just enough to wash the leaves.
EDITH LORING FULLERTON,
In *How to Make a Vegetable Garden.*

WHETHER or not it will pay to irrigate is a problem which varies widely according to character and drainage of the soil, topography, crops, temperatures, costs, time of application, water supply, quantity of water needed, relation of irrigation to tillage methods and, perhaps most of all with the differences (timeliness and untimeliness) of rainfall.

In arid regions irrigation is a necessity. It includes dams, reservoirs, canals, ditches, flumes, furrows and other features that have to do with supplying, conducting and applying water to immense tracts of land at about ground level of the fields. Usually it involves large scale engineering problems

IRRIGATION

beyond the scope of this book. The only application of these methods with which I have had experience is the furrow system by which attempts were made before the advent of overhead irrigation, to grow strawberries in Michigan. As both the texture and the quality of the fruit were impaired the trials were abandoned. On Long Island I have used the overhead system in strawberry, raspberry, and vegetable growing with gratifying results, mainly because it assures

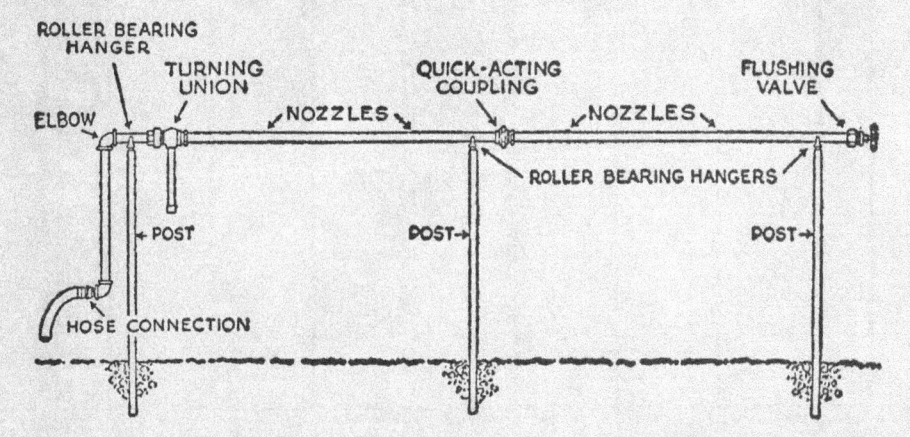

Fig. 28. Portable or temporary overhead irrigation system.

better control and economy of water than does the furrow system.

Some form of overhead irrigation is applicable to every conceivable condition.* It may be either stationary or portable, hand operated or self-operating, laid on or (preferably) raised above the surface to suit local conditions. Its first cost in any case is not much greater than an equivalent of hose and nozzles but it will last almost indefinitely, whereas hose usually must be replaced in two or three years. (Fig. 28.)

The essential feature of the various systems is the specially constructed "nozzle." The one I have throws a stream $\frac{1}{32}''$ in diameter. When the water pressure is turned on full and when there is no wind the minute drops into which the streams breaks fall about 40' away from the nozzle. Under

*Part of what follows is quoted from the Author's *Modern Guide to Successful Gardening*.

greater pressure 50' is often reached. By changing the pressure, reducing the flow or turning the nozzle at various angles and in nearly opposite directions breadths of 80' to 100' may be covered. For a small additional outlay a water motor operated by the current of water will do the turning automatically.

To irrigate vegetables and berries the best way is to place

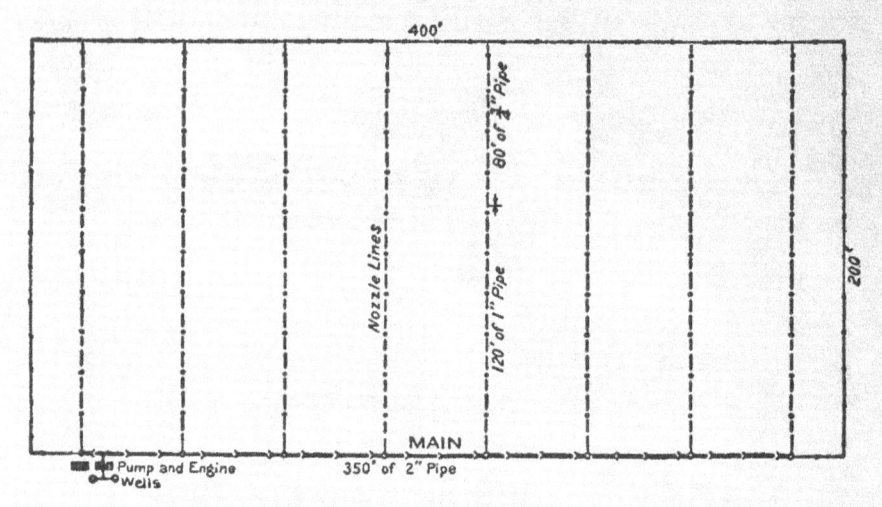

Fig. 29. Typical lay-out of overhead irrigation system.

straight lines of pipe 50' to 70' or 80' apart depending upon the pressure and with nozzles at 4' intervals. They may be laid on the ground but will work better if raised. When only a foot or so above ground they are inconspicuous and easy to step over, the iron post supports cost less than do those 6' high, are less likely than tall ones to be shifted by winds, more convenient and easily removed and replaced for plowing or digging. Tall ones, which are easy to walk under, throw the water farther, so are often better than low ones, especially for low pressures of water. Manufacturers supply wheels or rollers to facilitate turning the pipes.

For small gardens one line may be made to serve by having it in readily portable sections. Each line may be connected with a hydrant by hose or have its permanent supply pipe with a gate valve and a turning union to control the water distribution. For convenience, however, a water motor

IRRIGATION

is far better than a handle because the whole area will be sure to get an even distribution of water.

Overhead irrigation in some form has the great advantage over all other styles in its applicability to every type of soil, every elevation, every size of garden or field. As the water is evenly distributed in minute drops it sinks slowly in the soil without puddling or baking and neither seeds nor plants are injured. Also the gentle showers cleanse foliage and encourage healthful, vigorous development.

In view of all these advantages, low cost, durability of the equipment, positive insurance against damage to plants by dry weather and time saving such a system will effect, you should certainly provide overhead irrigation, choice being made after considering the specific merits of several styles. If your conditions suggest any difficulty do not hesitate to ask manufacturers or their agents for suggestions as to the best type of apparatus and equipment to use. When you do so be sure to give as complete data as possible so these people may understand the whole problem. It is to their interest as well as yours to advise you correctly.

The most significant experiments in irrigation that have come to my attention are those two-, to six-year ones reported by Erwin and Haber in Iowa Experiment Station Bulletin 308, from which the following statements have been condensed. The data should furnish a basis upon which estimates as to costs and benefits may be calculated.

Where the normal annual precipitation is ample for the production of vegetable crops irrigation must be regarded as crop insurance. The question is, then, how often do drouths injurious to vegetable crops occur and how much may irrigation increase yields?

From the standpoint of moisture supply the summer months are usually the critical time for vegetable production. A study of United States Weather Bureau records of a typical locality shows that summer drouths usually occur on an average of once in three years, that the range of duration is 15 to 45 days, the average 26 days and that the average deficiency of rainfall during drouths is 2.87".

The moisture supply in the subsoil at the beginning of a drouth has, of course, an important bearing on the effect of the drouth. [Hence the importance of drainage and deep preparation of the soil—Chapters 18 and 26.] It is generally accepted that a minimum application for vegetable production of ¼" of water, either from rainfall or irrigation is, in most instances, not the most profitable. In dry seasons at least double this amount may be applied to advantage.

The number of ten-day periods in the summer months having less than ¼" of rainfall is suggestive of the possible

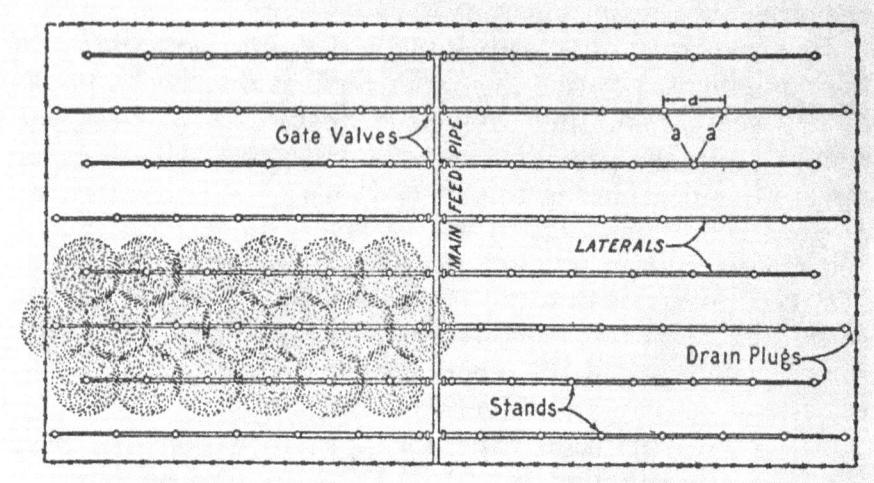

Fig. 30. Circular spray irrigation system showing staggered, equilateral arrangement at a, a, a, to assure best distribution of water.

need for supplementary irrigation. Four such periods occurred in each of four out of five years but only one in the fifth.

High temperatures (90° or above) which usually accompany summer drouths deplete the supply of soil moisture and increase the water requirements of plants. Hence, even in seasons of normal rainfall irrigation may be an important factor in tiding over temporary periods of high temperature. That such conditions are frequent is shown by the weather records which cover seven consecutive years. These or even higher temperatures are not necessarily detrimental except as they are accompanied by deficiency of moisture supply.

The chief sources of water for irrigation are ponds, reservoirs, artesian wells and municipal water supplies.

IRRIGATION

The size of the plants and the amount of foliage influence the desirable amount of water to supply. Shallow rooted crops, such as radishes and onions, usually require more frequent applications of water than do tomatoes or carrots.

Light and frequent watering is inadvisable; applications of less than ¼" is considered enough for seed-beds and young vegetables and from ½" to 1" for maturing crops.

An "acre-inch" (1" of water over an acre), requires 27,152 gallons or approximately 110 tons.

Overhead irrigation (Fig. 28) may serve as an aid to cultural practices in the following ways: 1. To tide over periods of drouth which might otherwise wipe out or materially reduce the yield. 2. To make it possible, through the control of moisture conditions, to produce larger yields. 3. To insure better quality of vegetables by making possible continuous growth, which is necessary to high quality. 4. To obtain a good start for vegetables transplanted into dry soil. 5. To obtain prompt germination at sowing time, especially of succession crops, which, if delayed, may miss the higher markets or be damaged by frost. 6. To make possible the preparation of ground which, without irrigation, would be too dry to plow or pulverize. 7. To protect certain crops from light frost injury. [Irrigation lines may also be used to apply fertilizers in solution and to wash undissolved fertilizers off foliage.]

Though the foregoing points represent the main advantages of overhead irrigation they do not necessarily mean that irrigation will pay. Against these advantages must be set the cost of irrigation, before a fair judgment of economic value may be reached.

Circular spray systems distribute water from nozzles fixed to the tops of upright equidistant pipes distributed uniformly through the field. They are supplied with water through the uprights, placed 30' to 50' apart on underground pipe lines. As the nozzles are placed at the corners of equilateral triangles (Fig. 30) they assure a more even distribution of water than if placed at the corners of squares or rectangles. Each lateral pipe line and sometimes each nozzle is controlled by a valve. The system is largely confined to light soil that takes water rapidly and to crops that will not be

FIVE ACRES

injured by coarse sprays. It seems more particularly adapted to irrigate fruit than vegetables and is used largely in citrus orchards with nozzles higher than the trees.

When irrigating it is important to moisten the ground thoroughly to the bottom of the root zone each time. On clay soils the application must be slow to prevent puddling and to allow the water to seep well into the ground. This demands more frequent turning of the lines than for light soils; hence the advantage of automatic turning. Nozzles

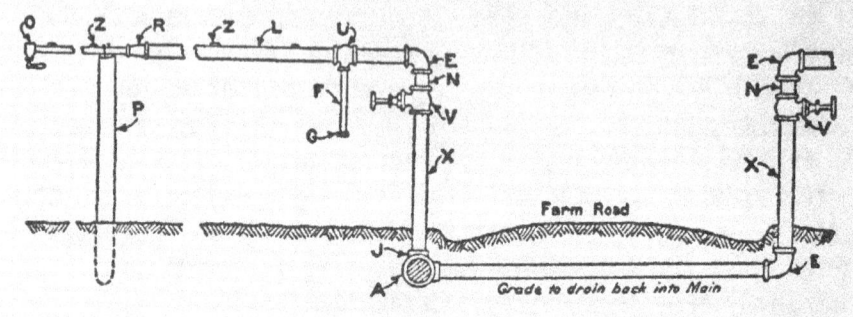

Fig. 31. Typical fittings for overhead irrigation system. A, Underground main. J, Tee with vertical and side openings. X, Upright. V, Valve. N, Nipple. E, Elbow. U, Turning union. F, Handle for turning irrigation line. G, Cap on handle. L, Nozzle line. Z, Brass nozzles. R, Reducer connection to smaller pipe. O, Cap or flushing cock. P, Post support.

smaller than usual are also an advantage in such cases. The ground should not be allowed to become dry between waterings and then soaked, but kept moist, not wet, all the time.

The actual time of operation varies widely from year to year due to variations of rainfall and the length of the growing seasons. Systems of one to four acres in the Middle Atlantic States may require pumping averages of about 100 hours annually; in the Middle West, probably 150 or more hours. Gasolene and oil to operate a three-horse-power engine would be about 7¢ an hour, or $7. for 100 hours.

As soon as the irrigation season ends the machinery should be overhauled, the pipes thoroughly drained before winter, and repairs made when necessary. Before starting operations each spring the pipes should be thoroughly flushed out to get rid of any loose rust particles. Repairs and overhauling should not cost more than $10. annually for a system of one to four or five acres. If properly handled the engine and the

IRRIGATION

pump should last for 12 to 15 years and the piping still longer.

As adequate presentation of construction, water sources, engines, motors, pumps, pressures and other details would demand excessive space the reader is referred to U. S. Department of Agriculture Farmers' Bulletin 1529.

The initial outlay for equipment for the Iowa experiments, exclusive of pump and power, was estimated at $400. Interest on the investment at 6% would be $24. an acre annually and the annual charge-off for depreciation (20-year period) at 5% would be $20. The item of repairs is practically nil. Aside from an occasional adjustment of the water motors which turn the sprinkler lines, there were no repairs. The cost of water per acre-inch was arbitrarily placed at 10¢ per 1,000 gallons or $2.72 per acre-inch. By readjusting the above figures to fit any local conditions, a grower may determine his irrigation costs for a given season.

There are no hard and fast rules by which to decide when to irrigate. An examination of the soil is helpful in determining moisture conditions; but the best guide is the plant itself. Checking of plant growth is a sign that moisture is needed; wilting is another sign.

Temporary wilting such as affects plants on hot days, is not necessarily a symptom of moisture deficiency. Plants will recover from such wilting if sufficient water is in the soil. But if they remain flaccid until early morning they indicate permanent wilting, so water should be applied promptly; for if the wilting is allowed to continue, the quality and succulence of the crop are likely to be seriously impaired and the yield greatly reduced.

The question is often asked: Is it injurious to irrigate while the sun is shining? There need be no hesitation in watering plants at any time they need water. But in overhead irrigation less water is lost by evaporation at night because of the lower temperature and higher humidity. [In windless weather also the loss is less than when high winds prevail.]

Irrigation experiments with porous, but tight-weave, canvas hose (Fig. 32) in growing potatoes, strawberries, small fruits and "garden truck," conducted by O. E. Robey of the Michigan Experiment Station (Extension Bulletin 133) have proved that the method is practical, profitable and useful in meeting special situations where soil conditions, crop, ground contour or capital investment would put other methods out of the question. The following statements condensed from the bulletin are of interest:

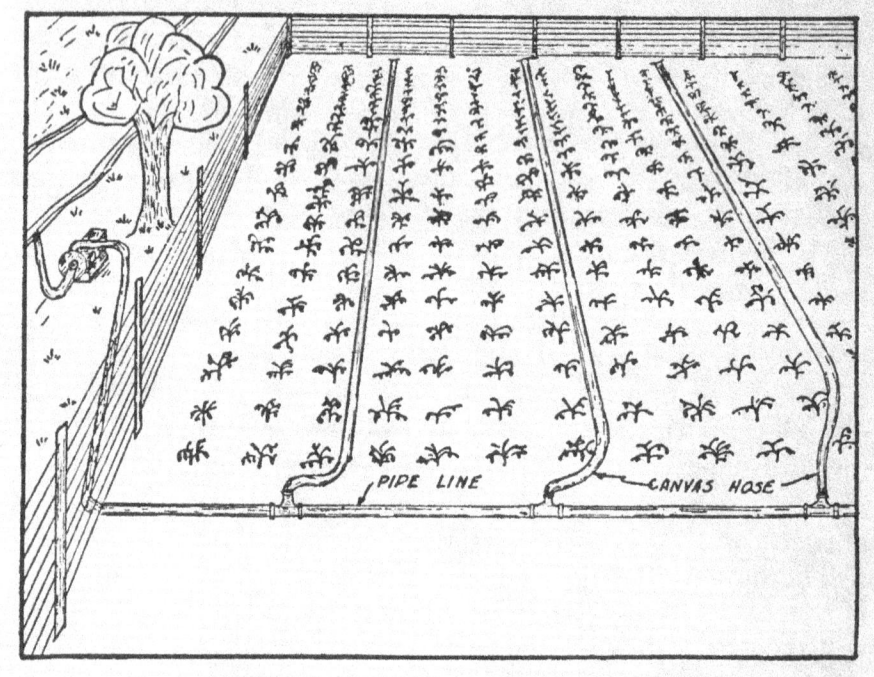

Fig. 32. Typical lay-out of porous hose irrigation system.

When the hose (closed at the end remote from the supply pipe) becomes distended by slight pressure the water oozes out through pores in small drops, not as sprays or streams, thus preventing erosion of the soil and wetting of the foliage.

The supply pipe is located, preferably, at the higher side of the field and a pressure of 15 to 20 pounds to the square inch maintained in the porous lines. Where the lift is not too great old fire hose will serve as temporary supply pipe, otherwise iron pipe will be necessary. Though the canvas hose may run up hill, a better distribution may be secured by having it run down. Lengths of over 600′ have worked well, especially when heavy weight canvas is used near the source and lighter at the distal and when the current is up grade and the reverse when it is down.

Though the hose in some of the experiments has lasted three years its durability has not been fully determined. Its life may be prolonged by treating with a solution of one gallon asphalt paint, ½ pint each of kerosene and gasolene thoroughly stirred before applying either with a brush or by soaking and running through a clothes-wringer to squeeze out the excess. It must be dried for 24 hours at least before using.

In use the hose is merely laid between the crop rows, the water turned on until enough has been applied, then moved to the next place. Soil conditions and methods of tillage will decide the width of effective distribution.

20

FROST DAMAGE PREVENTION

A GOOD deal remains to be learned about orchard heating, and even after the general principles are better understood than they are today, there will always be need of careful adjusting of the methods to the requirements of each particular orchard.

FRANK A. WAUGH,
In *The American Peach Orchard.*

THE killing of plants and plant parts by frost may often be prevented by simple, inexpensive, easily applied means. Before we discuss these, however, let us see what frost is and how it acts.

Frost is the term which indicates the conversion of a liquid into a solid by the reduction of temperature. Though this definition covers all cases such as the solidification of molten iron and other metals it is popularly understood to mean the formation of ice from water at the temperature of 32° F.

As air cools its power to hold water in vapor form decreases until it deposits more or less in tiny drops (dew) on objects cooler than itself, such as foliage. The temperatures at which this deposition occurs depend upon the proportion of water vapor in the air at the time. This varies as the cooling proceeds. During summer the dew point, as the deposition or saturation temperature is called, is often above 60°; in winter often below zero.

When the dew point is below the freezing point the water vapor condenses on still cooler surfaces in the form of fine particles of ice, which because they reflect the sunlight and appear white we call hoar or white frost. Often a similar appearance or "false hoar frost" occurs when the temperature is several degrees above freezing point. Usually this is due to the way the light is reflected from the dew.

The condensation of water vapor tends to check the fall of temperature because what is called "latent heat" in this vapor is returned to the air which, scientific instruments prove, becomes measurably warmer. Thus within variable limits the deposition of dew protects plants from frost damage as the temperature approaches or in some cases even goes below the freezing point. This accounts for many escapes of plants that would have been killed by frost had the dew point been below instead of above the freezing point. We can often take advantage of this phenomenon and save our plants, as explained further on.

Plants vary in their resistance to frost damage according to their origin and their condition. Those which originate in a cold climate (apple, cabbage) naturally are strongly resistant; those from a warm climate (orange, tomato) are weak and easily destroyed. Between these extremes are many intermediate grades. Those plants that, in a given locality, live through the northern or alpine winter in spite of deep freezing of the soil are called hardy in that locality; those that succumb to the slightest frost are tender, and those between are variously classed as half-hardy, semi-hardy, half-tender and semi-tender (fig, French artichoke).

Tender plants are usually injured more or less when the temperature continues at 32°, 31°, or 30° for several hours, especially when bright sunshine or quickly warming air strikes them in early morning. Few if any of them can stand lower degrees than these for even a short time; half-hardy plants often survive temperatures of 20°, but seldom lower degrees.

In all these cases plant condition plays an important part; for plants that have made rapid growth, have soft, immature tissues and are full of water, are far less resistant than those which have developed slowly, have denser, stockier growth and are less full (perhaps even in need of) water. This statement applies to hardy as well as tender plants; for even trees normally hardy in a given locality may be winter-killed because they made a late, sappy growth which did not ripen or which was full of water when cold weather arrived.

The mere deposition of frost on the surface of foliage does not necessarily indicate that the plants have been killed or even damaged. But when the air is too dry for dew to be deposited they may be frozen by a dry wind, or on a clear night without the deposition of either dew or hoar frost. In such cases damage is due to freezing of the water inside the plant and the consequent rupture of the tissues. When the sun shines on tissues thus injured the internal ice melts, the leaves have no chance to mend the broken cells, so the leaves droop, wilt and turn black; hence the term "black frost."

Anything that will prevent the fall of temperature to or below the freezing point and anything that will shield the plants from direct sunshine while they are still frozen, covered with hoar frost or severely chilled will help ward off damage or even save plants that would otherwise die or be seriously checked in their development. A wind that springs up in the evening, clouds that appear during the night or early morning, or a rain that follows a frost will often either prevent the freezing or check the thawing process and thus save the plants. All these phenomena of Nature are, of course, beyond our control but we may imitate them as outlined further on.

We can largely control the development and therefore the hardiness of our plants in several ways. For instance, during spring we can avoid over-feeding our young plants with stimulant fertilizers such as manure and nitrate of soda and also avoid giving them excess of water. Both these tend to make sappy growth easily killed by frost. On the other hand, by keeping the plants cool, almost cold ("hardening them off") as they approach the time for transplanting to the open ground we can increase the hardiness of hardy, semi-hardy and even tender plants. Plants so prepared will stand cold snaps whereas those of the same species not so inured would probably be killed or so chilled that they would "sulk" for several weeks before recovering or renewing a normal rate of growth.

Similarly we may prevent winter injury of hardy trees, shrubs and vines by supplying ample water during summer and early autumn, withholding it later, avoiding applications of stimulant fertilizers and manures from midsummer forward, counteracting any excess of these by liberal dressings of potash and phosphoric acid during early fall (Chapter 28) or by sowing buckwheat in July or rye in September or both these together in July. As these crops grow they remove excess water and nitrogenous plant food from the soil and develop plants which when plowed or dug into the ground in spring return the plant food and their own bodies to form humus.

Fortunately we can predict accurately enough for practical purposes when to expect frost. The daily forecasts by the United States Weather Bureau give suggestions as to the general weather to expect; but we can make our own observations and predictions. Local conditions influence temperature. For instance, a near-by body of water, such as a lake, the sea, a wide or a deep river, or even a large pond affects the rate at which air temperature changes. In spring, because the water is cold, it keeps the air also cold and thus more or less retards plant development. In autumn the reverse effect occurs; the water being warm not only warms the air but fills it with water vapor thus warding off frost.

Open and flat country and small villages are more likely to suffer from late spring and early autumn frosts than are large cities and their near-by suburbs because, in the former, heat loss by radiation into space is more rapid in clear, clean air and under cloudless skies than in the latter where the air is filled with smoke and dust and where the fires in countless houses, factories and other buildings directly raise the temperature.

Dark colored, sandy and well drained soils absorb and hold more sun heat than do light colored, clayey and poorly drained ones so are less likely to be frosty. Other conditions being equal, southern and eastern slopes are also warmer than western and northern ones because they more quickly absorb the sun's rays. Though this favors earliness of plant development it often makes the growing of certain fruits (apricot, peach, Japanese plum, more especially) precarious or impossible because the flowers are encouraged to open so early that spring frost kills them and thus prevents fruit production, though not usually killing the trees.

We can discover for ourselves that cold air, being heavier than warm air, flows like water from high to low ground and "settles" in hollows or "pockets" unless it can drain to still lower levels or be driven out by wind; that frosts are much more likely to occur when the air is still, the sky clear and the stars brilliant than when there is wind or clouds, especially when the former is strong and the latter cover the whole sky. The direction and force of the wind also help in making a local forecast. One that blows strongly from the north is far more likely to bring cool or cold weather than one from the south, just as one from the east is likely to bring clouds and rain and one from the west clear skies and colder weather. The rate at which the barometer rises also helps because it indicates the approach of clear weather and, if rapid, also of cold weather.

An unusually warm spell is almost sure to be followed by a cooler or cold one because our general weather moves in prodigious waves from southwest to northeast across the country. Hence a light frost following a warm spell is likely to do more damage than an even more severe one following cool weather. For this reason we should be on our guard when one of these warm spells occurs in spring—be ready to protect our seedlings, newly transplanted plants and the flowers on our fruit trees and bushes.

When the sky is cloudy, when there is fog or even when a haze occurs during or toward evening, frost is less likely to occur than when the night is clear because these conditions of moisture in the air prevent loss of heat from the earth.

A reliable sign of approaching frost is the rate at which the temperature falls during the late afternoon and early evening. Starting with 50° or less, clear skies and no wind, a fall of 2° or more an hour between four and eight o'clock usually indicates that freezing temperatures will be reached before morning unless clouds or winds develop or unless we do something to prevent frost.

In a small way individual plants may be protected by inverted flower pots, peach baskets, and other receptacles placed over them, by newspapers spread and held in place by stones or clods of earth. A more convenient adaptation of this way is to use a light screen of burlap mounted on a frame placed over the plants or beds. These all tend to hold the heat around the plants.

Smouldering fires which produce abundant smoke and steam form artificial clouds which check radiation in the same way as do true clouds. When the air is still the smoke spreads out evenly and proves effective as a protection nearly as far as the clouds extend. This method is infeasible where the smoke would prove objectionable to neighbors. Numerous small, bright fires of wood, coal or (preferably) oil are used extensively by commercial growers of fruit and vegetables to heat the air. They are less useful in small areas than the methods already presented and those that follow.

The most generally feasible method is to fill the air with water vapor in one of the following ways: Stirring the soil with the wheelhoe or the cultivator toward evening to expose an increased surface of damp earth; sprinkling the plants, the ground and the adjacent area with a hose nozzle that breaks up the water into small drops, or using an over-head irrigation system for this purpose. The water evaporates and as the vapor condenses it liberates latent heat and thus checks the cooling process.

Freezing of the ground may injure even established trees, shrubs and vines of some kinds, so anything that will reduce the depth of frost penetration or prevent alternate freezing and thawing will tend to prevent such injury.

Experiment has proved that under a sod freezing reached a depth of 8" whereas in an adjacent sodless area it reached 18". Peach trees on the sod ground made healthy, uniform growth whereas in the sodless soil they were slow to start, had many dead branches and made poor development.

In another experiment, just before winter a few forkfuls

FROST DAMAGE PREVENTION

of manure or shovelfuls of soil or peat were banked around the trunks of exceptionally vigorous peach trees with the result that every tree so treated came through the winter without injury whereas a few not banked all died.

In order to have extra early beans, corn, melons and cucumbers I have often sown seed much earlier than was locally popular, thus risking frost damage. When no frost occurred I was ahead of competitors and when it did come I usually saved the plants by one of these methods. I have never thus risked transplanting eggplants, tomato or pepper plants because, even though not frozen, they "sulk" if chilled and start fruiting late.

In case you have not protected your plants and a frost has occurred during the night you may be able to save them, even those covered with frost or whose tissues have been ruptured by freezing of their sap, provided you will spray them with cold water as soon as possible after dawn or before sunrise and also shield them from direct sunlight after the sun appears until they have thawed out and apparently resumed normal activity. Better keep them so shielded until between ten o'clock and noon. The most conveniently applied shield is the screen mounted on a frame with short legs.

When water freezes it swells and lifts the crust of frozen earth above the unfrozen ground below. As it does so in autumn and early winter it also lifts shallow rooted plants, roots and all. When it thaws the soil settles back but the plants do not. They are left with more or less root exposed. Each succeeding freeze lifts them some more and each thaw leaves them farther out of the ground with the result that they dry and die. Hence the importance of applying a mulch in the fall.

In the spring equally fatal results may follow unmulched plants because when the surface thaws above a lower layer of still frozen earth the thawed layer settles and when it later freezes and lifts it breaks the roots of small plants by pulling them. Hence, again, the importance of a mulch; for beneath a sufficient layer of such loose material heaving and settling are reduced to a minimum and thawing of the ground proceeds from below upward until the mulched soil has thawed out and thus eliminated danger of root breakage.

21

LIVE STOCK

THE animals of the farm should be regarded as living factories that are continuously converting their feed into products useful to man. A fact of great economic importance is that a large part of the food they consume is of such character that humans cannot directly utilize it themselves.

HENRY AND MORRISON,
In *Feeds and Feeding*.

WHETHER or not to keep live stock of any kind on the small farm is a question to be decided only after the pros, but especially the cons, have been carefully considered. Perhaps the most serious objection to the majority is that most animals require space that can usually be more profitably employed on such small places.

In many cases there is less objection to a team of horses, than to other live stock (except hogs) provided that the animals are to be used a large part of the time for plowing, hauling, cultivating, and other work. The "pleasure horse," especially in these days of autos, has no place on the small farm; his care and the area he requires, to say nothing of the cost of his keep, may be more profitably devoted to other purposes.

Even the work horse is open to the objection that though he may be a good worker there is not enough work to keep him busy a sufficiently large proportion of the time, he must be fed, groomed, bedded and watered almost as diligently while idle as when at work and given time-consuming exercise in order to keep him in good health. So far as the small farm is concerned heavy work such as plowing may be done more economically with a rented team or a tractor or even by a hired plowman or tractor owner, and light work mostly with a garden tractor or a wheelhoe. Hauling is more expeditiously and cheaply done with a truck.

A cow will require a minimum of an acre of pasture to support her, even though she will get much of the waste from vegetable and other crops. In addition to this, more or less area will be needed to supply her with fodder even when hay and grain are bought. On a ten-acre farm one acre in forage means 10% of the gross area; on one of five acres, 20%! These percentages are far too high to warrant keeping the usual cow for her milk and manure. The amount of milk needed by the family might better be bought so as to release her supporting land for the production of definitely profitable fruits or vegetables.

The cow is objectionable from another standpoint; she positively must be tended and milked at least twice daily, thus tying down someone to this duty. Neglect or postponement of milking may be fraught with danger of impairment of her ability as a milker, if not to the animal herself. Furthermore she should always be milked by the same person, not a series of experimenters who differ in their natures and abilities as milkers, for thus she may acquire bad habits.

On a farm large enough to warrant keeping a cow and where she may be properly managed she is a highly desirable animal to have; for if a good one she should not only produce ample milk for drinking, cookery, butter and pot cheese for the family and manure of special value in vegetable growing but profitably consume large quantities of unsalable produce and waste, thus converting worthless material into profit.

Though many a grade cow is a good milker, expecially if her sire be a dairy breed bull, pure bred ones of the dairy breeds are more likely to be reliable, so preference should be given to them. Among dairymen who supply city markets the Holstein is the favorite because she is noted for abundant production; but the quality of the milk is distinctly poor when compared with that of other dairy breeds. The Holsteins are also poor foragers. When turned out to graze the pasturage must be good or their yields of milk will decrease.

The Channel Island breeds (Jersey, Alderney and Guernsey) are famous for the richness of their milk but they require more care in stabling, dieting and handling than other cattle. Of these the Guernsey is probably the hardiest and most easily managed.

I am prejudiced in favor of the Ayrshire because for more than a century my relatives have been breeders of it. The animals are hardy, active, wonderful foragers, abundant producers of excellent milk which though it is less rich than that of the Jersey and less copious than that of the Holstein it is, in my estimation, more palatable than that of either. The best cow I have ever owned was a pure bred Ayrshire for which I paid (not to any relative!) only $75. and which supplied my family with all the milk, cream, butter, and pot cheese we needed, besides furnishing all the whole milk two other families wanted.

Sheep have no place on the small farm. As they require grazing range cheaper, rougher, untillable land should be devoted to them. Moreover they should be kept in moderate to large sized flocks in order to have enough high grade wool to interest buyers. Similar comments apply to lamb raising for meat.

Where enough untillable grass land exists on the small farm it is more sensible to keep a milch goat or two than a cow or sheep. In fact, a goat will thrive where a sheep would starve. As the animal is hardy it will largely shift for itself except in winter when only the most ordinary shelter and feeding care are necessary. Its milk is highly nutritious and is specially noted for cheese making.

The one domestic quadruped that best fits the conditions of the small farm is the hog. He requires smaller area than any other—merely a pen and some range, with shade and a good wallowing place. Except to provide him such items, good and ample feed, he demands no unreasonable time or attention. He is the most wonderful of all domestic animals as a converter of waste and worthless fruit and vegetables, even weeds into profitable meat. Hence his pet name, mortgage lifter!

It is not profitable on the small farm to breed pigs but it is highly so to buy young ones in the spring when only a few weeks old, feed them until fall and then convert them into pork, sausage, headcheese, scrapple, liverwurst, pickled pig's-feet, and other toothsome delicacies. By purchasing in spring and butchering or selling in the fall the farmer may be relieved of the care and cost of winter feeding—four to six months.

Rabbits and Belgian hares, kept not as pets but for meat,

are often profitable where most of the feed is produced on the place and where marketing conditions are favorable. They require good housing, exercise paddocks and preferably summer range on clover or alfalfa. During winter they need good hay with regulative rations of cabbage and other vegetables. As they are prolific breeders and grow rapidly they soon begin to return profits to the man who takes good care of them.

Personally I believe that no farm is complete without a dog, not necessarily a pure bred animal but one well trained and obedient. Among the 15 or 20 I have owned since boyhood the worst happened to be pure bred—poultry killers, garden destroyers, cow chasers and thieves. But their records are not mentioned to condemn good breeding! The best was a 50-50 Airedale-Collie—powerful, intelligent, well trained (thanks to his former owner, not myself!), a splendid watchdog and protector of wife and playfellow of the children, wonderful ratter and woodchuck destroyer and without a fault. No wonder he was stolen!

Though I strongly advocate having such a dog I as insistently urge that he be not allowed to feed and house himself as best he can but be given his own special, comfortable sleeping quarters, his own feeding and drinking vessels and that he be well fed at regular intervals. Surely he is entitled to expect these. "The laborer is worthy of his reward."

As strongly do I sponsor the cat. Nothing equals him (especially her) as a preventive and cure of mouse and rat trouble. But, more than the dog, the cat needs protection when outdoors. As the law gives hunters the right to shoot any cat seen in hunting territory he should be taught while still young to wear a conspicuous, fairly stiff, leather collar, but loose enough to slip over his head in an emergency. A bell is a good thing to add to the collar because he can ring it when he wants to go out or come in the house. (At least, ours does!) It will not interfere with mouse and rat hunting as he soon learns how to keep it mute! Good feeding will remove the need of killing birds. Training while young and whipping him across the face and head with any dead bird *he has just caught*, will usually cure him unless the habit has been allowed to form.

Properly managed and properly located live stock of almost every class except scrubs may be made to pay, but the small, general farm is usually not the place to keep them; also it is not the place, (few places are!) in which to indulge in "freak" animals such as skunks, frogs, muskrats, groundhogs, foxes, ornamental fish, snakes, alligators, guinea-pigs, squabs, songbirds, *et al.* So when tempted to "take a flier" in these or others, first consider the natural adaptability of the property to the special kind you have in mind; second, remember that no matter who declares "there's money in it," even though he can prove it with reliable records, nevertheless you are justified in concluding that if you yield and indulge you must be prepared to pay "fool tax" until you gain experience. You are much more likely to succeed with the ordinary domestic animals whose milk or meat you can surely sell than with freaks which by no stretch of the imagination can be considered as "staples."

22

POULTRY

THE best way to be successful with poultry is to start with a few hens, give them good care and comfortable quarters, and—keep both eyes open. By this I mean that one should study the matter in a practical way by familiarizing himself with the habits and requirements of his fowls, and observe the effects of different kinds of food on them.

EBEN E. REXFORD,
In *The Making of a Home*.

No matter how small or large your farm, you should keep hens. The outlay for good stock, equipment and feed and the amount of time and work to manage them properly are small in comparison with the advantages. Chief of these are consumption of much food (for instance, kitchen and table scraps) which, barring the friendly aid of a pig, might be otherwise wasted, the destruction of countless insects, the constant supply of eggs whose freshness cannot be impeached, an occasional chicken dinner and the production of appreciable quantities of highly concentrated manure.

Whether or not to keep a large enough flock to supply regular customers, or a road stand, is a question which thousands of small farmers have decided affirmatively to their profit. It must be said, however, that many failures have resulted from attempts to conduct the business on a commercial scale before the fundamentals have been mastered and before perhaps thousands of dollars have been spent in the hope of developing an egg farm to supply the city market. For the majority of people the profitable limit seems to be about 100 hens, more or less, the reason usually being that the owner can give these his personal attention without neglecting other features of his place; whereas large numbers demand hired help.

Under specially favorable conditions other branches of poultry keeping may be profitable but unless these conditions are assured they will almost certainly not pay. For instance, the presence of a lake or a river may lead you to think that duck and goose raising would be a good venture simply because these birds, being waterfowl, would forage for much of their food. But if you allow ducklings or goslings to swim there while they are small you may wonder what becomes of them until you perhaps see one pulled under water by a snapping turtle, a pike, or an otter, or some other voracious creature. My own small flock was reduced 75% in a mill race before I learned this fact! But I was only a boy then!

Commercial duck growing is a highly specialized business in which many single farms raise from 10,000 to 50,000 ducklings each season and place them on the market within 10 or 12 weeks of hatching. The only water these ducklings are allowed to have is what they drink. None but the breeding stock is allowed to swim.

Goose growing is not similarly specialized because, apparently, geese cannot be so closely confined but must have ample grass range on which to graze. Relatively small flocks are kept on pasture with no access to lake or river until the goslings are at least half grown, though they may have a small pond in which to swim.

Turkeys, unless reared by modern methods, are the most disappointing perhaps of all poultry to raise because the poults are delicate until after they have passed a stage of development called "shooting the red." The turkey hen is the worst fool of a mother imaginable. Unless she is confined in the morning and during rainy weather she will lead her brood through wet grass with the results that they get chilled and so weakened that many, if not all of them die. Then, too, the young turks are subject to diseases which are highly fatal. So turkey raising, though profitable to the specialist and the raisers of large flocks under modern methods, is a branch of poultry raising advisable for the small farmer to avoid.

Guinea fowls are the least tamable and most independent of all poultry. Once their wings are developed they will fly over anything—a barn, for instance! They largely take care of themselves and their young, are wonderful foragers, alert "watch dogs" giving alarm on the slightest suggestion of danger, and the young ones are delicious when they come to the table. Guineas are therefore in good demand at relatively high prices, but they must have free range, must be educated to feed at headquarters and to roost where they will be safe from prowlers.

Pigeons raised for squabs are also profitable when well managed but are often a nuisance because of the mess they make on roofs and walks. Like all other classes of poultry, except hens, they are for the specialist to keep—if the object is profit. For supplying the home table, however, a few pairs in a properly made cote should produce enough squabs to add a pleasing variety to the family menu.

Before you decide to raise poultry of any kind be sure you understand the importance of having stock bred to meet the purpose for which you plan to keep it. If all you care for is pretty fowls to stroll across the lawn choose among the many beautiful, ornamental fowls of large or bantam breeds. They will play their roles well without overburdening you with eggs or their carcasses. If you want abundance of eggs avoid mongrels and poorly bred fowls as you would the plague as they will "eat their heads off" without yielding an equivalent of anything; if you want poultry for meat, choose Brahmas, Jersey Giants, Langshams, Cochins or some other "meat breed"; if you want eggs regardless of meat, choose among the Mediterranean breeds—Leghorns, Spanish, Minorcas, Anconas, Andalusians; and if you want both meat and eggs pick among the English, American and Australian breeds—Rhode Island Reds, Plymouth Rocks, Orpingtons, Wyandottes and others.

Hatching chicks by hens, though still practised on many farms, is so inconvenient and the results are so unsatisfactory, especially from loss of chicks, that artificial hatching has largely replaced it, but small, lamp incubators, which were highly popular between 1890 and 1915, and still used, are being rapidly replaced by custom hatcheries which, in the aggregate, produce millions of "day-old chicks" and ship them successfully and satisfactorily by express, airplane and parcel post for hundreds of miles.

As a man who has raised chicks and ducklings with hens, owned and operated a lamp incubator and a home-made brooder and who has also reared day-old chicks, let me assure you that the hen method has nothing to commend it, that the lamp incubator is satisfactory if you intend to raise several broods in a season from your own hens' eggs, but that neither of these methods of hatching is as satisfactory as the day-old chick method.

The outstanding advantages of this method are great saving of time and labor over both the other methods, replacing mongrel and other inferior stock with pure bred and bred-to-lay strains, the practicability of hatching pullets early enough in spring to start laying in early fall so as to command high egg prices, to say nothing of the avoidance of investing money in equipment of adequate size to meet such requirements.

Artificial brooding avoids many of the disadvantages of natural brooding, especially fatalities due to lice, exposure to unfavorable weather, and the time, labor and annoyance incident to hen brooding. By means of stove brooders in portable colony houses it is as easy to manage flocks of 300 chicks (the maximum satisfactory number) as a small fraction of this number with hens, and to be surer of developing more uniform pullets.

Improved methods of feeding, largely the results of experimental work, have made poultry production both more efficient and more economical. Especially have these methods affected chick development and egg production, both of which have been enhanced and made more successful. Egg production has also been profitably increased by using electric light to lengthen the hens' "working day" to 12 or 14 hours between the equinoxes of September and March, through encouraging the fowls to consume larger quantities of food.

Opportunities to make egg production profitable, whether the eggs are sold in the open market, through local stores, roadside stands, or to personal customers, include selling eggs according to grade as to size, weight, shape, cleanliness, absence of blood spots, infertility, and freshness. The quality of the product may be maintained by sanitation, cleanliness of the poultry houses, runs and nests, gathering the crop at least twice a day in autumn, winter and spring, four times during warm weather, keeping them cool and away from direct sunlight to avoid deterioration, discarding all males except for breeding purposes so as to prevent fertilization and consequent rapid deterioration, and marketing at least twice weekly—preferably oftener.

On general farms the poultry flock is usually the most neglected, even abused, feature. In spring chicks are hatched haphazard by hens and suffer heavy losses through parasites, disease, accident, and destruction by hawks, weasels, owls, foxes, hogs and other enemies. During summer the flock roams at will, picks up most of its food—insects, worms, waste grain and weed seeds. In winter, being badly housed, cold and wet reduce its number and its production of eggs. Under such mismanagement eggs are plentiful only when prices are low and scarce when they are high. Hence there is little or no money in such "farm poultry." Yet no branch of farming is capable of making such quick turn-over of capital or such high profits, as may be seen from the following paragraph, the data in which are culled from Research Bulletin 83 of the Wisconsin Experiment Station which investigated costs and practises of poultry raising on Wisconsin farms.

Only 8 of the 22 farm flocks investigated produced more than 48 eggs a year per hen. The average cost in these cases was 25.4¢ a dozen. Eight other flocks, whose hens ranged from 24 to 48 eggs did so at a cost of 32.4¢ a dozen; and the remaining 6 flocks laid less than 24 eggs a hen at an average cost of 99¢ a dozen! When the value of the meat is credited to these various flocks the costs of eggs are reduced. In one case, where credit for the meat sold and used on the farm was not given credit the cost of eggs was 16.3¢ a dozen, but when credit was given it was reduced to 3.2¢ a dozen. On another farm where egg cost was 19.1¢ the net cost was 4.8¢ less than nothing (!) when meat sold and used was credited.

The data further indicate that the most liberally fed flocks produced most eggs. Seven flocks received only 40¢ worth of feed a head a year and produced about 36 eggs a hen, whereas 7 others which received 70¢ worth of feed averaged 54 eggs a head. Feed, however, is not solely responsible for such gains; credit is also due to better care.

In Massachusetts, studies showed that "each additional dozen eggs per bird increased the labor income by 35¢ . . . $1. more per bird was made with flocks averaging 154 eggs than with flocks laying 120 eggs; also average egg production per bird was the most important factor influencing profit per bird."

Ohio flocks which laid 180 or more eggs a bird a year cost $2.51 a bird more than flocks which laid less than 100 eggs a bird a year; *but* the high-producing group returned $4.83 more in cash receipts than the lowest-producing group.

California statistical studies "indicated an increase of 153% in net profits between two groups of layers, one producing 122 eggs [per hen] per annum and the other 163 eggs. Such an increase is much greater in proportion than the 33.6% in average number of eggs produced per hen. In other words, as production increases income over feed cost rises more rapidly."

Among the various factors that influence returns from poultry are prices of feed and poultry products (eggs and meat), size of flock, productiveness of the fowls and diversification. Where general purpose fowls are kept the proportion of income is usually about 40% for meat and 60% for eggs; but where egg production (especially by White Leghorn fowls) is the main aim the figures may range from 70% to 90% for eggs.

Egg production may be increased by systematic breeding, careful culling and correct feeding. For instance, at the New York State College of Agriculture, White Leghorns of a highly prolific strain averaged about 180 eggs a year whereas those of a "low" strain averaged 120. In a 5-year test the value of eggs in the former case was 60% higher than in the latter and yielded approximately $2. a hen more above feed costs, even though the cost of feeding the former was slightly greater than that of the latter.

One characteristic of poor production seems to be lateness of maturity; that is, fowls that require relatively long time to mature are usually inferior as layers to those that mature most rapidly. Experiments conducted by the Department of Agriculture discovered that Leghorn pullets of quick maturing strains had records of 224 eggs for those that started laying in September, 226 for those that began in October, whereas those that started in January laid only 86 eggs. Rhode Island Reds that started in October and November respectively averaged 212 and 209 eggs, whereas those that began in January laid only 161.

The quick maturing Reds made a profit of about $2.28 a fowl more than did the slow ones. This was due to both the number and the higher price of eggs, for 22% of the eggs were laid between October and December inclusive when prices were highest. Feed cost for these early maturing pullets was only about 18¢ a fowl more than for late maturing ones.

As the most feasible way to increase profits is by increased production the improvement of fecundity is fundamental, but the season of the year (September to January) is also highly important. In developing flocks with this end in view nothing must be left to chance. In the former case trapnesting the year around and testing each generation of descendants; in the latter, earliness of hatching and steady development of sturdiness, are success requisites. The five essential factors in breeding toward these ends have been discovered by trap-nesting; namely, early maturity, steady winter laying, high rate of production, elimination of broodiness and persistency of egg-laying.

As application of these principles by means of trap-nesting is tedious and insistent few other than breeders of bred-to-lay fowls and the large producers of eggs for market systematically undertake the work. Most poultrymen and farmers find it more convenient, economical and expedient to buy hatching eggs or day-old chicks from reputable breeders. Such stock is well worth its somewhat higher price over ordinary stock because when properly managed it will produce larger numbers of eggs and higher profits, especially when large size is characteristic of the eggs laid.

In Michigan a recent survey was conducted to determine the cost to produce pullets from baby stage up to 24 weeks of age. Thirty-seven farmers completed coöperative records with the following results: *

The average net cost per pullet was 69¢; on the 10 most efficient farms, 37¢; on the 10 highest cost farms, $1.09. The average number of chicks at the start was 638. Mortality at 12 weeks was 11.3%; at 24 weeks, 15.7%. Average cost per chick at 12 weeks was 35.2¢ which was 1.6¢ less than their estimated meat prices. Broilers were usually sold at about 13 weeks and weighed 2.2 pounds. Their average sale price was 20¢ a pound. At 24 weeks the average number of pullets was 38 out of 100 chicks; low-cost farms had 41 and high cost ones 34.

Pullets of the light breeds averaged 3.4 pounds at 24 weeks on 25 farms; heavy breeds, 4.4 on 12 farms. Of the total costs at 24 weeks, feed constituted 42.2%; chicks, 22.2%; labor, 16.3%; equipment and brooder charge, 7% and other costs, 12.3%. Average feed consumption of mash and scratch was 5.1 pounds per pound of poultry produced. The 10 most efficient pullet producers used 4.2 pounds; the high cost farms, 6.1 pounds. The feed cost $1.61 and $1.81 per hundred respectively.

In "clean chick raising" campaigns conducted in many states individual poultry raisers have lowered the percentage of mortality to less than 10% as against 25% to 35% as

* Mich. Quar. Bull., Vol. XIV, No. 3.

chicks are ordinarily raised. By thus maintaining sanitary conditions they have not only had healthier flocks but increased the egg-yield in at least one coöperative group of flocks by an average of 40 eggs to the pullet.

There is no reason why you may not benefit by applying the same types of care that these campaigns have adopted; namely, clean chicks, clean incubators, clean brooders, clean ground, clean litter, clean feed, clean hoppers, clean water, clean management, clean laying houses—in two words *cleanliness throughout.* Let your slogan be: "Health sticks to clean chicks," for it is true.

Though costs and receipts necessarily vary with locality instances may prove helpful and suggestive of what may be expected, especially when they represent averages of considerable numbers of farms and fowls. For this reason the following figures are presented.

The Arizona Experiment Station reported (Bulletin 145) the average receipts and expenses of egg production on 44 farms as follows: Total expenses, 100%; feed, 41.2%; labor, 22.5%; replacements, 14.3%; inventory depreciation, 8.2%; interest, 4.7%; auto, 2.5%; upkeep of buildings, 1.9%; taxes, 0.3%; lighting, 0.3%; miscellaneous, 4.1%. Total receipts, 100%; eggs, 90.7%; hens sold, 6.3%; flock appreciation, 2.3%; miscellaneous, 0.7%. The cost of eggs per dozen: Average, 25.1%; 161 or more eggs, 19.3%; from 141 to 160, 23.6%; from 121 to 140, 26.9%; from 101 to 120, 26.8%; 100 or less, 52.4%.

In Oregon, 441 flocks totaling 271,337 hens produced 3,864,162 dozen eggs in a three-years' test, averaged 615 hens to the flock and produced an average of 171 eggs a hen. The averages of costs and receipts are as follows: Operator's labor, 14.6%; unpaid family labor, 3.3%; hired labor, 0.7%; total man labor, 18.6%; total feed, 54.2%; miscellaneous, 18.5%; depreciation, 3.8%; interest, 4.6%; gross cost, 100%. Receipts other than eggs, 16.7%; net cost (three-year average) 83.3%.

Though the figures as to market receipts and prices of eggs compiled by the Department of Agriculture are admittedly imperfect as to yields because they take no account of eggs used for hatching and for farm consumption or of prices paid in personal customer trade; they show that market receipts and yields are greatest during April and May and least during November and December. They also show that market prices are lowest from April to June and highest from November to December. Thus they help support the statement that every producer should aim to have his greatest egg production between September and February because prices always average higher during that half of the year than during the other half. The figures show egg receipts and prices for a ten-year average in the five principal markets—Boston, New York, Philadelphia, Chicago and San Francisco.

23

BEES

FEW persons think of bee-keeping as a business. The ordinary conception is that of a diversion, a side line on the farm, or a harmless pursuit for old men. Perhaps ninety percent of those keeping bees may be included in one of these classes, of which a very large number come under the head of keeping bees as a diversion.
FRANK C. PELLETT,
In *Productive Bee-Keeping*.

IN my estimation, no branch of agriculture is of such absorbing interest as bee-keeping. Having kept bees of my own I know something of the marvelous origin and coöperation of the colony not only in gathering nectar and in manufacturing honey but in making "bee-bread" from pollen, beeswax from honey, "propolis" or "bee-glue" from the waxy exuditions of flower buds, and of the "royal jelly" on which the developing queens are fed. I also know something of the profit that usually accompanies good management and the loss that surely follows carelessness and neglect.

The highly satisfactory experience of sitting in the shade on a hot summer day and watching throngs of workers flying between the hives and the basswood trees or the white, crimson, alsike or sweet white clover (or melilot) fields during a "honey flow" and realizing that soon there will be another "super" full of well filled "sections" that sell on sight to personal patrons is another experience that I have had.

But from having worked on two "bee farms" I also know that commercial bee-keeping is a man-size job—one that demands far greater ability, patience, tact, and attention to details than any other branch of "live stock" breeding, rearing or management, to say nothing of crop production. I therefore can indorse the statement of the late Professor J. H. Comstock, the famous entomologist and bee-keeper, that "any man who can make $1,500 out of bees has ability to make at least $2,000 out of something else!"

So this chapter is written to warn you not to go in for *commercial* bee-keeping until after you have served your apprenticeship with one colony, then with others developed from it; for perhaps in no branch of agriculture is it so important to learn to creep before you try to walk.

Before you start with even one colony first make sure that you are willing to be punctilious in attention to details, to do the thing that must be done when it should be done. If not, don't attempt bee-keeping, for you will almost certainly fail. As the nectar flow of one plant species lasts for only a few days it is essential to have everything in readiness so the bees may make maximum amounts of honey during such times. By attention to details the careful man will often get a profitable quantity whereas the careless one in the same locality with the same equipment will have little or nothing.

By starting with only one colony, preferably in early spring, the necessary equipment will cost only a few dollars and by following directions in good books on bee-keeping you may end the season not only with honey to sell but with two or more colonies developed from the first one. Then by assuring proper conditions, especially in having abundant stores of honey, the colonies should pass the winter in safety and be prepared to double or triple their number as well as increase the honey production average during the second season. And so on! The greatest increase of which I know was made by Dr. C. C. Miller, one of the most noted bee-keepers, who in a single season increased nine colonies to 56!

BEES

The first colony with the small necessary equipment and extra hives should enable you to teach yourself in beekeeping and make a profit at the same time.

Low prices or low yields of honey may not affect you as a small bee-keeper who combines honey production with your other work so as to use your time profitably. Local markets and personal customers assure the readiest way for you to maintain profits. Reduction of costs is another way to increase net returns. Yet the lowest cost per pound does

Fig. 33. Display greenhouse for sale of plants.

not necessarily mean the greatest profit. Amount of business and total production also figure. Yields vary from 30 to more than 200 pounds of extracted honey per colony, or one to seven cases of comb honey per colony, with three as a high average.

Comb honey should be produced only in localities particularly adapted to this class—where there is regular, abundant, rapid flow of white honey such as basswood, white, alsike or white sweet clover (*Melilotus alba*) and where there is a minimum of off-color honey and propolis. In all other regions it is more profitable to produce extracted honey unless the local market can be developed for the dark comb honey, such as buckwheat. Often this can be done because a local dark honey sells well where produced, even in preference to white honey shipped in.

A man with a maximum of 250 colonies may busy himself

profitably in other directions, provided the other lines do not exact work at "peak times" of bee activity—brief summer periods. To handle the work properly as much work as possible (e. g., hive, frame and section assembling, foundation making and placing) should be done during the previous winter so only the actual, necessary handling of the bees may be left for summer.

Coöperative work with the United States Department of Agriculture in Wyoming with bee-keepers and more than 25,000 colonies showed that the total cost of handling the bees ranged from $2.04 to $12.50 a colony! At 50¢ an hour the highest labor cost was $4.47, the lowest, $1.09. Average net incomes were $4.65 a colony. This figure included labor of operator at 50¢ an hour, interest on the investment at 6% and average prices of 6¾¢ a pound for honey and 27¢ for beeswax. Hour-returns for labor and management ranged from $1.61 to $11.78. The net income ranged from a loss (four out of 39 bee-keepers) up to a profit of $16. a colony. Costs for extracted honey were as low as 5.4¢ a colony but averaged 7¢ a pound for apiaries smaller than 400 colonies. Good yields are usually characterized by low costs and vice versa.

24

GREENHOUSES

[A PLANT grower] just starting in business may be compelled by lack of means to commence upon a small scale. While he would find a lean-to house the cheapest to erect—provided he built it against a south wall of a building—the excess of cost for a span-roof house would be so slight, and the results obtained so much greater that he would be wise in selecting that form of a house. The size must be determined by the business to be done, but for most purposes a house twenty feet in width is preferable to anything narrower, and an enterprising [plant grower] should be able to utilize one that is fifty feet long.

LEVI R. TAFT,
In *Greenhouse Construction.*

ONE of the most striking developments of gardening in recent years is that of the small greenhouse. So great has been the demand that leading greenhouse construction companies have developed standard sizes, sectional parts and unit features, the materials for which they can ship without delay.

Gardeners benefit in many ways through this standardization. For instance, the cost of construction is wonderfully reduced, 1, through the mere standardization itself and the consequent reduced cost of making the parts; 2, through the manufacturers' purchases of raw materials in large quantities and favorable sizes thus eliminating or greatly reducing waste; 3, through the construction of such parts by less skilled labor than that required to make special sizes; 4, by the making of these parts during slack seasons, thus eliminating over-time high rates; 5, by making it possible for a purchaser to put the parts together with unskilled labor when he thinks that thereby he may save money.

On this last point, however, the saving is often more speculative or fancied than real, mainly because skilled men know exactly what to do first, next and last, thus making up the difference in the cost of their time by requiring less time than unskilled men to do the same work. Moreover, they are sure to do the work correctly and thus give the owner more confidence in the job than would be the case when he does the bossing.

One of the most important advantages of buying certain standard styles and widths is that these are made in sections. Thus a beginner or a person short of money may start with two or, preferably, three sections and add others as he gains confidence through experience or as his finances improve and he is able to meet his obligations. The application of this sectional idea is somewhat similar to that of the sectional bookcase which years ago proved its worth.

Many a gardener has built a greenhouse say 15' or even only 10' wide by 25' long only to discover too late that he might have bought a standard size 18' by 25', ready to assemble, of far better material and for less money, to say nothing of greater convenience of working in such a house, the certainty that it would be adequately warmed and ventilated and that should he wish to enlarge he can do so conveniently, economically and harmoniously.

Favorite widths for even-span greenhouses have not become such empirically or through accident; but through the experience of men who do greenhouse work. These widths have been decided upon because of three factors—the "reach" of the men who work with the plants, the widths of the benches and the necessary walks between benches or beds. Where plants are grown in flower pots placed on the benches men less than 5¾' tall find side bench widths 30" wide or wider and center bench widths 5' tiring because of the reach from front to back or to center, respectively. Where plants are grown in "solid beds" instead of flower pots and where tall men are employed the side benches are sometimes as much as 3' wide and the centers 6'; but even under these conditions such widths are not nearly as popular as the ones first mentioned. In commercial greenhouses walks are sometimes less than 2' wide.

By using bench and walk widths as a basis of measurement the greenhouse construction companies have decided upon standard widths. Among the favorites are 15', 18' and 25'. Within reasonable limits the length of any house may be as many units as desired, but not less than two, preferable three or more for the sake of convenience of operation and arrangement. To illustrate: a standard house 25' long would consist of two sections, each 12½' long, a 50' one of four such sections or two sections of 25' each.

Though the sectional idea and standard widths insure excellent advantages the greenhouse construction companies have not stopped here. Their experience has proved that in thousands of cases beginners have started with one small greenhouse, gained experience during a year or more and then added other sections, often of a different type. So now whole series of units are available for combining to form not only easily operated parts but harmonious combinations of units attractive to the eye as well as being easily managed.

Such being the case the beginner may not only avoid making countless mistakes in construction but, knowing that he may make harmonious additions to his first small greenhouse he can at the start deliberately plan for a complete lay-out properly adapted to the entire available area at his disposal and toward which he may work, unit by unit, with the knowledge that as each is added the outfit will be both pleasing to look at and well adapted to its various uses.

Though greenhouses are made in many "fancy" designs and though they vary considerably as to dimensions, by far the most popular forms are the even-span greenhouse and the lean-to conservatory. The main reasons for this popularity are lower cost, adaptability both to plant growth and to adjacent architecture and convenience of arrangement.

The even-span, as its name implies, is a greenhouse which, if cut from apex to ground and from end to end would present a right and a left half. Such houses may stand alone, be connected with a head-house, a garage, a residence or another greenhouse usually at the north end. Houses of this form are usually and preferably placed with their ends respectively north and south so the sun will give as much light to one side as to the other as the day advances. Under such conditions plants, theoretically, develop more evenly than if they receive light from only one side as in the case of the lean-to conservatory. This advantage is perhaps more fancied than real because plants properly managed do well in all styles of correctly built greenhouses.

The lean-to greenhouse, "just a lengthwise half" of an even-span house, is usually built against an already existing wall, either of a residence, a garage, another greenhouse or some other building. It may be and usually is much narrower than an even-span house, often less than 10'. Whenever possible it should face south. A northern exposure does not favor plant growth, though for propagation purposes it is often preferred by people who grow large quantities of plants from cuttings.

With certain exceptions, such as special propagation and lean-to houses, the standard even-span greenhouses 15' or 18' wide will prove a better investment than any narrower size. This is because narrower sizes (which are also lower in height) cost more instead of less proportionately to build and are harder to operate during changeable weather because of the relatively small volume of air they contain. The air is affected by outdoor temperature and the fluctuations of intermittent sunshine and clouds, to say nothing of wind. For the same reasons the 25' house is still more economical, relatively, to construct and convenient to operate.

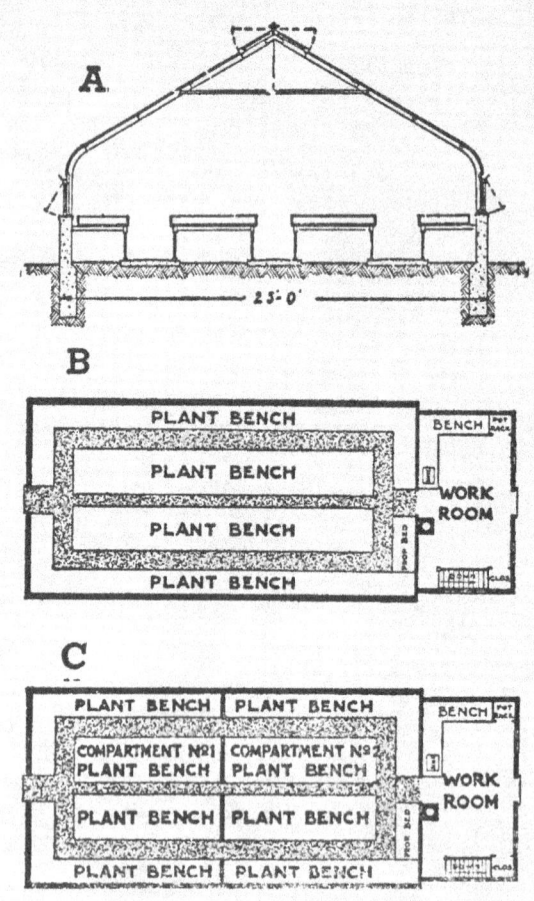

Fig. 34. Cross section and two floor plans of even-span display greenhouse.

It has a far larger available proportionate growing area than has the greenhouse of any smaller width and its increased height not only makes it easier to manage during variations of outdoor temperatures and air movements but permits the growing of much taller plants.

A superficial comparison of these sizes shows that whereas

a standard greenhouse 18' x 50' feet contains 900 square feet of ground area, one 25' x 40' has 1,000 square feet—100 square feet more—yet the latter has advantages as well as appearance which make it more attractive to own, whether or not the whole length be built at one time or at different times. It is the more adaptable size. These statements gain support from the fact that the greenhouse companies report the steadily increasing popularity of the 25' house as compared with the 18' size.

One of the most striking and significant advantages of owning a greenhouse built by a construction company, especially one of standard size, is that the company stakes its reputation upon the materials it supplies and the skill of the men it sends to do the erection. The materials are what many years of experience have proved best, each for its purpose. Most important, perhaps, of all is the selection of cypress for the wooden parts. This wood withstands decay even under the exceptionally trying conditions of dampness and heat characteristic of greenhouses. It also retains it shape astoundingly well. Such wood is rarely carried in suitable sizes or properly seasoned by local lumber companies so the man who wants to build his greenhouse of local materials is only too often tempted to substitute some far less desirable wood. The result is that sooner or later the greenhouse roof will warp, leak, break glass and decay.

Another advantage of standard greenhouses is that guesswork is eliminated. This has special application to the heating. Each construction company has worked out the heating requirements of each of its standard size houses for each general temperature required by the plant species to be grown. It supplies a heater not merely capable of warming the house under ordinary winter conditions but under severe tests such as blizzards during zero weather. This fact is of special comfort to the owner who has valuable plants or a crop to keep growing vigorously.

Adequate ventilation is scarcely less important than heating, for without it the plants may be "cooked," even when the weather outside is below freezing. Unless the ventilating system is properly proportioned to the area of the roof and the volume of the house it may be inadequate to keep the temperature and the humidity favorable to plant development. Moreover the ventilating apparatus must be so constructed as to be easily operated regardless of outdoor conditions.

But why go into further structural details? Why not close the case in favor of standard greenhouse units by saying that the greenhouse construction companies present prospective patrons various propositions. For instance, two companies with whose business methods I have long been familiar, and presumably other leaders in the field, offer to supply all materials, do all the work of erection and to turn over the greenhouse complete, ready for plant occupancy. Undoubtedly this is the best offer because the companies assume the entire responsibility and place their own skilled men in charge of the work, even when local laborers may be hired to do certain rough work under the supervision of the company's foreman.

This offer may be modified by the patron who may wish to employ his local masons to do the masonry and laborers to do the rough work. In such cases the company supplies plans and specifications for the patron to follow.

Where a patron wishes to do all the work with his own men the company will supply all the materials cut and marked to fit—everything ready to assemble according to plans supplied. This plan appeals most to men far from the company's supply depot because it obviates traveling expenses.

Whether or not the prospective owner of a greenhouse builds the house himself or has it erected by a construction company it will be greatly to his interest to familiarize himself with the principles of greenhouse construction, equipment and management before taking up discussion of a purchase with a company's representative. Thereby he can know beforehand what to look for and what to avoid and can the more intelligently pick and choose what will suit his local conditions as well as himself.

Though each construction company publishes worthwhile pamphlets which show pictures of many styles, present the pleasing features of greenhouse ownership and discuss structural features these leaflets usually say too

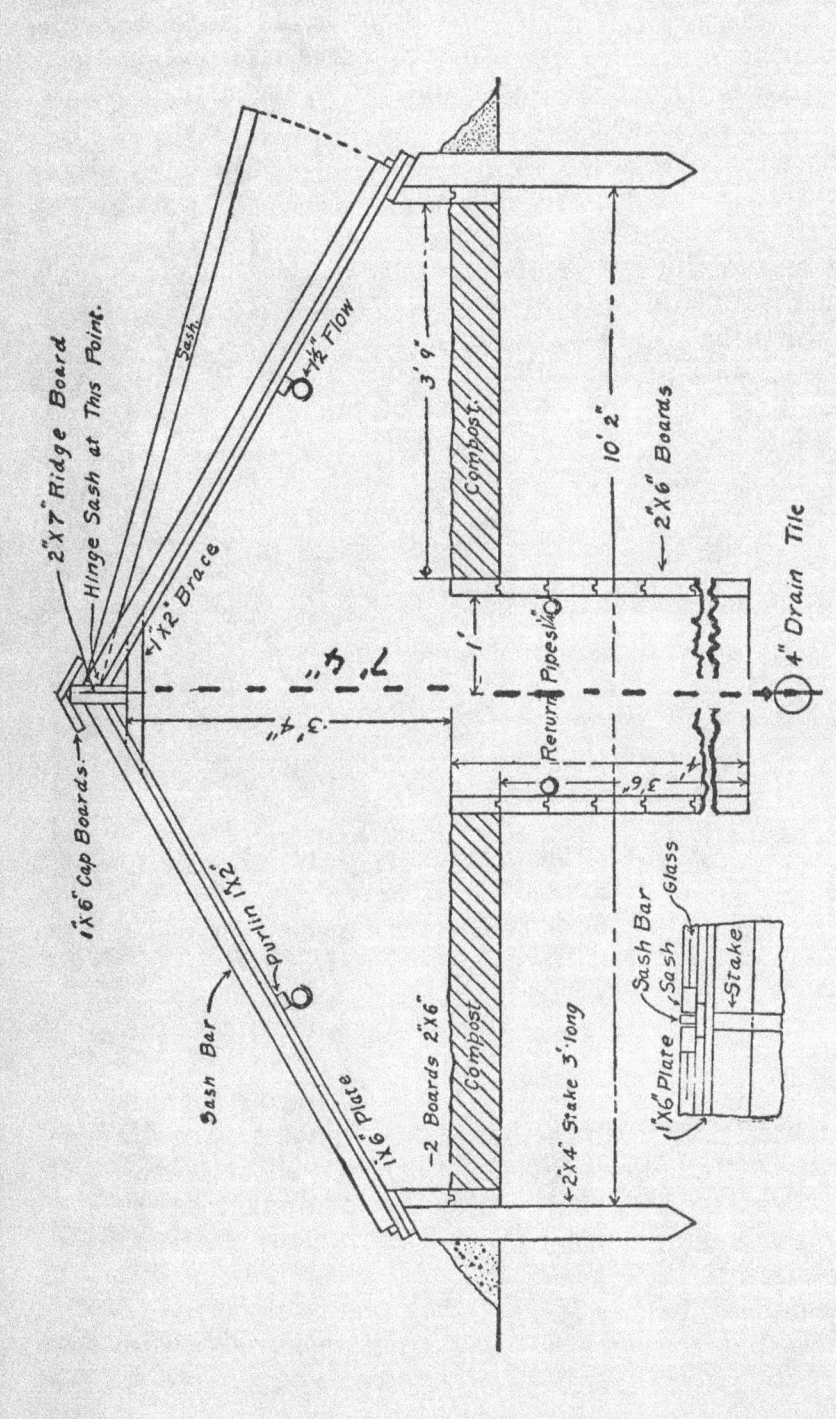

Fig. 35. Greenhouse built with hotbed sash. Walls and walk sunk below ground level. Economical to heat because little wall is exposed to the cold.

little about operation principles and details to be of much help to the owner of any specific greenhouse. They are good to have and to study but far more information than they give is needed by the would-be or the actual owner.

At best greenhouse literature is meagre. The recently published books which have come under my notice have been disappointing in various ways. So far as I have been able to discover there are no recent or available government or experiment station bulletins on this subject, though Farmers' Bulletin 1318 of the Department of Agriculture is well worth having. Two of the standard books, *Greenhouse Construction* and *Greenhouse Management* by Professor L. R. Taft, have long been out of print. Though written in the late nineties they are well worth picking up at second-hand book stores. *Greenhouses, Their Construction and Equipment* by Professor W. J. Wright emphasizes conditions at the time of its publication and is still available. It also is well worth its price.

25

COLDFRAMES AND HOTBEDS

As the name implies, coldframes are sash-covered frames without heat. The application of heat at once transforms them into hotbeds. While coldframes can hardly be accorded the dignity of forcing structures, yet they play an important part in the protection of plants in autumn and spring, as well as during the winter.

LEE C. CORBETT,
In *Garden Farming*.

IN no one direction has interest in gardening been so characterized as in the increasing numbers of coldframes and hotbeds employed by amateurs as well as commercial men. Under modern construction, these adjuncts have many striking advantages among which the following are sure to make appeal.

By means of hotbeds (Fig. 38) plants of desired flower and vegetable varieties may be started weeks or even months

COLDFRAMES AND HOTBEDS 137

before they could be sown outdoors. Instead of relying upon other sources for supplies of seedlings or potted plants, one's own seed may be started in the hotbed, pricked-out in coldframes, inured to the weather ("hardened off") and when outdoor conditions are favorable, transplanted to the garden.

By starting hardy perennials and biennials several weeks before outdoor sowing would be safe the plants may often be

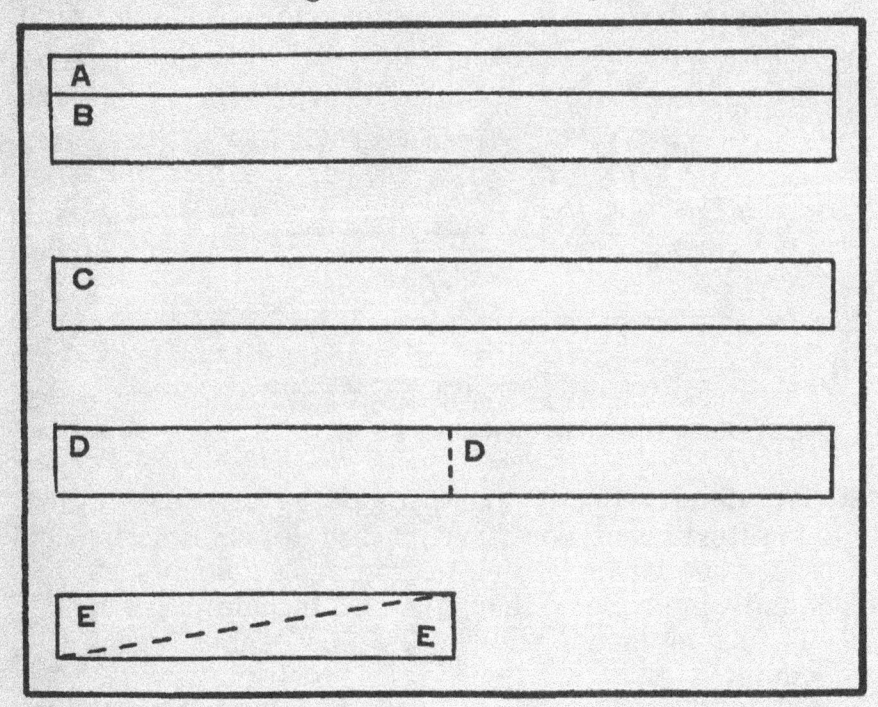

Fig. 36. Planks for hotbed frame. A, B, C, D, 12′ long, 2″ thick; A, 6″ wide; others, 12″ wide; AB, rear wall, C, front; DD and EE, ends to be nailed together with cleats to make slope from rear to front. Saw D and E on dotted lines. Bolt sides and ends to corner posts of 2″ x 4″ scantling. (See Fig. 37.)

made to bloom during the first season instead of having to wait until the second year. Similarly cuttings of many perennials, roses, and other flowering subjects may be started in a hotbed and advanced by the same stages to the open ground.

The coldframe alone may likewise be used for slower-rooting varieties. One great advantage of this practise is that you may propagate a personal stock of favorite plants and thus be sure that the young ones are true to name.

With a hotbed or a coldframe the difficulties and disappointments often incident to starting minute seeded plants may be wholly avoided; for when a shower threatens the beds may be protected by sash and the danger of washing out or burying either the seeds or the seedlings in mud obviated.

Risks of starting tender subjects too early or too late outdoors may be avoided. When seed of such varieties is sown too early in the open the weather may be so cold and wet it may decay or the seedlings be nipped by a tardy spring frost; when too late, the plants may meet unfavorable summer con-

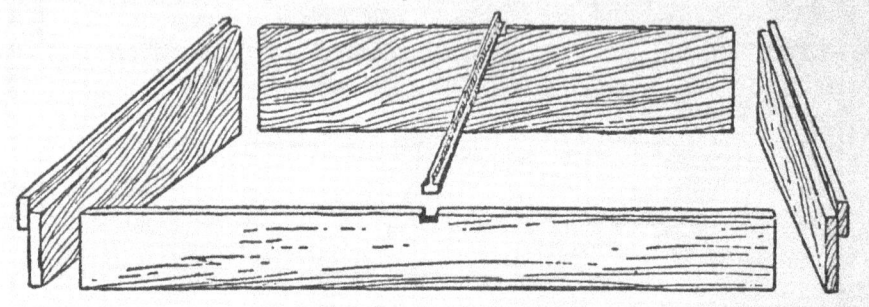

Fig. 37. Hotbed frame ready to assemble with corner posts and bolts.

ditions, develop poorly and perhaps be destroyed by an autumn frost before they have reached desired development. This is of special application to growing cantaloupes, cucumbers and watermelons, though it also includes other tender plants such as tomato, eggplant, pepper, dahlia, canna and geranium.

In localities where blight destroys watermelons, cantaloupes and cucumbers, hotbeds and coldframes also enable the gardener to avoid this disease or to ward off its effects until after the fruits have been ripened; for by starting the plants on inverted sods, in flower pots or other convenient receptacles to favor transplanting they may be kept in the frames and readily sprayed, dusted or fumigated to kill cucumber beetles and squash bugs which not only feed on them but spread blight infection from plant to plant.

Even when not growing plants, coldframes are admirable for extending the ripening season of tomatoes that would otherwise be spoiled by an early fall frost. The fruits that show pink may be gathered when falling temperature threatens damage, placed on deep layers of straw in the frames and covered with sash whenever the weather is wet or cold. The rate of ripening may be accelerated to a week or so by keeping the sash on the frames during the day, thus raising the temperature; or it may be delayed by leaving them off whenever the weather will permit without risk of frost-bite. Thus the tomato season may be extended to Thanksgiving Day or even later. The frames may also be used to store such late vegetables as celery, endive, cauliflower, Brussels sprouts and witloof chicory (so-called "French endive") and for carrying semi-hardy plants such as chrysanthemums over winter.

In spite of such an array of advantages, many people hesitate to have either a hotbed or a coldframe for fear their lack of experience or their conditions may induce failure. If you are one of these cast your fears aside, for in spite of both inexperience and lack of ideal conditions thousands of other novices have plunged and succeeded beyond their most sanguine hopes. Of course, the nearer you can approach ideal conditions the more likely are you to succeed. These are a southern exposure (or next best a southeastern) and protection from high winds, especially from the north and the west, by a windbreak such as woods, tall hedges, or buildings.

Modern, standard, commercial (3' x 6') sash and (6' x 12') hotbed frames (Fig. 38) are made so accurately and of such desirable materials (cypress or redwood), are so easy to duplicate and cost so little that they should always be preferred to home-made ones. The 6' x 12' frame is especially desirable to start with because it is more convenient to operate (having a larger volume of air) than a smaller one. If desired it may be divided by a partition across the middle so half may be used as a coldframe and the other as a hotbed. Being assembled by bolts and angle irons it may be readily taken apart and stored flat or on end in small space or moved from one property to another, thus appealing to renters. While "knocked down" it may be easily cleaned and painted and thus made to last for many years.

For a coldframe no excavation is usually made; for a hotbed the area to be excavated should be at least 6" wider and longer than the frame so there may be plenty of space for the foundation. This may be made of stone, brick, concrete or 2" planks, preferably of "pecky" cypress, when such are available; otherwise clear cypress or some other wood that resists decay—locust, cedar, chestnut. The foundation should extend a few inches above the surface to increase the longevity of the frame itself. A post is at each corner and at 4' intervals on the sides of all wooden foundation frames.

The depth of excavation will depend upon the climate; it varies with the local frost line. From Maine to Minnesota

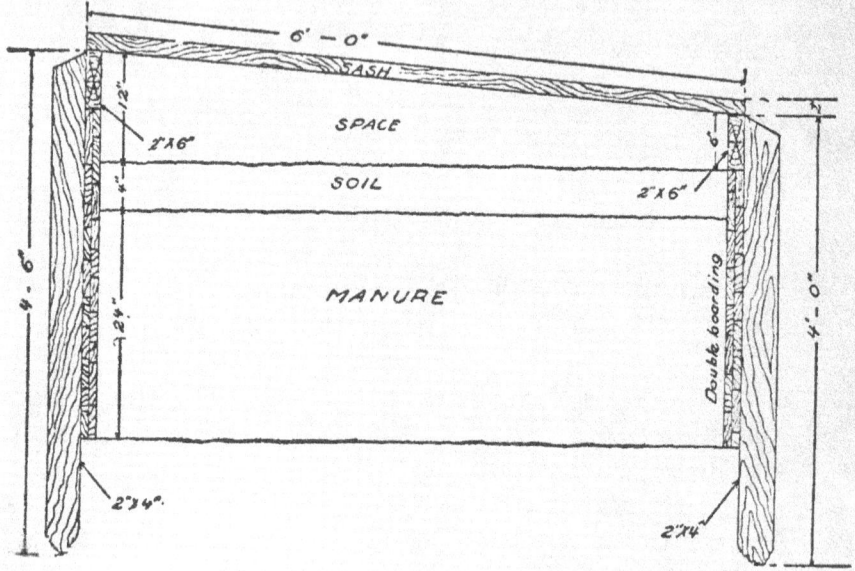

Fig. 38. Construction of manure-heated hotbed.

and northward 24" to 30" is favored; in southeastern New York, 18" to 24"; near Washington, 12". In the South the frame usually rests directly on the ground without excavation or foundation frame.

Though all commercial sash are 3' x 6', users have preferences and objections to various types. Some prefer light weight styles because easier to handle than heavy ones; others, the reverse because the frame, being sturdier, they claim glass breakage is less. For this same reason many growers prefer sash with four rows of panes rather than three. But, others point out, the three-row style casts less shadow.

Double glass sash have both proponents and opponents,

the former claiming protection equivalent to straw mats placed upon single glass sash, thus avoiding the work of laying and removing the mats (real work when wet or icy), giving the plants full sunlight and therefore better growing conditions than under single glass sash.

Opponents object to the greater weight of double glass sash, the 25% to 30% extra cost, the retention of moisture between the upper and the lower glass and consequently the encouragement of decay. They also declare that the accumulation of dirt between the panes results in shade and poorer growth of plants, and that all the advantages of double glass may be gained without any of the disadvantages by placing a second single glass sash on top of the first during cold weather. In any case it is advisable in northern localities to have protective mats (straw, burlap, felt or other material) and often matched board shutters to place over the sash when the weather is inclement, especially when windy.

The only difference between a coldframe and a hotbed is that heat in the former all comes from the sun. Hotbeds may be heated in several different ways. Fermenting material, especially fresh horse manure, though formerly popular is objectionable because of its present day scarcity, high cost, labor to prepare, short period of usefulness, fumes of fermentation (especially of ammonia), and the excess of attention as to ventilation that beds so heated require. It is inferior to all other means of heating hotbeds.

During the past few years electricity as a source of heat has been gaining rapidly in popularity because of its advantages over all other methods. Where electric current is available only an "outlet" is needed—such as used for the toaster. As proved by experiment stations (especially New York, Oregon, Minnesota, and Missouri), by various organizations as well as by electrical apparatus manufacturers (whose bulletins and advertising matter are available at little or no cost) the electric hotbed is free from all the objections to fermenting materials and is superior to steam and hot water. Among advantages thus determined the following are outstanding:

Electricity costs less to operate than do steam and hot water (approximately 50 kilowatt-hours to the sash during the growing season). It has probably longer life than either. When properly installed with thermostatic control it is safe and easy to operate, produces uniform heat at any desired temperature, demands a minimum of attention and eliminates the difficulty and work of preparing manure.

For temporary (especially rented properties) its convenience, adaptability, and portability make special appeal.

A coldframe may be converted into a hotbed or vice versa by merely turning a switch on or off. This saves time in starting seedlings because, regardless of weather conditions seeds may be sown later, seedlings grown faster and, thanks to the thermostat, sturdier plants may be produced than by any other system applicable to frames.

Electricity is clean and agreeable to work with and is always ready to use. As it may be turned on immediately after installation it saves loss of time incident to the use of manure whose heat must always be allowed to subside to less than 90° F. before seeds may be sown with safety. It is an economical substitute for manure when the cost of hauling and handling are included.

A hotbed 6' x 12' may be permanently electrified for about $5.; a manure bed must be made and the manure paid for fresh every year. Whereas an electrified bed may be heated every day in the year, one can be warmed by manure during only 6 to 8 weeks. When current is shut off a hotbed becomes a coldframe, so no extra coldframes are needed for hardening-off plants. Thus the work of shifting plants from one to the other may be avoided. Coldframes may be kept from freezing by hanging electric cables around the inside walls and turning on the current when needed.

Anybody can install an electrical heating system by attaching the heating "elements" and the thermostat as shown in Figure 39 and plugging to an outlet connected with the house wiring system. Regulation of the heat is done automatically by the thermostat which may be set to operate (shut off and re-connect) the current at any desired temperature. At the close of the season the system may be disconnected and stored along with the sash and the frames until the following spring.

COLDFRAMES AND HOTBEDS

Without going into detail it may be said that one amateur grew about 5,000 plants of 50 flower and vegetable varieties at an operating cost of only $1.60 for current (the rate being 10½¢ a kilowatt hour). The bed was equipped with wires, eight 25-watt vacuum bulbs and a thermostat for only $5. and half a day's time to install. The thermostat, set to operate at 50° F., maintained a temperature within "safe limits during sub-freezing weather," part of the time 16° F. above zero.

Hotbed heating by electricity is also done through thermal cables, either strung around the walls, laid near, upon or beneath the soil surface, or as plates placed beneath separate flats. Though styles of apparatus and application vary widely and are annually being improved, all are more satisfactory than manure, steam and hot water methods. Electricity has also proved advantageous for heating propagation benches and for forcing such vegetables as lettuce, tomatoes and cucumbers. In short, its convenience, economy, cleanliness and automatic control are what so strongly appeal to gardeners both amateur and professional, and account for its rapidly increasing popularity.

Various methods of applying the heat have been tried in several states but, so far as I know, not more than two had been compared until T. M. Currence conducted experiments with the four types illustrated in Figure 39 and detailed his findings in Bulletin 239 of the Minnesota Experiment Station. Among his conclusions the following have special significance to the man who wants to use electricity for heating his hotbeds or converting his coldframes into hotbeds. They are also of particular value because of the rigorous climate of the locality in which the work was done. This is far stronger proof that electricity is safe to install where cold weather is the rule than if the tests had been made farther south where the winters and springs are less severe.

For these experiments the hotbeds were built of 1" pine boards entirely above ground on a southern exposure and upon gravelly soil; they were lined with insulating material, covered with a waterproofing compound, the outside with roofing paper; in cold weather they were protected by ½" felt mats. All were equipped with meters and thermostats to measure the current and regulate the temperature, respectively.

Frame A was heated and lighted by four 100-watt Mazda lamps attached to a ½" pipe crosspiece about 12" above the soil; each lamp had a 6" shallow cone reflector. In Frame B, 120' of hotbed cable was placed in the soil to deliver 200 watts of heat and had 4 50-watt Mazda lamps suspended as in Frame A. In Frame C, 60' of cable was placed 6" below the soil surface to deliver about 400 watts of heat, and a supplementary 60' as a precaution to produce additional heat when necessary—though the occasion did not arise. In Frame D, 60' of cable was hung on the inside walls.

Efforts were made to adjust the thermostats so they would operate in a range of 50 to 55° F. but sunshine and the difficulty of making them operate at the same temperature caused—not serious—variations of the heat. During four weeks of winter weather the differences in current cost were not great.

The largest crops were grown in Frame A, doubtless because of the stimulating effect of the light. The temperature in that frame averaged more than 3° lower than in any of the others. Once when the temperature fell to 15° below zero the lowest temperature in any of the frames was 38° above zero. This indicates that when reasonable care is exercised in frame construction and the use of mat protectors, each method of applying current will protect plants from freezing. Frame D, with coil above ground, used least current in the winter test and produced better plants than did Frame C. Frame A produced somewhat uneven, rank growth. The unevenness could probably be prevented by better distribution of the lights. Combination of bottom and overhead heat was costly to operate but the plants made good growth. Much difficulty was found in regulating moisture in Frame C. When ventilation in this frame was reduced to a minimum in cold weather, moisture condensed on the glass and tended to keep the soil surface wet, and when properly ventilated the soil dried out rapidly, baked and current was wasted. Perhaps unfavorable moisture conditions accounted for poor growth of plants in this frame.

Each method has advantages and disadvantages for com-

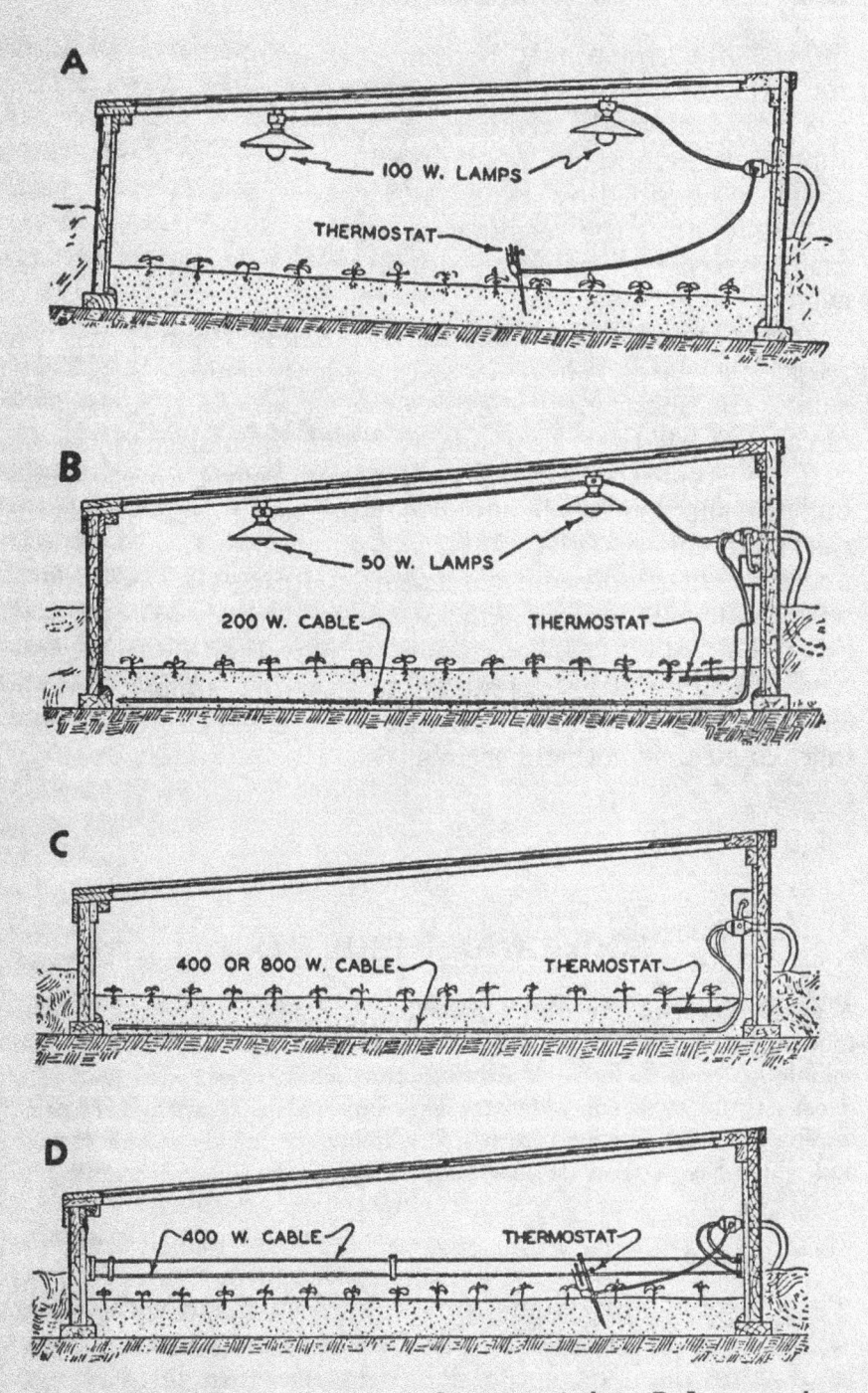

Fig. 39. Electric hotbed heating. A, Lamps used alone. B, Lamps and buried cable. C, Buried cable alone. D, Cable strung above soil.

mercial use. Apparently bottom heat should be least costly for seed germination; lights most desirable for rapid growth, especially in cloudy weather; combination of these two will supply widest range of adaptability; possibly best results might be obtained by using bottom heat to germinate seeds, and lights after the seedlings appear; 25-watt lamps, 4 to a sash, give good distribution of light and tend to eliminate uneven growth.

Lights would be desirable for beds smaller than 6' x 6'. To reduce the watts of heat necessitates lengthening the heating cable. Heating capacity of 200 watts would require 120' of cable, a length cumbersome to install in small area.

Commercial trials in Minnesota summarized in the same bulletin, supplement the findings in the above experiments by the statements that electricity makes a saving of 2 weeks over manure heated hotbeds; current used during the first month was 30 kilowatt-hours a sash; at 3¢ a kilowatt-hour (the local rate) the cost was 90¢ a sash in 2 cases (the third not comparable); all 3 growers plan to increase the number of their electric hotbeds—a sure sign that they recognize the advantages of electric hotbed heating.

26

SOILS AND THEIR CARE

A PERFECT soil is one which maintains a reserve supply of insoluble food material that cannot be washed away; which produces enough soluble material to feed the growing crop; which is so constituted that it can supply sufficient water to the crop; which is capable of maintaining the right temperature or of warming up quickly in the spring; and which has a structure that permits of proper root movement.

CHARLES W. STODDART,
In *Chemistry of Agriculture.*

IF you could start your farm on a piece of land with which man has never interfered and from the start could work as Nature does you could doubtless maintain fertility and crop production indefinitely at a high standard. The probability is, however, that the ground with which you have to deal has been so injured by man's abuse that its ability to grow desirable plants is at a low ebb. This need not discourage you because in a surprisingly short time you may easily bring it back to its pristine condition and can even improve on this to an astonishing degree.

Soils may be classified according to their principal components—clay, sand and innumerable combinations of these with vegetable matter (humus). Clay—blue, red or yellow —though rich in various mineral elements is undesirable without modification by the presence of the other two components. It is sticky, hard to dig or plow, so dense that water and air penetrate it with difficulty and its surface bakes in dry sunny weather. Sand, the opposite extreme, contains far less actual plant food material, allows water to rush through, carrying with it and wasting soluble plant food material added by nature or man. Combinations of these extremes with vegetable matter form "loans" of many grades popularly known as heavy clay loam, light clay loam, sandy loam, etc.

Loams are more desirable for plant growing than are either clay or sand because, on the one hand, they are easier to work, more porous and less likely to bake than clay and, on the other hand, are more retentive of moisture and plant food than sand. When your soil happens to approach the characteristics of one or the other of these extremes modify it by adding the opposite extreme to it; that is, "lighten" a clay by adding sand and make a sand "heavier" by adding clay. With each class of soil it is easier to increase the water-holding capacity by adding vegetable matter through applications of manure or by plowing under green manures or cover crops. (Chapter 29.)

Clay may also be lightened by plowing under a two-horse load of fresh horse manure to each 2,500 square feet of area in late fall, leaving the clods and furrows unbroken just as turned up by the plow so frost will break them, adding a 1" layer of sifted coal ashes during winter and, in spring, giving a surface dressing of lime (about a pound to 10 square feet, 250 to 300 pounds to the acre).

Texture is the principal characteristic that determines

the value of a soil for trucking and strawberry growing. This is mainly due to the size and the quantity of sand particles. When coarse, the soil is "quick" because it drains rapidly, warms up quickly and permits early sowing. Such soils are warm all season long and thus favor early crop maturity. Because of their open texture they require large quantities of humus, lavish feeding and irrigation.

Medium sandy loams are not quite so early but retain water and plant food better and are more productive. Fine sandy loams, though later than the preceding are usually best for summer vegetables and strawberries. For latest crops silty and clayey loams are often most valuable of all. Their fertility is also more easily and economically maintained.

If your money crops are to be largely vegetables and strawberries, as they should be, you will do well to remember that sandy loams have the following merits: Earliness and warmness especially in spring, earliness of tillage in spring and lateness in autumn, earliness of tillage after rains, quickness of fertilizer action, economical cost of tillage, ease of transplanting in them, facility of root crop harvest, smoothness of root crops, ease of cleaning and preparing root crops for sale, relatively less compaction of soil during harvest, particularly when the soil is wet and adaptability to efficient overhead irrigation.

Humus is the residue of decomposed organic matter (principally vegetable) contained in the soil. It is of great importance to the fertility of soils because it increases water-absorbing and retaining power, lessens soil tenacity, increases heat absorption of the sun's rays, seizes, holds, and develops plant food from the air and the mineral compounds of the soil, supports the life of creatures (including worms, grubs and bacteria) that live in and modify the soil, adds its elements to the supply of plant food, and on all these scores aids plant nutrition.

Nature supplies her soils with humus by the decay of fallen leaves, branches and the manures and dead bodies of animals of all kinds. When we upset her methods by cultivation we must supply humus in some other way or our soil will be so impoverished she cannot grow good plants for us. We can supply the loss of humus by liberal dressings of manure, muck, peat, leaf mold or other vegetable material or by growing green manures. (Chapter 29.)

When leaves and other forest litter decay they become "leaf mold," a material highly valued by flower growers but, as a rule, not so much by vegetable growers and farmers. The annual deposit of leaves ("needles") and other waste in populous pine woods is about a ton and in hardwoods probably twice as much to the acre. Not only does this gradually change to humus but it adds appreciable quantities of plant food to the soil.

In the 1928 Department of Agriculture *Year Book*, W. R. Mattoon gives the analysis of a ton as including 12.1 pounds of ammonia, 2.8 of phosphoric acid and 3.9 of potash. At an average of 1 pound of "dry matter" to the square foot, an acre (minus the area occupied by tree trunks) would contain at least 18 tons which, at the above figures and at prevailing prices, would be worth $39.20 for ammonia, $2.52 for phosphoric acid and $4.21 for potash, a total of $43.93.

Mr. Mattoon also reports that a farmer who applied the rakings of one acre of heavy oak woods to an acre of field crop land during 3 years secured yields of corn and cotton which increased his returns by an annual average of $16.15 over a similar acre not so treated.

Though these fertilizer and return figures are impressive they do not tell the whole story because leaf mold is noted for its high water-holding properties. When added to soil it acts as a sponge—absorbs rainfall, checks downward seepage, and yields moisture during long periods to plant roots. For these reasons, where it can be obtained at little or no cost but the gathering, it is valuable to use as bedding or litter in stables and poultry houses, as mulching material for strawberries and bush fruits, as a supplemental manure or as a source of humus in compost piles.

In a sketchy way the following paragraphs suggest how color may help even a novice to appraise the crop-raising value of soils.

Color has long been relied upon as an index of soil value for crop production. In a general way it proves the presence or absence of desirable and undesirable compounds and also suggests the composition of the most important soil ingredient, the clay. However, color, *per se,* is of small importance; though dark colored soils, especially black ones, when well drained, absorb larger proportions of sun heat than do light colored ones. (Chapter 18.) Hence, dark colored, well drained soils are earlier than light ones of otherwise similar composition and texture.

Though color differences of soils are due to differences of components it is not true that darker colored soils contain larger quantities of coloring matter than do light ones. Composition and combination account for many variations. Usually a black soil is rich and a dark soil productive in ratio to its darkness but black color is due to combinations of organic matter and lime, even in small amount (often less than 2% of lime).

Brown color generally indicates soil acidity, due to the presence of iron oxide. In such soils the organic matter, even though abundant, is not saturated with lime. When iron oxide is in the "free" state the soil is usually yellow when the quantity is small or red when it is large. However, the color is not primarily dependent upon the amount of oxide because the quantity of this material is fairly uniform in the clay of the surface soil, regardless of color. Clay, the finest divided material of soils, is formed by rock weathering. Its composition therefore varies with the nature of the original rocks. Red and brown soils are highly valued, less because of their iron oxide than because of their condition which this material (and color) indicate. For it suggests good drainage and other favorable growth conditions and proves the presence of abundant material which will both supply and retain plant food.

White and light colored soils are deficient in important components—organic matter and clay—and contain excess of sand. Hence they cannot absorb and retain water but permit such rapid drainage that the soluble components of manures and fertilizers rapidly disappear and are wasted in the drainage.

Where light colored spots appear in dark colored soils they prove that the soil there was water-logged and that the lime, organic matter, manganese, iron oxide and phosphoric acid have been leached out.

M. S. Anderson, chief chemist of the Bureau of Chemistry and Soils, has so strikingly presented certain functions of soils that I am somewhat condensing his article in the 1930 *Year Book* of the Department of Agriculture as follows:

Nature hoards some of her assets in an almost miserly manner. Even plant food may be kept in such closed vaults as to be only a long-time investment, bearing a very low rate of interest. Some of the coarser soil mineral grains may contain large potential supplies of fertilizing elements and yet the soil be of low fertility because of their lack of availability. Apparently little can be done to release them.

Another class or condition of plant food serves as a savings bank, whose deposits are subject to the "drafts" of a crop whether for the present or the future. The "colloidal material" (finely decomposed rock fragments combined with organic matter) appears to offer such banking facilities. It makes up the greater part of the clay in soils and differs otherwise from the coarser mineral grains than in particle size, besides its marked influence on the physical behavior of soil, its ability to serve as a plant food depository is most marked. Yet even its presence does not insure a rich plant food deposit. Its character varies widely, being the resultant of parent rock material, climate and vegetation. In the main, though sometimes badly impoverished by Nature, it is the soil's most valuable asset. Moreover, it is the plant food asset over which the grower has the greatest control. Little can be done significantly to alter the quantity of colloid in the soil, but much can be done to maintain or improve its quality.

A bank deposit can not be indefinitely drawn upon without making deposits to the credit of the account. Colloid material is the agency through which such credit may be effectively restored to the soil. This is possible because of the great holding power of the soil colloids for mineral constituents, apparently both mineral and organic portions of the colloid. When plant foods are added as fertilizers only part of the water-soluble material added may be taken up immediately by the crop. The rest may be held by the colloids in such a way that loss by drainage is slight. Thus the crop of a later season may share in the benefits, and the balance not used may accrue to the credit of plant food.

Not all plant food of soil colloids is "subject to check" by growing plants; some is "on time deposit," so additional requirements must be met before it is available for use. Organic matter often serves in this capacity. In addition to serving other beneficial purposes, it is a source of readily available plant food. Recent investigations have emphasized the availability of its plant-food constituents and its relatively high capacity for retaining plant food as compared with the mineral portion of the colloids of less fertile soils. Regardless of the kind of mineral colloid each load of manure added and each crop residue turned under becomes a "credit slip" to the plant-food account.

If a soil has a very low colloid content, as in the case of very sandy soils, banking facilities are not at hand, and valuable plant food may be lost. Even organic matter may be rapidly decomposed and much of its value washed away. Nitrates especially are leached out in the drainage water. Under such conditions plants must be fed in a hand-to-mouth manner without the expectancy of building a bank account for the future.

It is extremely difficult under certain conditions to build up the fertility of sandy soils, or even of some soils of finer texture containing certain kinds of colloidal material, beyond the necessities of a single season. Such soils for profitable use require frequent additions of quickly available plant food. Any attempt to treat them as store houses of plant food as in the case of most medium and heavy soils is likely to mean plant food wasted.

The loss of soil fertility due to sheet erosion is probably far greater than from gullying. Through it a thin layer of the most fertile soil is removed from the surface with each heavy rain. Because the material is thus removed gradually and because subsequent cultivation destroys all evidence of the erosion, the ill effects often go more or less unnoticed until after much damage has been done.

It has been shown by the experiments of R. I. Throck-

SOILS AND THEIR CARE

morton and F. L. Duley of the Kansas Experiment Station (Bulletin 260) that sheet erosion is greatly reduced when the land is kept covered with a crop as much of the time as possible. By using a cropping system that provides for a crop on the land most of the time, or at least during the seasons when the greatest erosion is likely to occur, much

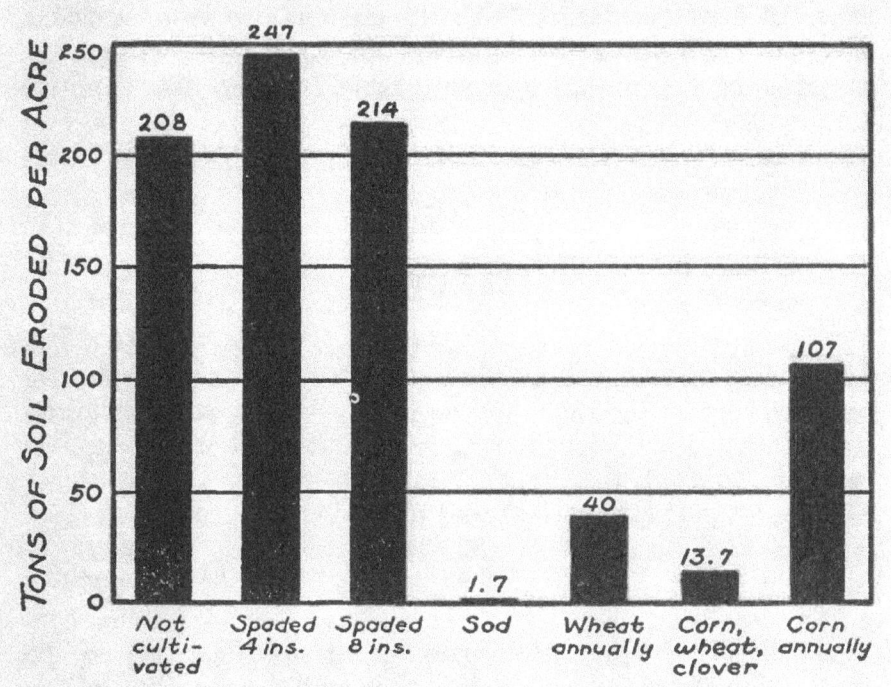

Fig. 40. Graphs to show relative annual losses of soil by sheet erosion under various treatments.

can be done to reduce the disastrous effects of sheet erosion. Cropping systems differ widely as to the time the land is covered with a crop.

Some crops give more effective protection than do others. (Fig. 40.) Small grain crops hold the soil far better than corn or other cultivated crops. Red clover forms a sod and protects the land more effectively than a crop like soy beans and thus gives not only more efficient but more extended protection from washing. Sweet clover or alfalfa used in the cropping system gives much the same protection as red clover.

On steepest lands, permanent grass pastures or meadows

should be used as much as possible since they form the most effective protection against erosion. The less sloping ground should be grown to small grain or kept in such rotations that the land will be protected by a crop at least three-fourths of the time. It is only on relatively level upland or bottom land that cultivated crops can be grown more or less constantly without serious loss from erosion. Even in such cases the land should be rotated with other crops since continuous cropping is seldom the best practise on any soil.

27

MANURES

IT IS the first business of the farmer to reduce the fertility of the soil, by removing the largest crops of which the soil is capable; but ultimate failure results for the land owner unless provision is made for restoring and maintaining productiveness. Every land owner should adopt for his land a system of farming that is permanent,—a system under which the land becomes better rather than poorer.
          CYRIL G. HOPKINS,
            In *Soil Fertility and Permanent Agriculture.*

UNQUESTIONABLY, the best of all fertilizers is stable manure because as it decays it supplies not only all its mineral elements but nitrogen and humus. It is in partially digested form and is full of micro-organisms which help to unlock other plant food held in the mineral compounds of the soil. Unfortunately it is scarcer and more costly than before the days of autos and tractors, though, for small garden work, it may be had in dried, pulverized form through garden supply stores or direct from the manufacturers.

Though drying concentrates these manures and kills the weed seeds most of them contain, it also destroys the bacteria contained in the fresh product, but this loss is offset by the almost odorlessness and convenience of applying the dried product.

Soils fertilized by manures and other natural products will grow better crops than those enriched with chemical fertilizers, unless these latter are supplemented by humus supplied in some way. But this remark must not be taken as condemning chemical fertilizers. (Chapter 28.)

Whether manure shall be applied while fresh, after rotting or in compost will depend upon various considerations. When the aim is to improve the physical condition of a heavy clay soil fresh manure will give best results, especially when plowed or dug under during autumn. It generally decays before spring and as it does its products of decomposition dissolve hitherto insoluble soil minerals. When the aim is to improve the physical condition of light, sandy soils well decayed manure applied in spring will give best results because its plant foods are more quickly available than those of fresh manure.

Decomposition is slow in heavy soils because the entrance of air and water is slower than in light ones. Hence in heavy soil no effect may be noticed for a year but the effects will last longer when they do start because there is less loss of plant food by leaching than in light sandy soils. Unless the season is dry and the supply of moisture scanty the plant food in fresh manures applied to sandy soils becomes available about as fast as the plants can take it up. If well decayed manure is applied to them its nitrogenous compounds may be washed out of the soil before the plants can utilize them. Clay soils do not waste food in this way but absorb and retain it. So though they require more work to get them in condition they are usually well worth the efforts made toward their improvement.

For ordinary vegetable crops a 2-horse load of manure to 2,500 square feet (50' x 50') is a liberal amount, though half this quantity will give fair results and twice as much will be best for crops grown for their foliage or stems—celery, spinach, cabbage, cauliflower, etc.

Always fresh and rotted manure should be applied to the surface *before* digging or plowing; dried, pulverized manures *afterwards* and then thoroughly raked or harrowed in the surface inch or so of soil. Liberal dressings of dried manures are: Sheep, 100 pounds to 1,000 square feet; poultry and pigeon, 75 pounds; horse, 100 to 150, cow, 150 to 200.

The crude materials upon which bacteria work consist of dead organic matter such as plant roots, leaves, stems, animal wastes and the bodies of dead animals. Higher plants cannot use these materials in their crude form; only after they have been converted into nitrates and other compounds as the final products of decay or their demolition by worms, bugs, molds, yeasts, bacteria and other forms of life, all of which play their parts in the process. When the quantity of crude material is abundant nitrate formation may be so great as to exceed the need of the growing crop; when deficient the crop may use it more rapidly than it is made and may even suffer because of this lack. To prevent such unfavorable results either nitrate must be supplied as fertilizer or the quantity of crude material in the soil must be increased.

As nitrates are developed from ammonia in the soil, applications of ammonium salts such as sulfate are often made—at the risk of making the soil more or less acid, unless lime is also added to neutralize the acid. When rotted manure is available it is better than ammonium salts because it supplies not only ammonia but considerable organic matter for the bacteria to work upon and for a much longer time. But as manure is scarce and often costly, green manures, especially legumes, are more available, cheaper and, when abundant, even more lasting in their effects. (Chapter 29.)

Unless soil conditions are favorable to decay—warm, moist and aërated—action will be slow. The physical condition of the soil must, therefore, be made favorable by such processes as plowing, harrowing and, where feasible and necessary, by irrigation. Organic matter such as manure and green manure plowed under when mature (as in straw and corn stalks) or when the soil is dry will decay much less quickly than if it is succulent and when the soil is warm and moist.

Until recently market gardeners near large cities usually considered 10 to 20 tons of stable manure to the acre annually necessary to produce good vegetables. Whether more or less would be profitable has been mainly guesswork, especially since the auto is steadily lessening supplies and

MANURES

increasing prices. Experience has suggested that though yields were increased they did not always pay, also that under certain conditions moderate applications of commercial fertilizer paid better than did larger ones. However, as such cases were unrelated they were unconvincing; so the Maryland Experiment Station undertook a series of comparative experiments with six leading truck crops (cabbage, potatoes, spinach, corn, tomatoes and peas) on a silt-loam soil of fairly good fertility underlaid with clay to determine if possible the profitable amounts of manure and fertilizers necessary to maintain fertility. After 13 years T. H. White and V. R. Boswell have tabulated and detailed results in bulletin 309 from which the following specially significant statements are quoted:

All treatments were beneficial insofar as yield is concerned, varying from a mean percentage increase of 123% for 500 pounds of fertilizer up to 312% for 6 tons of manure plus 750 pounds of fertilizer. Eight tons of manure yielded 19.3% more than four tons; 12 tons yielded 13.9% more than 8 tons and 37.08% more than 4 tons, the lower yield being taken as 100% in each case of these comparisons. Considering the increasing applications . . . successively larger applications gave successively, although not proportionately, larger yields. But such increases are not always accompanied by greater profits.

Applications tended not only to maintain but increase productivity. Though chemical fertilzers alone, without addition of organic matter, seemed to maintain or even increase productivity, actually considerable amounts of crop residues and green manure crops of weeds were plowed under. It is questionable whether fertilizer alone would have given such results on a lighter soil or under management in which no appreciable amount of organic matter was turned under.

Because of the scarcity and high price of manure it is necessary to depend on other materials to a large extent to maintain high yields. The experiments show that there was no significant difference in yield between 4 tons of manure vs. 500 pounds of fertilizer; between 8 tons of manure vs. 1,000 pounds of fertilizer; between 12 tons of manure vs

1,500 pounds of fertilizer; 2 tons plus 250 pounds fertilizer vs 8 tons manure; 2 tons plus 250 pounds vs 12 tons manure; 2 tons plus 250 pounds vs 1,000 pounds fertilizer; 4 tons plus 500 pounds vs 12 tons manure; or 4 tons plus 500 pounds vs 1,500 pounds fertilizer.

These results strongly emphasize the fact that under the conditions of the experiment: 1, Commercial fertilizers maintain yields as well as light to medium applications of manure and, 2, light applications of manure plus fertilizer are as effective as medium applications of manure alone or of fairly heavy applications of fertilizer alone. Furthermore, it was proved that certain manure-fertilizer combinations are superior to single treatments; for instance, 2 tons manure plus 250 pounds fertilizer yielded 33.19% more than 4 tons manure alone; 2 tons plus 250 pounds fertilizer yielded 38.50% more than 500 pounds fertilizer alone; 4 tons plus 500 pounds yielded 36.4% more than 8 tons alone; 4 tons plus 500 pounds yielded 19.17% more than 1,000 pounds fertilizer alone; 6 tons plus 750 pounds fertilizer yielded 32.8% more than 12 tons alone; 6 tons plus 750 pounds yielded 19.36% more than 1,500 pounds of fertilizer alone.

In emphasizing these results it must be remembered that the soil is a silty loam of fairly good native fertility and that the crop residues and some weed growth were turned under annually on all plots. The same amounts of manure might prove relatively more important on soils especially low in organic matter to which no plant residues were added.

After all, the experimenters conclude, the grower is interested in yields insofar as they increase his money (net) returns. They remark that the data show that manure produces desired yields, but does so at high cost if it must be bought. If manure is available at low cost, or produced on the farm, there is no question as to its value. If it must be bought at a high price and hauled a considerable distance, it appears that under conditions similar to those described, it can be replaced more profitably by crop residues and commercial fertilizers. To these may also be added green manures and cover crops. (See Chapter 29.)

28

COMMERCIAL FERTILIZERS

THE farmer . . . is often inclined to pay little attention to the forms in which the fertilizing elements are applied, even though he may employ sufficient quantities of a given mixed fertilizer to supply the proper quota of each element. As a matter of fact, the selection of a proper form or forms in which to supply the needed plant foods will, in many cases, determine the success of the application of a given formula to the crop. Too much care and attention cannot be given to this important question.

WILLIAM S. MYERS,
In *Food for Plants*.

IN soils that have long been cultivated, especially in those which have been mismanaged, one or more of three or four essential elements of plant nutrition may be deficient. Hence we must resort to commercial fertilizers to supply the needs of our crops.

Commercial fertilizers are of two classes; organic (of vegetable and animal origin) and inorganic (of mineral origin). The former, which include manures, play their chief role in improving the physical condition of the soil, though they also supply important amounts of plant food; the latter play less important or (some of them even negative) physical roles, but are useful for their plant foods. The former have the advantage of ready conversion into available plant foods and of harmlessness to foliage upon which they fall. Some of their characteristics are as follows:

Dried blood, dried fish and cottonseed meal are all rich in nitrogen but also contain more or less potash and phosphoric acid; ground bone, chiefly noted for its phosphorus, also contains more or less nitrogen and potash; tankage, also noted for its phosphorus, but varying considerably in the amounts because of different materials from which made—fish, garbage, sewage, etc.—also contains potash and nitrogen.

The plant food most likely to be lacking in the soil, nitrogen, is the growth maker. Though it is only one of the essential elements of plant growth it is of special importance because its compounds are deficient in most soils, costly to buy, easily lost by drainage, and yet easy to replace by inexpensive methods. It constitutes about 80% of the air, but not until it becomes combined with other elements in such forms as ammonia and nitrates can most cultivated plants use it. Most of these combined forms, together with varying quantities of other combinations in organic compounds, are formed in the upper layers of the soil where plant and animal wastes accumulate. Such supplies are reduced more or less rapidly by cultivation, mismanagement, seepage to lower levels, and by escape into the air in gaseous forms.

In the presence of sunlight and when supplied with water, plants are able to get all their necessary plant food except nitrogen from the air and the soil. But without nitrogen they can neither grow nor even live. Growth, repair and reproduction are all directly or indirectly dependent upon nitrogen. One of the best indexes of its quantity in the soil is the character of the foliage. When this is lush and dark green the supply is ample or even excessive; when puny and yellowish, it is deficient. Rank stems and leaves are more susceptible to disease, slower to reach maturity and less hardy than those of only moderate growth.

The reason why nitrogen is often lacking in soils is that, being highly soluble, it is readily washed out of the soil and lost in the drainage water. Hence it is best applied in frequent but small doses during the first half of the growing period of the crop to be fertilized by it, never late in the growing season of shrubs and trees because this would probably result in sappy growth that might not ripen before winter and would probably be killed by freezing weather.

In addition to the supplies from manure and other organic materials, nitrogen is obtained in several chemical forms of which nitrate of soda and sulfate of ammonia are best known. The former, which contains about 15% nitrogen is quickly available to plants because of its ready solubility; hence it should be applied only in small doses at a time but repeated at intervals of two to four weeks; the latter, which contains about 20% nitrogen, is less quickly

COMMERCIAL FERTILIZERS 161

dissolved and is "fixed" or "held" much better in the soil so is less likely to be lost by leaching. It is therefore of special value in sandy and other light soils and in wet seasons under which conditions nitrate of soda would be quickly lost.

Nitrate of soda tends to make and maintain soils in neutral or alkaline condition—favorable to vegetables and most ornamental plants; sulfate of ammonia tends to make soil acid and therefore unfavorable to these plants but favorable to blueberries, rhododendrons and various other acid-tolerant plants. The acid reaction of this fertilizer may be neutralized by applications of wood ashes or lime. (Chapter 30.)

Potash, the fiber maker, is often lacking in sandy soils and in soils which have grown root crops (turnips, carrots, beets, parsnips, etc.) for several to many years without its adequate replacement in fertilizers or manures. When deficient the stems and branches of plants are weak and spindling and easily broken by wind.

Potash is obtainable from four main sources—wood ashes (about 4%), muriate of potash (50%), sulfate of potash (48%) and kainit (12% to 16%). Ashes contain all the mineral elements of the plants burned to make them. In order to be most useful they must be stored and applied dry. It will usually not pay to buy them but it will pay to use what supplies are made on one's own place, dusting the material on the ground anywhere plants are to grow.

Reliance for a potash supply should be placed on the muriate or the sulfate (the former preferred) because these are high grade so need be bought and applied in only small amounts. Potash is not washed out of the soil in the drainage water but fixed or held by various materials in the ground. It may therefore be applied at any time of year when the ground is not frozen.

Phosphorus, the ripener, causes fruit and seed to ripen well. When lacking in the soil, crops may be slow to mature or may fail altogether. It is usually applied as superphosphate, basic (or Thomas) slag or pulverized phosphate rock ("floats").

Though superphosphate (formerly called acid phosphate)

has long been the leading phosphate fertilizer of America it is open to the objection that its analysis (17 to 20% of available phosphoric acid) is too low to meet the increasing demand for higher grade fertilizer mixtures; moreover, in humid weather it tends to cake because of its absorption of moisture from the air.

To offset these objections, manufacturers of mixed fertilizers add one or another of several concentrated materials to their goods. Doubtless the most popular of these is double superphosphate which is often marketed as "triple superphosphate" and "treble superphosphate."

In physical properties and effects double superphosphate is on a par with the best of other concentrated inorganic fertilizers such as nitrate of potash, phosphates of ammonium and potassium. When properly prepared it neither absorbs moisture, cakes in storage nor becomes sticky even in humid air. Hence it enhances the drilling properties of fertilizer mixtures to which it is added and therefore makes them flow more freely through fertilizer drills.

Its solubility in water being only a tenth as great as other concentrated phosphate fertilizers is of great advantage because it is far less likely to burn the crops to which it is applied. Burning is due to highly concentrated solutions of soluble salts either upon the foliage or in the soil. When it replaces superhposphate in mixtures no undesirable mechanical or physical changes occur. However, one caution is necessary: it should not be mixed with excessive amounts of lime, ground limestone, cyanamid, or other materials rich in lime. When mixed with potash or ammonia or other soluble sulfates a chemical reaction occurs and results in more or less caking, but when this has completed itself the mixture is readily pulverized without recurrence of caking.

Before applying any of these fertilizers it is advisable to know which one is needed and which not; for thus we may avoid wasting materials and money. The simplest way to do this is to divide the garden into strips of at least 10 feet width at right angles to its length and sow only one unmixed fertilizer on each alternate strip, leaving the others unfertilized. Then sow crops lengthwise of the garden so as to cross the strips. (Fig. 41.) When the crops grow the

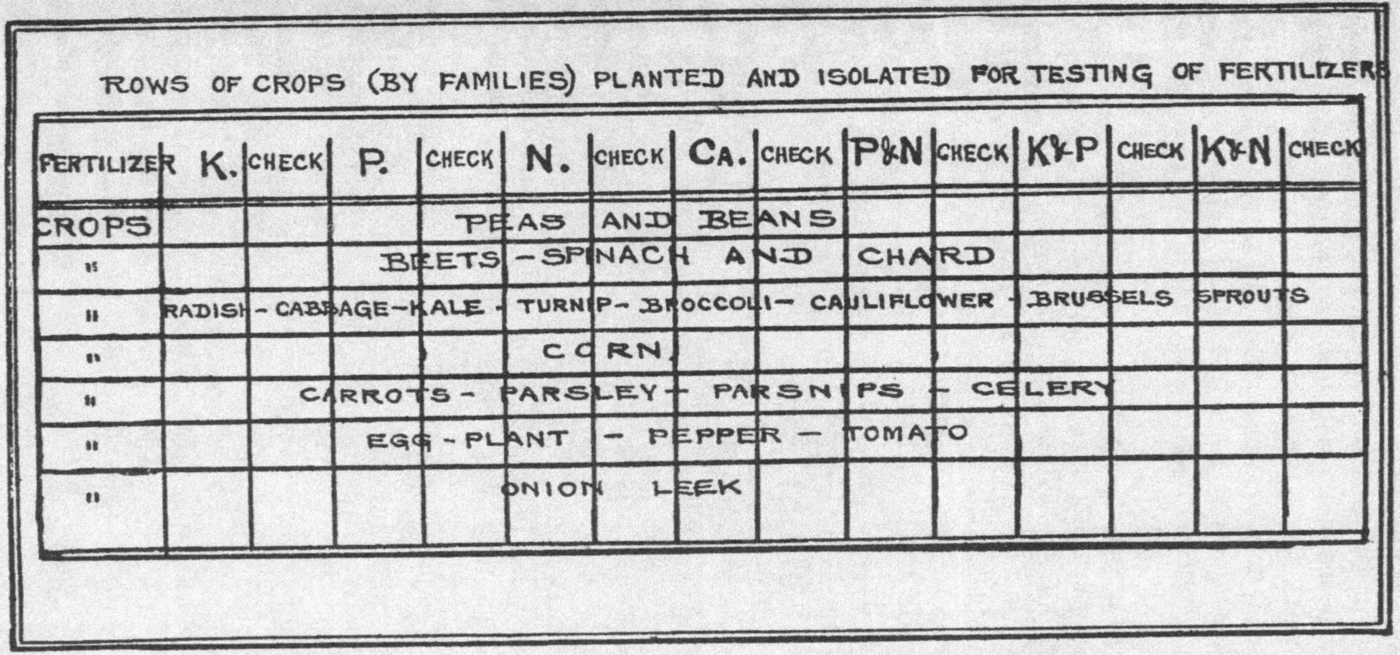

Fig. 41. Fertilizer tests to determine needs of soil. K, potash; N, Nitrogen; P, phosphorus (phosphoric acid); Ca, Calcium (lime).

differences of development will suggest what plant foods are lacking in the soil and therefore what ones to supply. As already noted yellowish foliage will indicate shortage of nitrogen; weak stems, lack of potash; and poor ripening of fruit and seed, lack of phosphorus.

There should be at least five strips, one to test the need for nitrogen, another for potash and a third for phosphorus with blank, unfertilized strips between. Perhaps the best fertilizers to apply in the experiment are nitrate of soda for nitrogen; superphosphate for phosphorus and muriate of potash for potash. When enough space is available other strips may be added as follows: Sulfate of ammonia and sulfate of potash with unfertilized strips between. If desired, combinations of two or even three of these five fertilizers may be made thus: Nitrate of soda and superphosphate; nitrate of soda and muriate of potash; superphosphate and muriate of potash, etc., but not two of the same kind such as nitrate of soda and sulfate of ammonia or muriate of potash and sulfate of potash.

It is not considered advisable to divide complete * fertilizers into successional applications because abundant experiments have proved that the best time to apply potash and phosphoric acid is shortly before seed is sown or plants transplanted. Spinach, lettuce, celery, cabbage, cauliflower and other leafy vegetables are generally stimulated by top dressings of quickly available nitrogenous fertilizers, especially during chilly, wet spells when the rate of growth is slow.

Vegetables whose fruits are the parts wanted—melons, cucumbers, eggplants and tomatoes—are often helped by an application of nitrate of soda just as the first flower buds develop. Root crops do not respond so strikingly to surface applications, doubtless because they forage more deeply than do the other crops mentioned. Top dressings usually range from 150 to 300 pounds to the acre, the lower amount being applied, generally, all at one time, the larger in two dressings. In cases of larger total applications it is advisable to make fractional ones at 2-week intervals.

* "Complete" fertilizers are "mixed goods" in which all three elements, nitrogen, potash and phosphorus, are included.

COMMERCIAL FERTILIZERS

A good general formula for an ordinary fertile soil well managed as already explained is made of high grade materials, in percents, as follows: Nitrate of soda, 5%; sulfate of ammonia, 10%; dried blood, 15%; muriate of potash (or sulfate), 15%; superphosphate or ground bone, 55%, (the former preferred where quick availability is desired, the latter where a long effect is wanted). Its application should be followed in the vegetable garden by one to three surface dressings of nitrate of soda at intervals of three or four weeks.

Fertilizer efficiency may be increased by assuring uniformity of distribution, placing it where it will be of greatest service to the seeds or plants without risk of burning to roots, and by using the best available fertilizers.

Though fertilizer distributing machines are time and labor savers they often fail to apply at least some kinds of fertilizer evenly—too little resulting in insufficiently fed plants; too much in burning and killing. Such irregularity may be due to inefficiency of the machine, either because of poor design or working parts, to improper mixing or segregation of the ingredients, to stickiness, caking, moisture absorption or other cause that interferes with drillability.

Separation or segregation in mixtures may be due to size or to the specific gravity of the individual grains, the rate and amount varying with the materials. Granular and dry materials drill much better than damp and finely pulverized ones. Deliquescent materials—those that, like nitrate of soda, absorb moisture from the air—become caky or sticky so that their drillability may vary from day to day and thus necessitate frequent adjustment of the drill.

Numerous experiments have proved that fertilizer gives best results when placed more or less locally to the seeds or plants rather than scattered promiscuously over the surface, but that this distance should be relatively distant, not in contact with the seed or plants.

Other experiments have proved that granulating both soluble and insoluble fertilizers not only improves drillability by reducing the tendency to cake and become sticky but prevents segregation of the ingredients.

The only way to judge the value of a commercial brand

of fertilizer is by its content of actual plant food. The laws of most states require that guaranteed analyses appear on the packages. When these are taken as the basis of calculation the values of various fertilizers can be determined and comparisons made to see which are the best ones to buy. Such calculations will usually show that pound for pound the concentrated or so-called high grade fertilizers, though naturally higher priced, are the more economical to buy, first because they contain larger percentages of the various plant foods and second because freight, cartage and application cost less for the smaller weights of material need be handled to obtain given quantities of plant food.

Concentrated fertilizers and proprietary brands, especially the new synthetic nitrogen products, must be applied to soils with more caution than the low grade goods. In no case should the manufacturer's recommendations be exceeded because these may be taken as maximum safe quantities. In order to avoid injury it will be advisable to apply less than the makers advise so as to be on the safe side.

29

GREEN MANURES AND COVER CROPS

A COVER crop is not a crop at all in the sense that it is to be removed from the land like grain or forage; neither should it be confused with a permanent sod of grass. The growing of cover crops in the late summer and fall presupposes some sort of cultivation during the spring and early summer.

ALBERT E. WILKINSON,
In *The Apple*.

GREEN manures are crops grown solely for the improvement of the soil. When sown toward the close of the season, either alone or among other crops as these are approaching maturity, they are often called cover crops because they are intended to cover the ground during winter and thus prevent loss of plant food through washing over the surface ("sheet erosion") or by seepage to lower levels and drainage. In the latter cases they are always dug or plowed under in early spring before they have made much growth. Otherwise they might become so woody they might decay slowly and thus, for a time, be a detriment to the soil.

Plants used for green manures are of two classes: Nitrogen gatherers, those that work over atmospheric nitrogen from the air in the soil; and nitrogen consumers, those that cannot perform this function but use what nitrogenous compounds are already in the soil. The former are generally the most important because they increase the supply of this essential element of plant growth—the most expensive to buy and the one most easily lost from the soil.

The principal nitrogen-gathering crops are clovers, vetches, peas, cowpeas, and soy beans; the consumers, buckwheat, rye, cowhorn and common turnips, and Dwarf Essex rape. When the former are dug or plowed under they add important quantities of nitrogenous material to the soil as well as other organic compounds; when the latter are similarly treated they add no plant food but what they took up from the soil—no nitrogen but what they were able to save from loss.

Often these crops are sown together with one or more of the first group so as both to save nitrogenous compounds already in the soil and to manufacture new supplies. One of the favorite combinations is rye and winter vetch; another buckweat and crimson clover. Sometimes all four of these are sown together in July after an early vegetable crop has been harvested, or even while the crop is still occupying the ground. In the former case the ground is dug or plowed and made fine before sowing; in the latter only the surface is made loose by a cultivator and the seed sown either broadcast or in drills between the rows of standing crops. Though buckwheat plants are killed by the first frost and though winter may kill crimson clover the vegetable matter these crops develop will be just as good as if alive when turned under. Rye and vetch which will probably live through the winter must be dug or plowed under before they get 8" high or the job will be difficult and the effects may not be as good as if the plants were more succulent.

When fresh or rotted manure is available it is highly advantageous to apply liberally just before a cover crop or a green manure crop is turned under because the bacteria these contain will help to break down the buried plants and thus make their plant food material more quickly available to the succeeding crops.

A green manure is measured for efficiency by its effect on the crops that follow. However, a failure should not always be laid to the crop since any one of many factors may be influential in the decomposition of the buried plants. For instance, most soil bacteria require a temperature of at least 65°F. to work properly. In spring the soil is usually too cold for their rapid work. Shortage of water in the soil also slows or even stops their action. Under most favorable conditions succulent green manures decompose in about a week but under lower temperatures and less moisture twice as long or longer.

Decomposition of green manure converts much of the plant matter into carbon dioxide which either passes into the air as such or combines with water in the soil to form carbonic acid. This acid converts more or less various soil minerals into forms which plants can utilize or which pass out of the soil in the drainage. Other parts of the green manures are converted into ammonia which remains in the soil and through bacterial action is converted into nitrates. Hence the larger the percentage of nitrogen in the plants buried the greater will be the quantity of nitrates. Hence also the high value of legumes—clovers, vetches, etc. Conversely, when the content of nitrogen is low (as in rye) the bacteria may use all of it for their growth. However, when these die and decay they give this nitrogen back to the soil.

When the quantity of nitrogen is very low, as in straw, harmful effects may result from an application because the bacteria may take their nitrogen from the soil and actually compete with the crop plants being grown.

During winter nitrate formation practically ceases, in early spring while the ground is cold and wet action is slow but during summer, *provided the soil is moist*, formation is greatest. From this it is evident that under ordinary condi-

GREEN MANURES AND COVER CROPS 169

tions losses are greatest while the ground is wet and bare during late fall, winter and early spring; least during summer when it is dry and also when liberally covered with a growing crop.

This indicates that in order to prevent losses of nitrates by seepage the management of the soil must be such as to prevent the accumulation of nitrates during summer and early autumn. One way to do this is to plan the crop rotation so that one will be in active growth while, or soon after, nitrate formation is most active. Should it not be convenient or profitable to do this with a "money crop," then a green manure or a cover crop may be added to the rotation so as to keep the surface covered with verdure. Such a crop will not only utilize the nitrates as formed or soon after and produce a lush growth, but will prevent their losses in drainage water.

Choice of crop will depend upon the time of year when the money crops will reach harvesting development. Some kinds such as cowpeas and buckwheat require warm or hot weather for their development and are tender to frost; others such as rye and winter vetch need cool weather and are resistant to frost damage. The time to sow need not necessarily be after the money crop has been harvested; it need not be postponed until the ground can be plowed and harrowed or dug and raked prior to sowing. On the contrary, it is not only often possible but advisable to sow the seed of the green manure or cover crop among and several weeks before the money crop will be ready for harvest. The amount of loss of plants due to tramping while harvesting the money crop will be more than offset by the avoidance of fitting the land for the green manure or cover crop and the extra saving of nitrates and other soluble plant foods in the soil.

Choice of the green manure or cover crop will also depend upon whether or not an increased supply of nitrates is desired in the soil. When such an increase is wanted the choice should be a nitrogen gatherer—some legume; when not, it should be a nitrogen consumer. For summer use cowpeas, soy beans, velvet beans and summer vetch are most popular among the legumes; for fall and winter, crimson clover, hairy or winter vetch and Canada field peas. The former are generally sown during late spring or early summer to be plowed under during late summer or early fall; crimson clover is sown during midsummer, winter vetch between July and September and Canada field peas during August or September.

Sweet clover or melilot has notable value as a green manure, especially on heavy soils, because of its deep rooting habit and the abundance of its foliage. However, if the soil is acid it may fail unless lime or superphosphate is applied shortly before seeding. Also it may fail if "unscarified" (machine scratched) seed is sown late—after the ground has become dry in spring. Such seed gives best results when sown in late fall or on the snow during winter. In these cases the plants get an earlier start than the weeds which they choke out. Scarified seed cannot be safely used in this way because it germinates too early. Like other legumes, the seed should be inoculated unless the soil has already grown this plant. When the plants are 8" or 10" high is the best time to plow them under for green manure.

Though the plants are grown for improving the soil they may need fertilizer to get a good start and to make a strong growth. When properly inoculated they get all the nitrogen they need from the air, so usually no nitrogenous fertilizer need be applied. Yet when the soil is suspected to be lacking in this element a little nitrate of soda or sulfate of ammonia may be applied, especially when seeding is delayed until May. This will help overcome the handicap. A dressing of 50 pounds to the acre may make a difference of 4,000 pounds or more of plant growth to the acre.

The fertilizers usually most needed by melilot are phosphates, even though sweet clover is famous for assimilating crude phosphates from the soil and converting them into other forms which become available when the plants decay. Yet if the soil is acid the plants seem to be unable to use the crude phosphates. Hence the advisability of applying lime or basic slag which are strongly alkaline and thus neutralize the acidity.

On sandy soils lacking in potash an application of muriate or sulfate of this element will often prove helpful.

30

LIME

INDIRECT fertilizers are of increasing interest and importance. Their intelligent use is attended with satisfactory results in crop growing, but blind, indiscriminating use may work harm to soil and crops and loss to farmers.

LUCIUS L. VAN SLYKE,
In *Fertilizers and Crop Production.*

VARIOUS materials are called "indirect" fertilizers or "amendments" because, though they may or may not contain plant food they are used chiefly to make changes in the soil and thus help plants to get food already present but in otherwise unavailable forms. The most important is lime.

Lime is an alkali, a chemical which has the power to combine with an acid to form a "salt." It is therefore used as a soil amendment chiefly for this purpose and thus to maintain the soil in "neutral" or slightly alkaline condition. If we were to use only natural manures our soils would continue favorable to plant growth much longer than when we use chemicals that contain acids (especially hydrochloric and sulfuric) which we apply whenever we fertilize with muriate or sulfate of potash or with sulfate of ammonia.

When we use such fertilizers we must sooner or later use lime to neutralize the acids they contain. The salts formed by these acids with the lime (chloride of lime and sulfate of lime—or gypsum), being highly soluble in water, are washed out of the soil in the drainage water, thus preventing damage these two powerful acids would do to soils and plants. Other acids are similarly neutralized by lime.

When land is cultivated and fertilized but yields only 75%, 50%, or 30% of a full crop it is like running an automobile with its wheels off the ground—it doesn't get anywhere! Soil, fertilizer, and water are all present and labor is expended but acidity prevents fruition.

Although experimental work in many states indicates that most vegetable crops do best on neutral or only slightly acid soils, the New Jersey Station felt that further information would be valuable, so field plots where lime has been used in varying amounts at five-year intervals for 20 years were chosen for gathering data. Two kinds of limestone were used—calcium and magnesium—both finely ground and applied broadcast shortly before seeding. Without going into the details of Bulletin 498, all of which were identical except for the lime, let it here suffice that these experiments prove that:

In many cases labor and fertilizer are wasted because productive capacity is distinctly lowered by the acid condition of the soil; that in every case the yields were much increased where lime was used; that beets and carrots were a failure on the unlimed plot, partly because acidity reduced germination; that poor germination of seeds due to soil acidity delays marketing or even results in loss of crop; that, because of its value in growing soil-improving crops, lime may have an important indirect effect, because most of these crops do best on soils only slightly acid; that certain soil organisms which play important parts in soil fertility are greatly favored by a soil well supplied with lime; that strongly acid soils may contain soluble aluminum compounds which are toxic to certain crop plants, whereas lime puts these compounds out of action; that heavy applications of superphosphate will do the same, but lime is more economical; and that when used in moderate amounts, there is little choice between the magnesian and the non-magnesian limestone.

Gypsum or land plaster cannot be used to neutralize acids because it is already a salt (sulfate of calcium). It may be used, however, for the other purposes for which lime is employed so it is safer to apply when a soil is already neutral or slightly alkaline and therefore not in need of the caustic form of lime.

Several fairly accurate apparatuses to determine whether a soil is acid, neutral or alkaline are for sale at garden supply stores. For practical purposes, however, the following ways will answer:

1. Shake a sample of soil with some rain or distilled water in a bottle, allow the sediment to settle, dip strips of druggists' red and blue litmus paper in the water and note any

LIME

change of color. If the red paper turns blue the soil is alkaline; if the blue paper turns red the soil is acid; if neither paper changes much it is neutral.

2. Instead of the litmus test add a teaspoonful of weak ammonia (not the "cloudy" household stuff but pure ammonia in water) to the muddy water. After standing over

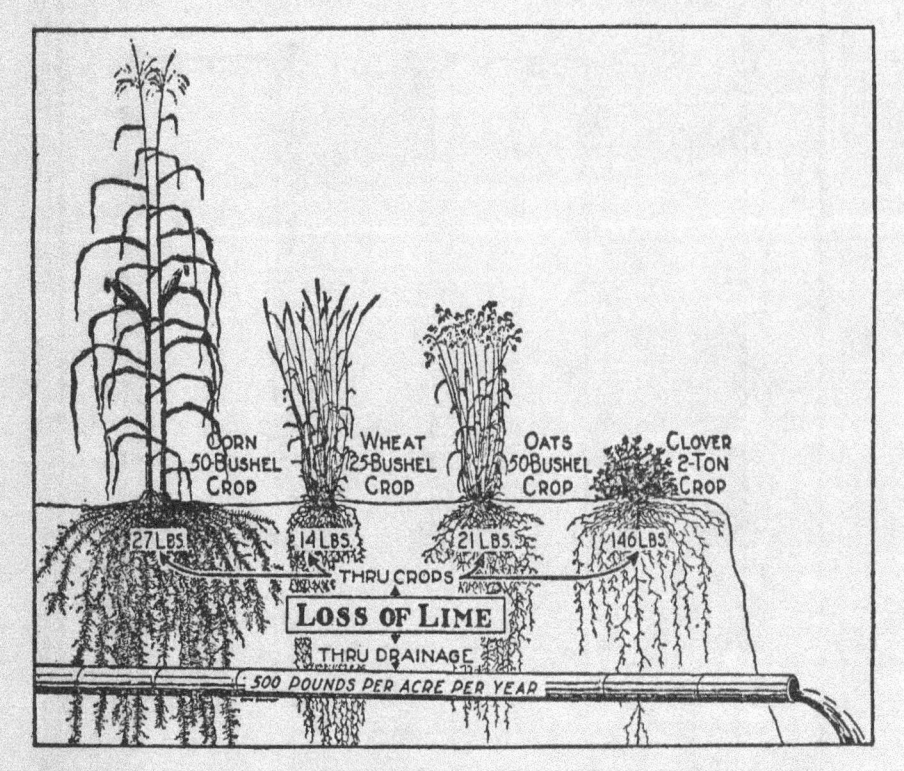

Fig. 42. Diagram to show losses of lime through crops and drainage.

night if the liquid has turned dark brown or black the soil is acid; if not it is neutral or alkaline.

Govern your applications of lime or gypsum by what several such samples of soil taken from various parts of the field indicate.

Because of its caustic nature lime must never be brought in direct contact with manure. Its chemical action drives the ammonia of the manure into the air. We need every bit of this ammonia for our plants so should try to prevent its loss by dusting the manure with some absorbent or by bury-

174 FIVE ACRES

ing it in the soil as soon as possible. Among the best absorbents are superphosphate, gypsum, dry muck and shredded peat moss. These may be scattered freely over the manure as made.

Wood ashes should be handled in the same way as lime

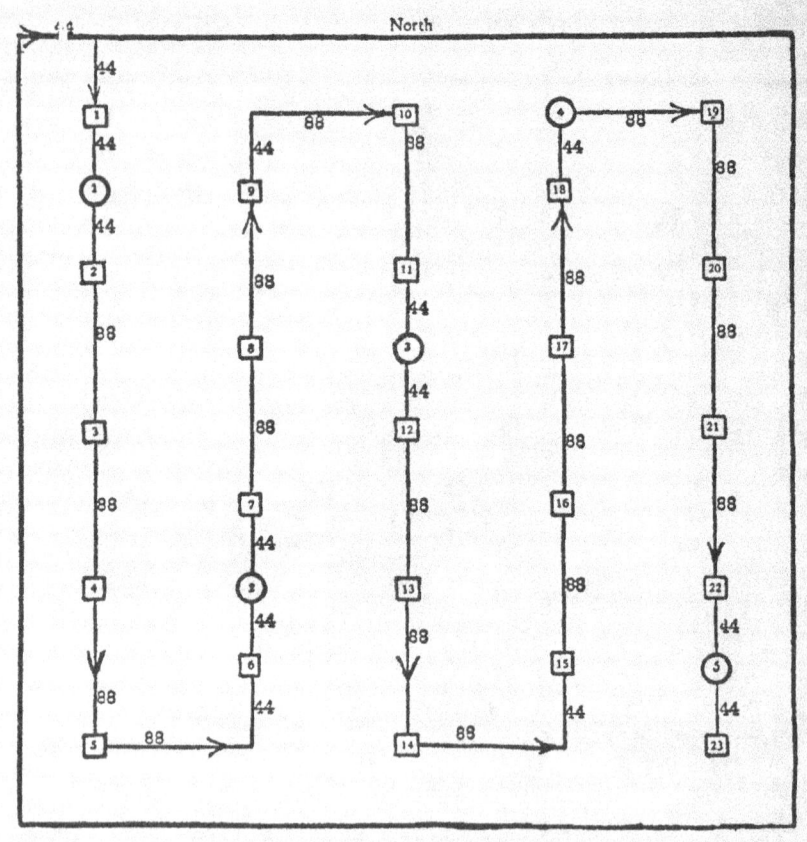

Fig. 43. Soil sample-taking to test for acidity. Squares indicate surface samples; circles those of subsoil.

because, though they contain all the mineral elements of plant food, they are noted for their potash and lime content and the effects of the latter upon soils. Never should they be mixed with manures.

As lime and gypsum always work downward in the soil from the level of their application it is important to apply them *after* the ground has been dug or plowed and to rake or harrow them in the surface. If they are applied before plowing they work downward from the level of the inverted soil and are therefore of no use in the surface layer. When these materials are to be used at the same time as manure the manure should be applied before plowing or digging and the lime or gypsum afterwards.

The amount of lime to apply will vary with the character of the soil and the degree of acidity. For light sandy loams an application of 1,000 to 2,000 pounds of pulverized limestone or 700 to 1,500 pounds of slaked lime (hydrate) to the acre should be ample as a dressing unless the soil is decidedly acid; then double the quantity may be applied. For a heavy clay, 1,500 to 5,000 or 6,000 pounds of pulverized limestone or 1,300 to 5,000 pounds of slaked lime to the acre are liberal applications. As a rule it will not be necessary to apply lime oftener than once in 3 to 6 years.

Of the several forms in which lime may be bought the kind generally known as hydrated or agricultural is the most convenient. It is sold in paper bags, is finely pulverized and passes through one's hand like melted butter. However, it may be more evenly applied by a fertilizer distributor.

31

COMPOST

THE compost heap can be made a prolific source of homemade manure. Every farm and garden should have one of sufficiently large proportion to take care of all refuse organic material about the place.
SAMUEL B. GREEN,
In *Vegetable Gardening*.

COMPOST plays such an important role and has so many uses that its manufacture is one of the fundamental processes of professional gardening. Whether your garden be large or small you should have a compost pile to supply the needs of your cold frames, hotbeds and greenhouses if no more.

Compost is specially prepared loam which contains all the elements of plant food added to it artificially. It is worked over for uniformity in composition and texture and is almost a fertilizer itself. If your soil is a heavy loam or a clay it will be greatly benefited if the compost contains a considerable proportion of sand, sifted anthracite coal ashes, muck, leaf mold, rotted sawdust, chaff from a straw stack, etc., all of which tend to lighten the soil with which mixed and when they decay to increase the content of humus.

One good way to make compost is as follows: In the autumn place wide boards an inch thick on edge so as to form a rectangle 6 or 8 feet wide. The length may be as desired. Fill this bin with fallen leaves well wetted and tramped down. On these place a 3" x 4" layer of sods upside down and close together. Next spread a 3" or 4" layer of manure, preferably from a cow stable, on top of the sod. Repeat these alternations of layers until the pile is 4' or 5' high. As it rises make each layer a little narrower than the one below so the completed pile in cross section will be like a letter A with a broad top. The finishing touch consists in covering the sloping sides with 2" or 3" of good soil well pounded and flattened with a spade to make it firm. Make the top of the pile about 3' across and hollow it out so as to form a basin 4" to 6" deep and nearly as long as the pile. Fill this with water in dry weather.

Besides the materials mentioned it is good to scatter ground bone, hardwood ashes, superphosphate, ground phosphate rock ("floats") tankage, dried blood, or other fertilizer or materials rich in plant food upon the various layers so as to enrich the final compost. Ground limestone or lime liberally sprinkled on each layer will help break down the vegetable matter and "sweeten" the final compost.

So prepared the compost will be of special value where the soil is somewhat acid as well as more or less depleted in plant food. Always add to the pile anything that will readily decay and thus make plant food—vegetable and animal refuse, weeds or garbage free from fats and oils.

The main reasons for making such a pile in the autumn are that the seeds of weeds in the manure and the sods will become moist enough to swell, be frozen and destroyed; that there is usually abundance of material at that season and that time is generally less fully occupied than in spring.

In case a pile cannot be conveniently made in the fall it may be made at any season when materials are available.

An autumn-made pile should stand until late the following summer when a sharp, square bladed spade should be used to slice the pile vertically downward so as to cut through all the layers. The sliced material is used to form another pile, the outer parts of the first being thrown on the inner part of the second to insure decay. When possible the pile may be allowed to stand for two years because the compost will "ripen" and be better than 1-year material.

All that is necessary in the final handling of such compost is to slice it vertically downward and pass the material through a mason's sieve with a ½" mesh to get rid of stones, clods and other debris.

32

CROPPING SYSTEMS

THERE is no one question of greater importance to the farming industry than that of soil fertility. In order that the industry may be successful, it is not enough to produce crops; it is necessary that the production shall result in a genuine profit. That is, it is not enough to produce crops which bring more than they cost in the way of labor and manures, without taking into consideration the effects of their growth upon the future productive capacity of the soil.

E. B. VOORHEES,
In *Fertilizers*.

No man can do his best when sick; neither can a soil! Under such conditions the character and quantity of both human and soil products are below par and continue to slump as long as unfavorable conditions last. In countless cases diagnosis shows that one or more fundamental principles of health have been or are being violated; for experience proves that when these are followed health and production are normal and characteristic.

The commonest way by which soils become "sick" is by being forced to produce the same crop or closely (botanically) related crops year after year without change. Like yourself, soils not only enjoy variety of diet but evidence their appreciation by increased and more profitable production. Hence has arisen "crop rotation," a general farming practise which consists in having from 2 to 5 unrelated crops occupy the land in prescribed sequence, usually during as many, often more years.

The advantages of rotation are: 1, Each crop of a well planned rotation leaves the soil in good physical condition for the next; 2, faults and disadvantages of one season's treatment may be corrected in the next; 3, plant foods may be maintained in better balance and supply than when the same or related crops are grown continuously; 4, plant diseases and insect enemies that live in the soil are starved out; 5, trouble from weeds is reduced; 6, labor is economized for various of the above reasons; 7, one or more of the crops (a green manure) maintains or helps to maintain the supply of humus in the soil and when this crop is a legume (e.g., clover), 8, the supply of nitrogen in the soil is supplemented or maintained at trifling cost. Rotation of crops implies and includes rotation of tillage, manuring and other cultural practises so the land benefits both directly and indirectly.

Actually, of course, soils do not become "sick." Unlike a human or an animal, they do not need to recuperate. The trouble with them is usually either that they leave an excess of certain by-products in the soil, due to a repetition of one crop (or of botanically related crops), or they have exhausted the available supplies of certain elements necessary to their welfare.

When such conditions occur in market gardening, as they often do, they are most easily and economically corrected by seeding down the area to grass for at least two years. By this practise a crop of hay (and aftermath) may be harvested, a sod may be plowed under to add humus to the soil and the physical texture of the soil improved.

Instead of resorting to this remedy it is better to divide the cropping area (when large enough to be worth while) into at least five equal parts each of which shall take its turn to produce a hay crop. Each of the other divisions should

1ST YEAR GROUP-1	1ST YEAR GROUP 2	1ST YEAR GROUP 3	1ST YEAR GROUP-4	1ST YEAR GROUP-5	1ST YEAR GROUP-6	1ST YEAR GROUP-7
2ND YEAR GROUP-2	2ND YEAR GROUP-3	2ND YEAR GROUP-4	2ND YEAR GROUP-5	2ND YEAR GROUP-6	2ND YEAR GROUP-7	2ND YEAR GROUP-1 →
3RD YEAR GROUP-3	3RD YEAR GROUP-4	3RD YEAR GROUP-5	3RD YEAR GROUP-6	3RD YEAR GROUP-7	3RD YEAR GROUP-1 →	
4TH YEAR GROUP-4	4TH YEAR GROUP-5	4TH YEAR GROUP-6	4TH YEAR GROUP-7	4TH YEAR GROUP-1 →		
5TH YEAR GROUP-5	5TH YEAR GROUP-6	5TH YEAR GROUP-7	5TH YEAR GROUP-1 →			
6TH YEAR GROUP-6	6TH YEAR GROUP-7	6TH YEAR GROUP-1 →				
7TH YEAR GROUP-7	7TH YEAR GROUP-1 →					
8TH YEAR GROUP-1						

Fig. 44. Rotation of vegetable crops according to botanical families. (See Fig. 41.)

be devoted consecutively to produce a root crop (turnips, beets), a fruit crop (tomatoes, cucumbers), a salad crop (lettuce, cabbage) and a seed crop (corn, peas). Where this practise is not feasible the seven-year plan illustrated in Figure 45 will provide a good substitute.

In addition to crop rotation, four other systems of cropping are practised to gain other ends. 1. *Marker cropping* consists in thinly sowing seed of a quick-sprouting and maturing crop in the same rows as a slow sprouting one (parsnip, carrot) or one whose seedlings are hard to see (beet, onion) when they first come up. The favorite marker is a forcing variety of radish. In less than a week (often in three or four days) the broad seed leaves of the radish appear and thus mark the location of the rows thus acting as guides for the hoe which should be started that early.

When the radish seeds are dropped about 2" apart the seedlings do not interfere with the permanent crop and when a quick maturing variety is sown the radishes should all have been gathered within a month, thus leaving the later crop in full possession. The practise is applicable on only a small scale because the seed of each crop must be sown separately by hand. When seed is sown by a drill marker cropping is not necessary because the roller of the drill presses down the soil above the rows and leaves an easily seen mark which serves as a guide for the wheelhoe that should be used a day or two after sowing.

2. *Partnership cropping* consists of growing two or more different crops from start to finish on the same area. Favorite crops so treated are pumpkins or winter squash with corn; pole beans with corn; dwarf peas, tomatoes or bush beans with strawberries. In this last case the vegetable plants are removed as soon as they have been cropped. The ground is then cultivated.

In one of my gardens I set strawberry plants 24" apart each way, planted corn between them in one direction, sowed pole beans to climb on the corn, bush beans between the rows and winter squash at 8' intervals. All did well except the squash which made a poor stand because of the heavy clay soil. The strawberries yielded abundantly the year following.

CROPPING SYSTEMS

3. *Companion cropping* consists in sowing alternate rows or setting alternate plants of two or more crops that require different lengths of time to mature. The short season crops are removed before the long season kinds need the whole space, thus neither interferes with the other so several crops may be harvested from the same area. To be successful it is essential that the companions mature at different seasons, that one be a catch crop or less valuable than the other, that

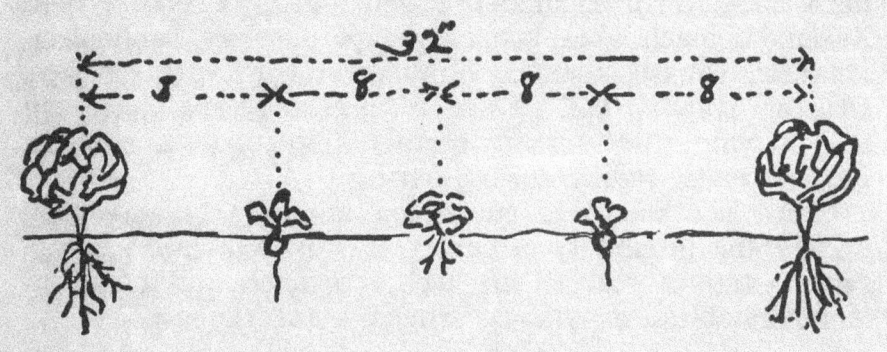

Fig. 45. Diagram of companion crops. Cabbage, outer rows; lettuce, center; radishes, between.

each requires the same tillage and plant food and that all are preferably of different botanical families so as to avoid insect and disease troubles and soil robbery of important plant food materials.

Favorite companion crop combinations are: Radishes between rows of beets, carrots or turnips; onion sets alternating with rows of onions sown as seed; onion sets and early cabbage or kohlrabi; parsnips and early carrots; leeks and early beets; bush beans and tomatoes or eggplants or peppers; cucumber or melon with bush beans; lettuce plants set between early cabbage plants or cabbage rows or both or also with radishes in continuous rows between the rows of cabbage and lettuce. In this last case the radishes are removed before the lettuce needs the space, then the lettuce is gathered and the cabbage left in full possession. This may then be followed by a succession crop.

4. *Succession cropping* consists in sowing or planting a second, a third or even a fourth crop in sequence on the same area during the same season. To operate it the maturing period of each crop must be short enough to allow its successor ample time to mature and, for reasons already given, preferably each crop should be of a different botanical family from its two or three predecessors.

Favorite succession crops are: *Earliest sown or planted*—round beet, peas, forcing carrot, lettuce, radish, early cabbage, peppergrass, mustard, spinach, kohlrabi, turnip, scallion, early potato, fetticus. *Second sown or planted*—bean, beet, corn, cucumber, melon, citron, eggplant, New Zealand spinach, okra, pepper, pumpkin, celery, cauliflower, kohlrabi, squash, tomato, rutabaga, turnip, late cabbage, Brussels sprouts, kale. *Third sowing or planting* (for fall use)—round beet, forcing carrot, winter radish, spinach, fetticus, cress, radish, turnip, mustard.

Crops not suited to succession cropping because they occupy the ground from spring to fall: Parsley, parsnip, sweet potato, chicory, salsify, leek, scorzonera and the perennial vegetables—asparagus, artichoke and rhubarb.

33

SOIL SURFACE MANAGEMENT

THE thoroughness with which the soil is prepared for planting determines, to a large extent, the cost of the after cultivation of the crop. . . . It also predetermines, to a very considerable extent, the stand of plants which will result from the use of good seed. Good seed upon poorly prepared soil will give an unsatisfactory and uneven stand of plants, while good seed upon thoroughly prepared land should give a perfect stand of plants.

LEE C. CORBETT,
In *Garden Farming.*

NEXT to developing and maintaining a good soil as outlined in Chapter 26, tillage is the most important feature of soil management because it largely influences both yield and quality of product. It modifies the physical structure of the soil, alters soil temperature, aërates the soil, makes conditions favorable to bacteria and other creatures

SOIL SURFACE MANAGEMENT

that change soil constituents, destroys weeds, prevents soil washing by rains and winter snows and covers manures, green manures, etc., by plowing and digging.

To attain these ends tillage must be done at the right time, in workman-like fashion and with the proper tools. Timeliness is of supreme importance because upon it success may depend more than upon any other factor. Soil conditions may be exactly right today to plow, dig, harrow, rake, hoe, or weed but tomorrow be unfavorable for any of these operations because of a rain and because wetness may continue for a week longer. Weeds will not be slow to take advantage of this opportunity to grow without molestation and they will be all the more difficult and costly to master in consequence. Apart from that, however, the delay may postpone seed sowing or plant setting until conditions are less favorable than would have been the case had the delay not occurred.

Thoroughness is gained first by using the right implement or tool and second by using it in the right way. Labor-saving implements generally do better work, more easily, cheaply and quickly than do old fashioned ones. So before you buy such be sure to determine its efficiency, adaptability to the work required, the ease with, and speed at which it may be operated, its durability and (lastly) its price.

Prior to plowing it is a good practise to use a disc or a cutaway harrow from end to end and when possible from side to side across the area so as to chop up the surface and combine the manure, green manure, weeds, etc., with it 2", 3" or deeper; for when this fined earth, manure, and other material is plowed under it mixes more intimately and with smaller air cavities than would be the case with large clods. Water connection between the subsoil and the surface will also be improved and the vegetable matter of the manure made to decay more readily. These remarks apply especially to spring and summer plowing.

You may plow (or dig) whenever the soil is in fit condition—autumn, spring or summer. In cold climates fall plowing (and digging) is especially useful in improving heavy soils, provided the furrow slices and clods are left unbroken during winter just as the plow turns them up; for thus rain and snow water pass into instead of over the ground, penetrate deeply and form a reservoir of larger capacity than if the ground were harrowed then. More important, however, the large, rough clods are broken apart by frost and made finer. Manure, green manure, weeds, etc., become decayed sooner than would the same materials plowed under in spring or summer, hence are more quickly available to the succeeding crop. The soil being more or less on edge, rough and therefore with larger surface, dries out more quickly than if it were flat, so spring harrowing and planting may be done sooner. Fall plowing may often be done when work is not pressing, thus reducing the rush work of the spring. Plowmen are then more likely to be available than in spring when everybody wants them. Fall plowing also exposes many hibernating insects to destructive agencies such as frost, birds and other animals.

In climates where the ground freezes little or not at all plowing in the fall is less desirable than in spring because plant food may be lost by leaching due to autumn and winter rains.

Whether in a cold or a warm climate wait to do spring plowing until the soil has lost its excess water either by drainage or by ordinary sun and air drying. To hasten drying it is advisable to avoid applying manure until just before plowing because this material when spread long before plowing becomes soaked with water, acts like a mulch and checks evaporation of water from the soil.

Harrowing follows plowing (except in the fall) to fit the soil for sowing and planting. Harrows are of many styles. At least two (preferably three) should be used, one for breaking down the furrow slices, the second for reducing the clods and the third for making the surface fine and smooth. The spike-tooth, one of the oldest styles, does poorer work than do many of the more recent inventions. Its teeth do not penetrate the soil deeply enough; they push aside or bury the clods instead of pulverizing them and it tends to bounce and thus miss working on patches of ground. Much better are spring-tooth styles, especially on stony ground because they level the furrow slices well, bring clods to the surface and break them up. Disc and cutaway harrows are even better as pulverizers, especially of soddy and clayey soils. When weighted heavily they cut almost as deeply as their axles. For this reason they are without rivals for preliminary harrowing of heavy soils.

To follow these implements the Acme and the Meeker harrows are especially valuable. The former pulverizes several inches deep, breaks clods missed by the other harrows and smooths the surface well; the latter starts where the former leaves off and makes as fine a surface as can be made by hand with the steel garden rake, crushing even the smallest clods and leaving a well smoothed seed-bed when its work is done.

Besides using the harrows best adapted to the soil and the work in hand successful harrowing depends upon the amount of water in the soil. When wet more harm than good may result because of puddling; when too dry a large proportion of the clods, particularly of clayey soils, will remain unbroken. One of the commonest errors is to leave the plowed furrows for several to many days before harrowing them. During this time the soil loses enormous quantities of water through evaporation and the clods, especially if stiff, bake more or less and cannot be broken by the harrow. Therefore, except for fall plowing, make this your invariable rule: Plow in the morning, continue until only so much is plowed as can surely be harrowed before the work day ends. That is, insist that all the ground plowed in the early part of the day be harrowed the same day—none left until the next.

In fitting the plowed ground for sowing or planting you will find the following order of procedure advisable for heavy clay soils: If a sod has been plowed under, the disc or the cutaway harrow will be better than the spring-tooth because the latter may bring lumps of sod to the surface. 1, Spring-tooth harrow, dragged parallel to the furrows; 2, disc or cutaway, also parallel; 3, disc or cutaway at right angles to the furrows; 4, spring tooth at right angles to the disc or cutaway work; 5, repetition of four if needed, but at right angles; if not needed then; 5, with Acme harrow at right angles to four; 6, finishing with the Meeker as often as needed to make a fine, smooth and even seed bed, each time being at right angles to the preceding one. If you have a choice between a team of horses and an equally powerful tractor choose the latter, especially if a "caterpillar" style is available, because tractors pack the soil more evenly than do horses.

When the area to be made into garden is too small to be plowed by team or tractor the work must be done by hand. Fortunately you can thus do a far better job than with power tools because you can give every cubic foot intimate personal attention.

Six tools are in popular use for garden digging; namely,

Fig. 46. Wrong and right ways to dig.

the long and the short handled shovel, the long and the short handled spade and the long and the short handled digging fork. The first three are incapable of doing as thorough a job as the last three. As the blades of the shovels are shorter than those of the spades and are curved they cannot be as deeply thrust in the soil; also only the curved tips reach maximum depths. Though the long handled spade and the long handled digging fork are powerful tools for prying up the soil clods their round handles make them hard to turn when inverting "spits" of soil. (A "spit" is a gardeners' term for the chunk of earth full depth of the spade blade and as wide both ways.)

The short handled spade and the short handled digging fork not only dig deeply when their blades or tines are thrust vertically in the ground but, because of their D-shaped handles, they are easy to turn over when dumping the spits of soil. Of these six tools the spading fork is the best for all soils except sandy ones which when dry may not lift out well but more or less slip between the tines. For such soils the spade is to be preferred.

The wrong way to use spades and digging forks is to thrust them in the ground more or less slantingly. (Fig. 46.) This demands more stooping than the correct way, is much harder work, does not properly bury the surface material (manure, weeds, etc.) and requires fully as much time as the right way.

The right way to use the spade and the fork is to thrust the blade or tines vertically full depth in the soil using the foot to bring the crosspiece at the top even with the surface, then to pry the spit up, invert it completely and finally break it small with blows from the tool. The advantages of this are that manure, weeds and other trash on the surface are completely buried.

Ordinary digging consists in inverting the soil only one spit deep. When the soil is well drained, either naturally or as described in Chapter 18 this is usually sufficient, though it is a good plan to deepen the whole area either by plow and subsoiler or by one of the various styles of trenching. In substitute trenching (the simplest but least satisfactory of trenching methods) the surface spit is dug in the usual way but the layer beneath this is loosened either by thrusting the fork deeply and at short intervals into it, by lifting up spits and letting them fall so hard as to break or by inverting them in the same way as the surface spits are turned upside down.

True trenching is safe to practise only where the soil is already deep either naturally or from previous good management. A trench two spits *deep* is dug at one end across the area to be trenched and the soil wheeled to the opposite end where it is laid in a long pile. Then a second trench is dug one spit deep next to the first and this top soil thrown in the bottom of the first trench. Before digging the lower spit of this second trench a 3" or 4" layer of well decayed manure, leaf mold or rich vegetable compost is spread on the inverted top spit. Then the lower spit of the second trench is dug and placed on the compost. This order repeats itself until the last trench has been dug when the soil wheeled from the first trench is shoveled in.

So-called "bastard" trenching is a safer method where the soil has not been improved. The surface soil two spits *across* (Fig. 47) and the full width of the area to be trenched is dug out and wheeled to the farther end where it is laid in a pile as long as the area is wide. Then a trench one spit

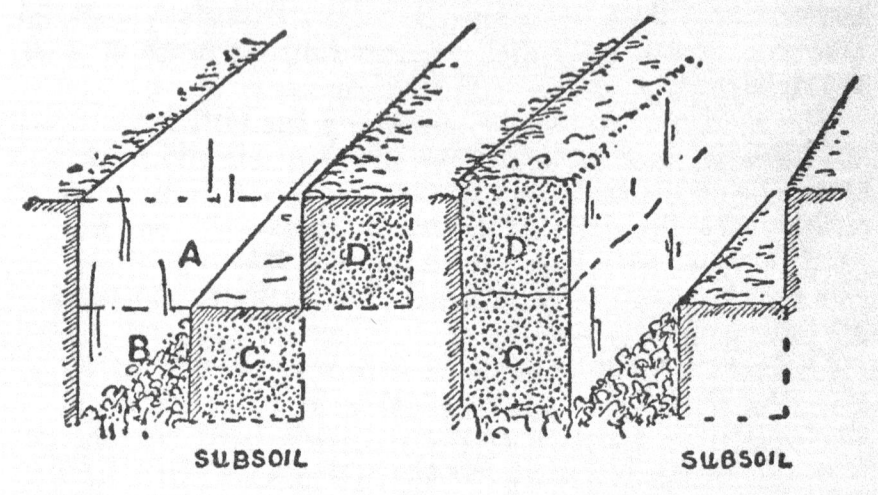

Fig. 47. Bastard trenching. Left, first move; right, completed work.

wide and one deep is dug in the subsoil (B), wheeled and piled parallel with but separate from that already placed there. Next, the subsoil (B) is broken with the spading fork (or a pickaxe if hard) and a 3" or 4" layer of vegetable matter worked into the loosened earth. Then the subsoil (C) is dug and placed on the loosened lower layer just described. Next, the top soil (D) is dug and placed at D, thus exposing the subsoil C which is dug and moved to C and the subsoil dressed with vegetable matter like B. This order of procedure continues until the farther end of the area is reached when the pile of subsoil wheeled from B is placed in the bottom of the trench, a layer of vegetable matter added and the surface soil taken from A placed on top of it. Bastard trenching is excellent as a preliminary process for one to three years prior to true trenching.

The advantages of trenching are so great that in private

gardens the work justifies itself; for the plants grown in trenched soil have at least double the feeding area that others have in soil prepared by ordinary digging or plowing. In most climates they require no irrigation because the trenched soil catches and holds far larger volumes of water than does soil ordinarily prepared. The results are that the plants thrive amazingly and in case of fruits and vegetables are of far superior quality to those grown in less carefully prepared ground.

Two great advantages that digging has over plowing are first that every cubic inch of soil may be given intimate personal attention and second that when properly done a larger proportion of moisture in the soil may be saved. To do the work properly break each spit with the fork as soon as you turn it over, by "spatting" it or thrusting the tines into it and twisting them or by doing both to break the clods. The best way to break the spit is to give it a sidewise blow so one tine after another will strike it and each take off a slice. In this way you will do a better job with less effort and with less risk of breaking the tool than if you whack the spit with all the tines at once. Furthermore you will leave the ground in better condition for the next operation —raking.

The steel rake takes up soil preparation where the harrow or the fork (or the spade) leaves off. When digging, always use it when you have dug four or five rows across the garden because you can thus stand on firm ground, avoid tramping on the soft, newly dug soil and, most important, do a thorough job and save the greatest amount of soil moisture. Always rake immediately after digging—never postpone this work until next day, or even from morning until afternoon.

Of all tools the rake is most often used improperly and carelessly but with which you can take most pride in using skillfully. Its primary mission is to break up and pulverize clods and remove stones and trash that later would interfere with cultivation and the well-being of plants. Second, it reduces hillocks and hollows to one uniform surface. Third, either turned upside down or right side up, it may be used to open little furrows for seed sowing. Fourth, when the handle is held almost vertically and the head patted upon and along the row it is excellent to firm the seed in the soil without packing too hard.

The proper way to use the rake is to start by a push and pull movement (not a downward whack) so as to hit only the tips of the clods with the teeth, then gradually work lower until the clods are completely broken and the teeth move freely their full depth through the soil without encountering obstacles, sods, weeds and other organic materials having been brought to the surface and thrown in the bottom of the trench, and the stones and other things that will not decay removed. Should there be small hollows in the surface, turn the rake upside down (teeth pointing upward) and use the back to draw loose surface soil from adjacent elevations. Then be sure to rake the elevations full depth of the teeth to get rid of any buried clods that may have been made nearer the surface by the removal of the soil. After the rake has finished, the garden will be ready for seed sowing and plant setting.

If your vegetable garden is smaller than 50 x 50 feet you will probably sow the seed by hand. But for this and larger areas and for quantities of vegetable seed larger than packets the seed drill will not only save a great deal of time and work but if you adjust and use it properly it will do a better job than can possibly be done by hand. It makes straight and narrow rows, opens the furrows, drops, covers and firms the seed, and marks the course of the next parallel row as you leisurely stroll across the garden. More even than all that, with its roller it leaves an easily seen mark from end to end of each row, thus enabling you to start cultivation and weed killing before the vegetable seeds appear. Still further, it confines the seedlings to a much narrower space than the hand can. Thus it favors the work of weed killing close to the seedlings and therefore reduces hand weeding.

These machines may be bought separately or as attachments to many styles of hand wheelhoes, the latter preferred for the small garden. As they sow one row and mark the position of the next, a taut guide line is needed for only the first row, the others being parallel.

SOIL SURFACE MANAGEMENT 191

Whether or not you buy a seed drill be sure to get at least a wheelhoe because it will save an enormous amount of work with the hand hoe and the rake and do it better.

When choosing a wheelhoe bear the following points in mind: Styles with high single wheels and broad tires are easier to run but more unsteady than those with small wheels

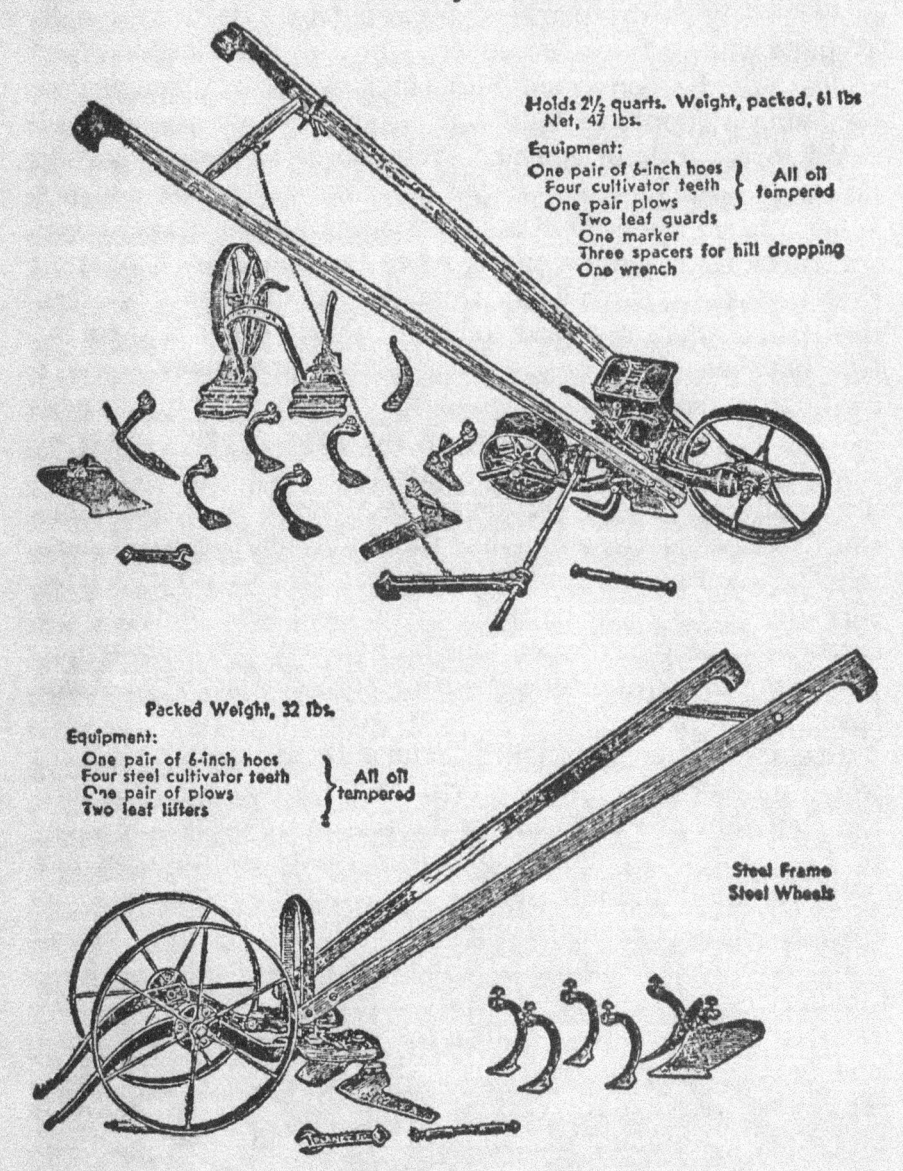

Holds 2½ quarts. Weight, packed, 61 lbs
Net, 47 lbs.

Equipment:
One pair of 6-inch hoes }
Four cultivator teeth   } All oil tempered
One pair plows   }
   Two leaf guards
   One marker
   Three spacers for hill dropping
   One wrench

Packed Weight, 32 lbs.

Equipment:
One pair of 6-inch hoes   }
Four steel cultivator teeth } All oil tempered
One pair of plows   }
Two leaf lifters

Steel Frame
Steel Wheels

Fig. 48. Wheelhoes. Seeder attached above; "sweeps" or "scuffle hoes" below.

and narrow tires and most of them have fewer attachments. Single-wheel styles will work in narrower spaces than 2-wheel styles but are less firm and therefore harder to guide. Though they work well between but not very close to the rows they can be used on only one side at a time. Two-wheeled styles can work both sides of the rows at once and more closely to the plants than would be safe to risk with a single-wheel style. Moreover, many of the double-wheel styles may be converted into single-wheel implements by removing a couple of bolts and attaching only one wheel.

Whatever style of wheelhoe you choose be sure to get the following attachments: *Right and left cultivator shovels.* When placed with their points together these make a fair substitute for a shallow plow; when the points are separated they will throw earth away from the row of plants between them; and when reversed they will throw earth toward the row between them or cover a furrow previously opened. *Cultivator teeth,* usually four with each machine. These tear up the packed soil between the rows. They cannot be run close to the rows without danger of injuring the plants. Two pairs of *sweeps (or scuffle hoes),* one long, the other short. When set close together these may be run within less than an inch of the seedlings without risk of injury. They slice the surface soil between or on each side of the rows, cutting ½" deep or deeper. One pair each of *three-prong and five-prong rakes,* the former for narrow, the latter for wider spaces between rows.

Besides these essential attachments are several others which may or may not be worth your having. For instance, the *plows* one of which throws the furrow one way, the other to both sides. Though convenient these are unnecessary when the right and left shovels are included. A pair of *small toothed three-gang cultivators* do better work close to the rows than do the ordinary cultivators because they turn up smaller clods. The three-gang discs are excellent to cultivate close to drill-sown rows before the seedlings appear and for one or two later cultivations—while the plants are small. Thus they destroy the earliest weeds close to the rows and so reduce the amount of hand work in weeding. An excellent attachment to substitute for the discs is a pair of *sweeps with extra high sides.* These not only cut the surface earth close to the plants or the row but prevent clods from being thrown over the seedlings. Another substitute is a *weeder attachment.* This is set so as to stir only the surface ½" in which it destroys millions of weed seedlings without reaching deeply enough to disturb the sown seeds over which, as well as between the rows, it is run.

You cannot over-estimate the importance of starting cultivation within three or four days of seed sowing or of repeating the work as soon after every rain as the soil crust has ceased to be sticky. If you do not have a wheelhoe or if the attachments you have are unsuited to get close to the rows of newly sown seed use the steel rake (or the wooden garden rake if the soil is light or sandy) over the entire area in the direction of the rows three or four days after sowing provided the soil is not too wet. The main object of doing this is to kill the tiny weed seedlings just starting to push through the soil. These always start to grow before the sown seeds are ready and thus may choke out the seedlings or force you to do a lot of finger and thumb weeding to save them. Have no fear that the first raking will destroy the seeds sown. Some may be brought to the surface and wasted but the chances are that you will have sown them too thickly anyway and would only have to thin out the excess plants.

This early cultivation with either wheelhoe or rake makes a loose surface which tends to check evaporation of water from the soil. Similarly you should cultivate at least once in 10 days (better every week) and after every rain that forms a crust, waiting, however, until the crust will break easily but before it begins to bake. As the plants grow and their foliage extends toward the centers between the rows you must narrow the scope of cultivation and when the leaves of any row begin to touch those of the next stop tillage altogether. The depth of cultivation should be only ½" close to the young seedlings but deeper a few inches away and deepest in the middles between the rows. As the season advances, however, it should be shallower as well as narrower than at first.

The hand hoe is by no means unnecessary. Even though you have a wheelhoe you will need to do more or less hand hoeing. However, you may lighten these labors by using smaller bladed tools than the ordinary rectangular style. Also various patterns do work superior to that of the common hoe. This hoe is too heavy for convenient use and too broad for deep penetration of the soil, for narrow inter-rows or spaces and close to delicate plants. Instead of it choose an "onion hoe," one with a narrow blade and a solid shank into the conical hollowed end. (Chapter 35.)

Of the many styles of narrow and pointed hoes that do

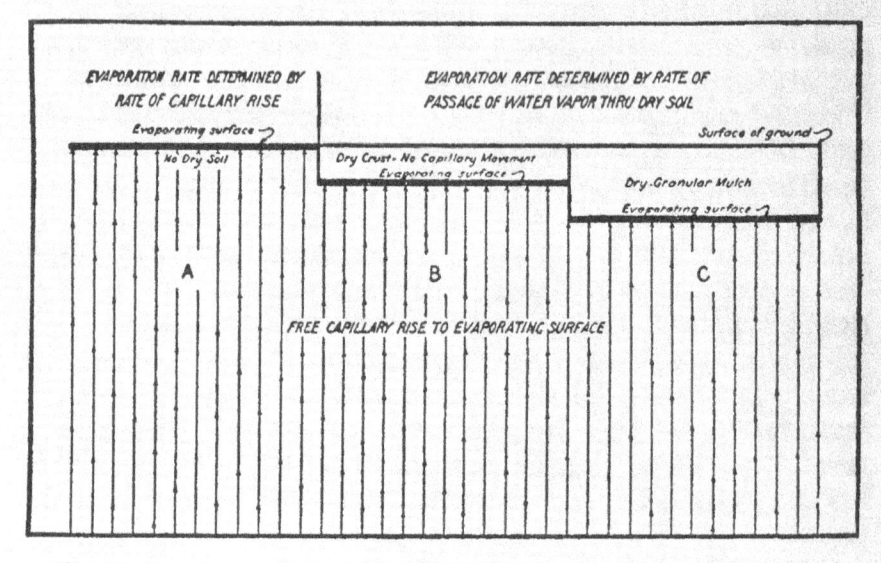

Fig. 49. How top soil condition affects evaporation. A, While soil is moist to the surface evaporation is free. B, C, When a dry layer is formed the evaporating surface is lowered so the rate of evaporation is reduced.

good work perhaps the Warren or heart-shaped hoe is the most generally useful. Its point opens furrows for seeding and the "ears" close them. It is also better adapted than the common hoe to dig shallow holes for transplanting, etc. Besides these pointed hoes are the scuffle (Dutch or push) hoes or scarifiers which merely slice the surface and which, being operated as one walks backward, leave an untrampled surface. They are specially useful after plants have grown too large to admit of using the wheelhoe, for maintaining a shallow dust mulch and controling weeds toward the close of the growing season.

Hand weeding is nowadays far less laborious than formerly. The tools and methods already discussed have reduced it and so have the many styles of hand weeders. My own favorite, the Hazeltine, is made of a piece of flat steel bent somewhat like a crooked finger and with one continuous cutting surface divided into five parts—a 1-inch end for close work among seedlings, a 2-inch "end joint" for thinning plants in the row, and a 4-inch "middle joint," also for thinning to this distance and for quick cutting of the surface beside the rows. The speed and skill with which this tool can be worked is astonishing and satisfying.

Lang's weeder is somewhat similar to the Hazeltine but is much smaller and has a strap over the fingers to hold it when not in use, thus leaving the hands free without having to drop it and then pick it up.

Another excellent tool for thinning and transplanting is the "onion weeder" which somewhat resembles a triangular hoe with a relatively broad blade and only a single hand hold.

The Excelsior weeder, like a hand with five stout, stubby but bent fingers and the Eureka, with three bent, stiff wires, are more useful for loosening the surface soil near plants and for scarifying than for actual weeding, though when used before weeds appear they kill countless seedlings.

If your garden is very small the Norcross cultivator (or weeder) in a limited way will take the place of the wheelhoe. It consists of five adjustable bent steel prongs with flattened, blade-like tips which readily penetrate and loosen the soil between the rows. When the center blade is removed the tool may straddle the rows of seedlings and weed or cultivate on each side. By sliding the top plate the width may be varied to suit the work in hand.

The long-handled Hazeltine weeder is not to be compared with the short-handled one or with the narrow-bladed hoe because it cannot be operated so accurately and is awkward when used as a hoe. As a loosener of the surface inch of soil and as a weed seedling killer it is also inferior to the "Speedy Weeder and Scarifier," a tool with a two-edged narrow blade that may be used to cut either shallow or rather deeply by either pushing or pulling it through the soil.

34

WEEDS

If the definition that a weed is a plant out of place, is accepted, almost any plant may become a weed. On the other hand, perhaps almost every weed may, in some way, become an economic plant.

P. H. ROLFS,
In *Subtropical Vegetable Gardening.*

WHEN Solomon "went by the field of the slothful" and noticed that weeds "covered the face thereof" he "received instruction"; but as he failed to record *what* he learned, later observers have been forced to draw their own conclusions. The result is endless books on weeds! Most of these place the emphasis where it does not belong—on the weeds!

Good farmers are not interested in weeds, but in good farming. They are not the people who fear that the Canada or the Russian thistle or any other "bad weed" will put them out of business. On the contrary, they are so intent upon growing profitable crops and improving their land that weeds just have no chance to become annoying.

Solomon noticed that the weeds thrived in the field of *"the slothful."* If he had been a better observer, or perhaps I should say recorder, he might also have included *the mismanager;* for weeds gain and maintain the upper hand probably just as often and as extensively where the hard working farmer follows unwise methods as where the man is lazy or shiftless. Clean fields are characteristic of good farming; weedy ones of poor farming.

So far as weeds are concerned good farming stresses prevention rather than cure, though when cure is necessary it acts promptly and rigorously. To know how weeds are disseminated and propagated is to be fore-warned and fore-armed. No control can be exercised over the transportation of weed seeds by wind, water, birds and animals. It is often also impossible to avoid their introduction in hay, straw or manure brought from outside the farm. But there is no excuse for allowing them to ripen their seeds on the place. If they have not been destroyed by cultivation they should be cut or pulled not later than their flowering stage. Similarly, in these days of state and government seed analyses there is no excuse for buying bulk seed, especially of grasses and clovers, without *knowing* it to be pure, or at least what proportion of weed seeds it contains and therefore what enemies and how many you are paying for when you admit them to your place and thus lay the foundation for future trouble!

If you will keep clearly in mind that, except in special cases, weeds are crops which do not pay for themselves but, on the contrary, reduce the profit on desired crops with which they compete—if you will keep this in mind you will better understand the importance and practise of rational farming methods. About the only ways in which weeds may be of benefit are to occupy fallow ground and thus prevent washing of the soil and leaching of plant food, to save plant food that might otherwise be wasted, to add humus to the soil when plowed under, and to provide a "costless" cover crop for orchards in which they are allowed to grow after midsummer. This last is decidedly inferior to a sown cover crop because it is likely to produce and spread weed seeds that give trouble in subsequent years.

Weeds are responsible for losses in many ways, especially in impairment of real estate value due to the appearance of the property; to robbery of plant food taken from the soil; to the cost of eradication—money wasted in work and reduced yield of crops among which they grew; to the lowering of value of "seed crops" such as clover, grain and grass; to the cost of removing their seeds from those of seed crops; to propagation of plant diseases and the harboring of insects that attack related crops; and to possible injury to live stock.

Though weeds are grouped as annuals, biennials and perennials they may all be destroyed most easily during the first week of their existence. Raking the surface half inch or inch of soil is enough to dislodge and kill them by exposing their roots to sun and air. In small gardens the steel hand rake or the wheelhoe set with "sweeps," rake teeth or 3-gang cultivators is highly effective; on a field scale, the horse-drawn weeder does excellent work. These tools when run in the direction of the rows (yes, and over the newly sown rows themselves!) two or three days after sowing will destroy millions of weed seedlings without appreciably damaging the sown seeds. The reason is that the weed seeds germinate sooner than the sown ones because they are moist; whereas the crop seeds are dry so must absorb water before they can sprout. After the crop plants appear cultivation can be given only between the rows.

The larger that weeds are allowed to grow the more labor will be needed to destroy them, the greater will be the damage they do and the more likely will they be to go to seed. When they do scatter their seed the trouble increases; for, to quote an old adage, "one year's seeding, makes seven years' weeding!" Mowing and hand pulling when in bloom will prevent this disaster so far as annual and biennial weeds are concerned. Perennials, however, usually bob up serenely the following year so must be fought in other ways.

Some species of perennials, such as quack-grass, Canada thistle and milkweed that spread by underground stems can be mastered only by constant cultivation or by "smothering." Both of these prevent the formation of foliage without which the plants will starve to death. In a small way smothering may be done with tar-paper spread and anchored over the infested areas. On a large scale I have found hemp sown broadcast and thickly to be satisfactory.

Vine weeds such as hedge bindweed cannot be mastered in this way because the plants would climb to the light and thus continue to live. Persistent and frequent cultivation is the "main strength and stupidity" way that I have always followed when I had this pest to deal with, but one that "listens good" is to plow the infested area in late August or early September and immediately sow rye and winter vetch broadcast thickly for a winter cover crop. In spring plow the rye when 6" or 8" high, harrow the ground and seed to mustard or Dwarf Essex rape to be turned under when about as tall for a sowing of buckwheat. Should hedge bindweed vines appear among the buckwheat this crop is to be plowed and the land again seeded to buckwheat, rye and vetch; if not, the buckwheat may be allowed to mature. The cost of seed and of planting all these crops will be more than offset, first by starving out the bindweed and making the land safe for cropping and second by incorporating large quantities of vegetable matter in the soil. (Chapter 29.)

With due respect to all the above practises and methods, the greatest means of preventing weed troubles on a field scale is crop rotation. (Chapter 32.) When land is sown to the same crop year after year the weeds that naturally thrive under the same conditions become entrenched pests. Also when hay lands begin to fail wild carrot, oxeye daisy, black-eyed susan and buttercup hasten the process until the grass is starved and the weeds are in full possession. However, crop rotation that includes at least one smother crop such as grass or clover or a mixture of both followed by inter-tilled crops such as corn, potatoes or turnips, the weeds become conspicuous because of their fewness—just one here and there!

Other ways by which weeds may be kept under control are: 1, changes in tillage practises—deeper or shallower plowing, harrowing or cultivating with different types of tools or at different times of year; 2, frequent stirring of bare soil surfaces with harrow or rake; 3, similar treatment of crops in rows; 4, hand pulling of weeds missed by tillage tools; 5, plowing land promptly after harvesting a crop and either replanting or sowing a cover crop; 6, use of sheep and goats to destroy weeds and brush and hogs to root out perennial tubers and rootstocks; 7, by letting "your light so shine before men that they, seeing your good works" may follow your example, reduce the number of weeds on their places and thus have less seed to re-stock their places and yours!

35

TOOLS

OUR aim should be not how much strain our strength can stand, but how great we can make that strength. With such an aim we shall incidentally and naturally find ourselves accomplishing more work than if we aimed at the work itself. Moreover, when such ideals are attained, work instead of turning into drudgery tends to turn into play, and the hue of life seem to turn from dull gray to the bright tints of well-remembered childhood.

FISHER AND FISK,
In *How to Live.*

ONE of the greatest satisfactions you can have in farming and gardening is to use the right tool for each kind of work. Just as auto mechanics have special tools for saving time so you can have a special tool for each kind of garden work. Such tools lighten and speed up the work and thus reduce both the time and the labor as well as add to the pleasure of doing the work well.

Though a poor workman quarrels with his tools he is to blame for having chosen poor ones. On the other hand, though a skilled workman may be able to do a good job with poor tools he is the man who never buys any but the best because he knows the satisfaction of owning, to say nothing of using good ones. No one better than he knows that a good tool keeps its temper and thus helps him to control his. If anyone should have good tools he is the owner of the small farm.

Suppose that good tools do cost a little more than inferior ones, the difference is, or should be, more than offset by higher quality, keener edge, better service, longer life, and the increased satisfaction these characteristics insure. Good tools properly cared for will often last many times as long as cheaper ones of the same kind. They will wear out rather than give out. Their initial cost is therefore less than two, three or more of the cheaper kind that might have been necessary to buy during the same period. By the time the good tools will have worn out their prices will have been forgotten, but the satisfaction they will have given during a term of years will always be pleasant to remember.

One of my trowels will illustrate these points. At the time I bought it other trowels were selling at 25¢ to 50¢ but I paid 75¢ (!) for mine because it was made of a better steel and because its blade and handle are all of one piece of metal, the handle being expanded to form a hollow cone in which a short piece of hardwood is fitted and fastened with a pin. Because of the quality of the metal and the construction this tool can neither pull apart nor be broken by any effort that a man can make when using it as a trowel should be used. During the 17 years I have used this tool other trowels of several styles have either pulled apart or broken, but this one, though its blade has worn down about a third, is still my favorite.

Most cheap trowels are made by thrusting a metal shank into a wooden handle which, being thus weakened, soon either pulls apart or breaks. A still cheaper style is made of one piece of metal pressed into trowel-shape so part forms the blade, and part the handle. All specimens of this style that I have seen are weak where handle ends and blade begins. They are therefore useful for only the lightest kinds of work.

The reason why I have discussed these trowels so fully is because two of the methods of construction are used in making various other hand tools, especially rakes, hoes and weeders. When these are made by the conical construction plan it is impossible to pull them apart and almost impossible to break the shank by ordinary use, because the tools are strongest at the point where greatest strength is needed. But in the other style, when the shank is tapered and thrust into the wooden handle, the wood is always weakened at the point it should be strongest and it is almost certain to work loose.

Attempts to prevent such pulling apart and breakage by placing a ferule where metal and wood meet is at best a makeshift way to offset a weakness. The only way to prevent the wood and the metal pulling apart is to use a ferule four or more inches long for rakes, hoes, etc.—much more for spades—and to pass a metal pin not only through it and the wood but also through the flattened metal shank of the tool. This forms a permanent union and positively prevents separation of the various parts. On tools of equal size such construction is weaker than the cone style because the amount of wood is reduced.

Another objection to this style is that manufacturing costs are higher than for common construction; hence selling prices must be higher. However, such costs are offset by the greater value and strength of such tools; for manufacturers are careful to put better metal as well as workmanship into tools of this type than into the cheaper styles. Of the three styles, other things being equal, the cone type is the strongest though not necessarily the most costly. Spades, shovels, forks and some other large hand tools are made after this pattern. Like all other tools so constructed they cost a trifle more than the other kind but are well worth the difference.

Much of the economy and success as well as pleasure of gardening is derived through adequate tool equipment; yet only too often people who supply only their own tables with home grown fruits and vegetables manage to get along with poorer and fewer tools than they really need. If they think at all about the matter they silence their better judgment by declaring that their gardens bring them no money, therefore they will put as little as possible into them. This applies as much to fertilizers, manures, seeds, plants and work as to tools. All mistakes!

These people fail to realize that they can doubtless make several times as much profit as market gardeners do out of the same crops; also that unlike these commercial growers they have, or should have, no waste; for they can gather each crop as soon as ready and sell it—all of it—without deterioration of quality to the best market of the world—their own home tables! If they will credit their gardens with only the cost of equal amounts of store stuff at market prices, even though they disregard the freshness which warrants higher prices, they should more than justify their outlay for adequate equipment of tools and accessories and plant food to assure abundant yields.

Both because tools are costly and because keeping them in good condition enhances the satisfaction in using them, tools should be kept in proper places. Next to having them exposed to the weather the worst place to keep them is probably a cellar because of the almost invariable dampness and the certainty of rust. Some of mine have suffered in this way because at first I had no other place to keep them. Several of my friends store their tools beneath their verandas, the floors of which are high enough above ground to serve as shed roofs. Other friends have tool rooms or sheds or use their garages. All such places are good, provided they are dry and airy.

One point seldom emphasized is that an owner is more likely to take care of good tools than of poor ones. A case that came under my observation illustrates this. Years ago I visited a Maryland farmer who had just bought a transplanting machine for, I think, $150. He kept it in a shed and maintained it (like his other tools) in good condition by timely care. About 12 years later when I again visited him the machine was still giving him good service. During these years, he told me, one of his neighbors had had three transplanters whose terms of usefulness had been shortened because the sky was their only coverlet!

No matter how or where stored, tools should always be cleaned after being used, especially before being put in place. The rough earth may be scraped off with a flat stick and the finer particles rubbed off with burlap or some other coarse fabric. Whether or not the tool is soon to be used it is a good practise, especially in damp weather and before winter sets in, to rub all unpainted metal parts with an oily rag before hanging it up. Thus a thin film of oil is smeared on the surfaces and rust is prevented. The oil drained from the crank case of an auto is good enough.

It is an excellent plan to overhaul the tools at least once a year, preferably as soon as the gardening season has closed. At this time worn and broken parts may be discovered and replaced. It is also a favorable time to paint the wooden and metal parts that do not come in actual contact with the soil. Instead of sticking to the colors applied by the manufacturers it is highly desirable to use only one color or a combination of colors so that all the tools one owns will have personal distinction. Whatever color scheme is adopted it should be such as will "stand out" well—orange, red, yellow or light blue—against the natural colors of the garden. Poor ones are green, brown, black and dark blue. Branding is also a good idea. Branding irons cost little but last "forever" and their brands last as long as the tools.

As part of the overhauling the tools should be sharpened; for much of the efficiency of spades, hoes, scythes, sickles, lawn mowers and pruning tools depends upon the keenness of their working parts. With the large tools, sharpening is best done by an emery wheel or a flat file. These are too rough for the finer edged tools such as knives and shears. An oil stone is best for them. After each tool is sharpened it should be rubbed with the oily rag, especially along the newly sharpened parts.

One of the greatest ways to waste time is to store a miscellaneous assortment of hand tools all huddled in a dark corner; for when so mixed desired ones are hard to find. In order to save time every individual tool should have its own place. Among the many satisfactory ways to arrange long handled garden tools are racks with holes bored for the handles to stand in erect; double hooks for spades, shovels and forks; and single hooks to hold up the handles of such tools as lawn mowers, wheelhoes, and wheelbarrows. These last must be provided with a cleat on the floor to prevent their escaping from their moorings. Small tools such as shears, trowels, dibbles and weeders may be assigned places on the wall and held by various home-made devices. The one place not to keep them is a drawer, unless each tool has its own especial compartment. The drawer should be well ventilated to prevent rust.

In order to make a quick check-up of tools to see that each is in its place and thus save time, perhaps also money, is to paint on the wall, in the cupboard and the drawer the exact position and form of each tool in some color that contrasts with that of the tool. When the design is made larger than the tool it will show at a glance whether or not the tool is in place so that, if necessary, search may be made for it.

Many, but not all, of the garden supply stores carry well chosen stocks of standard and nationally advertised tools. Thus they may be of far more distinguished service to beginners than most local stores that handle only limited lists of ordinary equipment; for there varied assortments may be examined before purchases are made.

From what has already been written the importance of this personal selection can hardly be over-estimated; for it is greatly to be regretted that clerks, especially the "extras" employed during the rush season are not always as well posted as they should be on gardening operations or on the quality of materials, construction, workmanship, advantages and often positive disadvantages of many tools they offer for sale. It is therefore advisable not to place too implicit confidence in the statements of the salesman. To avoid such risks it is a good plan to buy tools during the off season for then the clerk is probably one regularly employed in the department and therefore likely to be well posted.

Some points to bear in mind with respect to certain tools and implements not already discussed are well worth considering. For instance, avoid combination tools such as rake and hoe mounted back to back and attached to the one handle. Though the salesman may declare that you get more for your money because you are buying two tools for the price of one you will really be getting less and will be adding to your work when using them; for not only is there no sense in toting one while you are using the other (!) but such tools are usually made of metal so poor that they are likely to break as well as do inferior work.

Still worse is the double edged pruning saw. It is not only too light for heavy work and too heavy for light work but with it you are almost sure to do more damage than with any other tool; for it will often tear the bark of trees in unexpected places and even the hand that uses it!

By keeping the points discussed in mind it should be safe for anyone to trust himself in a garden supply store to choose the tools and other necessary items listed in the appendix.

Every year new tools, modifications and improvements on old ones are introduced. Seldom are they found in the garden supply stores sooner than the third year after they are first advertised because it usually takes about that long for the public to prove their worth. Until this has been done the stores are loath to list them. So it is advisable to be on the lookout for such tools as advertised in the farm and garden magazines and to try those that promise to be labor-, or time-savers.

To judge by my experience it pays to buy at least some of them. In any case it is well to send for circulars and catalogues offered by makers or introducers so as to be the better prepared to yield to the temptation to buy!

36

RE-MAKING A NEGLECTED ORCHARD

THE loss due to insect attack may be reduced materially by the adoption of proper methods of prevention and control. In many cases, the program to adopt involves no direct fighting, such as spraying, but simply the shaping of farm, garden, or orchard practise along lines unfavorable to the insects concerned—such matters as rotation of crops, or cleaning fields of weeds. Today's warfare against insects strives towards prevention as well as cure.

WALTER C. O'KANE,
In *Injurious Insects.*

WHEN a new farm is to be developed from an old one an important problem likely to arise is how to treat the fruit trees already existing on the property. As a rule they have been systematically neglected, abused or mismanaged for years and thus have supplied conditions for the spread of pests such as San José scale, borers, blight, canker and kindred enemies. As consequences the trees are full of diseased, dying and dead wood, interfering branches and watersprouts. The lack of cultivation has also permitted bushes and rank weeds to dispute possession of the ground with the trees and such vines as poison ivy and Virginia creeper have embraced the opportunity to clamber up the trunks to light and air.

On one property bought several years ago by a city man was an orchard which would fit the above description. The trees had been neglected though not maltreated, and the

RE-MAKING A NEGLECTED ORCHARD 207

place was so choked with undergrowth that his friends suggested he cut down the trees and save the berry bushes! During the previous five years this orchard had produced an average of less than half a barrel of apples to the tree!

Fortunately the new owner called in a competent orchardist who prescribed treatment which amply paid for being put in practise; for during the following six years (after renovation) the orchard averaged three barrels annually to the tree. Thus an orchard which had never paid the former owner became one of the most profitable areas on the place.

In another case "the trees were in such bad condition" that the advisability of cutting them down and planting a "real orchard" was considered seriously; but "it was finally decided to see what could be done to bring them back into usefulness." The first two years after revamping the trees averaged less than a peck of fruit each; but during the next four years they bore about two barrels to the tree as an annual average. As the fruit sold at an average of $4. a barrel it paid a fair profit on the investment of time in renovation.

Such cases indicate what may be done with apparently almost hopeless fruit trees. They also emphasize the importance of securing the services of a competent orchardist to advise whether or not it will pay to attempt reclamation not only of individual trees but of orchards; for often though individual trees may be brought back to usefulness there may be so few in a given area that the cost of annual management—on account of the area to be handled—might be disproportionate to the possible money returns. Nothing but careful examination of both the individual trees and of the area as a whole by a man familiar with orcharding can decide what would be advisable in any specific case.

Such cases as these must be kept distinct from those dealt with by the tree surgeon. The deciding factor in an orchard problem should always be the money cost and return; whereas tree surgery cases are influenced by sentimental, pictorial, historical or similar reasons—not at all by the sordid question "will it pay in dollars and cents?"

Before calling in such an orchard specialist you, as owner or prospective buyer of a property may apply the following suggestions to your problem so as to form your own opinion as to whether or not it may pay to attempt reclamation. After having formed this opinion will be soon enough to seek outside advice. Even where you may decide upon destruction of the trees and the planting of new ones, you should usually not take action until after consultation because the specialist may discover advantages which may have escaped your observation. On the other hand, should you think favorably of reclamation the specialist may advise destruction because of disadvantages that you have not noted!

When decision is made to reclaim, the specialist can not only indicate how the work with the trees should be done, but how the whole area should be managed. In fact, he can prescribe a schedule of treatment to cover a series of years. Thus his advice should enable you to improve the condition of the trees by pruning and spraying, to strengthen the growth, avoid trouble by cultivation and how to handle superior crops of fruit.

In the same ways as he makes his superficial (or preliminary) survey you may well consider the following points: Neglected peach trees older than five years are usually so badly infested with borers, yellows, little peach, rosette or other pests that it will not pay to work them over. They had usually better be pulled out and the ground rejuvenated by green manuring for at least two years before again being devoted to orchard. However, don't be too hasty! On one farm that I bought there were about a dozen peach trees "on their last legs" and at least 10 years old. As they had been planted for home use and were of choice varieties that covered a season of two months I overhauled them. The result was they supplied the family until the new orchard started to bear, by which time only about half of them were left. They were not worth saving for "making money" but they prevented our buying poorer quality peaches from the neighboring commercial orchards.

Cherries, plums and crab-apples cost so much to pick and the former two are so affected by brown rot, black knot and curculio that usually only enough trees should be saved to supply the needs of the family and of others living on the place.

RE-MAKING A NEGLECTED ORCHARD

Some varieties of pears are so subject to fire-blight that they are soon killed. Among others are some which may be so full of branches killed by this disease that when they are removed there will be little left. But so great is the vitality of pear trees that new wood may often be made to develop merely by cutting out the dead branches. Therefore, trees that are known to be of good varieties may often be saved to supply home needs for several to many years.

However, when there are many pear trees on a property, especially when they appear to be vigorous and free from blight they are probably Kieffer, the poorest variety of the list still carried by nurserymen. This variety may be identified by the branches which, in bearing trees bend in more or less pronounced arcs—much more than do most other varieties. Generally the trees bear heavily every year and the fruit finds a fair sale at moderate prices so they may be worth saving as a source of revenue, not primarily as a means of enhancing the pleasures of the table, even though the canned fruit is passable—when there is none else!

Whenever a branch affected by fire-blight is removed both the tool and the wound should be well sterilized by wiping with a rag or a sponge kept wet with a solution of corrosive sublimate (mercuric chloride) made by dissolving a tablet of the chemical in a pint of warm water. The dissolving must be done in a glass, crockery, or wooden vessel, not in any metal receptacle. Because of its highly poisonous character both the solution and the supply of tablets should be kept under lock and key to prevent children and animals accidentally getting it. Also it is a good plan to keep an emetic and a supply of dried egg-white, powdered milk or other albumenoid within handy reach, the former to cause vomiting, the latter to neutralize the poison after vomiting has been induced. Skim milk and fresh egg-white are also good. In every case call a doctor immediately!

The one outstanding fruit likely to be found in orchard blocks and to be profitable is the apple, especially if the varieties are few, in goodly number and in commercial demand. So far as home needs are concerned it is better to have 10 or even 20 varieties which ripen more or less successionally and are in season from midsummer until April,

May or even June. The first things to determine, therefore, are what varieties predominate, what others are included in the list and the variety name of each individual tree.

To supply the home it may be worth while to allow trees to live even though it will not pay to do much work on them or when they are the few survivors of a formerly much more populous orchard. But from the standpoint of income, individual trees that have badly decayed trunks and limbs are so likely to break down that it will usually not pay to do much work on them. Often it would be more profitable to cut such trees down and convert their trunks and main limbs into fireplace wood.

When an orchard has many blanks after the removal of such decrepit specimens, even though it may be feasible to rejuvenate the remaining trees the number left may be too small to permit of profitable returns from the area; for the annual cost of management must be considered in terms of area rather than in number of trees. The cost of managing a 40% stand is more than twice that of handling a 100% stand.

In cases of partial stand it will not pay to replant vacancies because many of the new trees will fail through impoverished soil, because those new trees that might survive would probably require 10 years to reach profitable age by which time many of the old trees might have disappeared. This would leave a lot of blanks to be filled with still younger trees. An orchard so planted would be both unsightly and unsatisfactory to manage.

Individual trees that may be readily worked over to other varieties are those which have sound trunks and limbs even though many of the branches may be interfering and there may be quantities of watersprouts. The presence of these latter in abundance is a good index that the trees are vigorous and that the soil is at least fairly well supplied with plant food.

Trees with numerous dead branches are not necessarily doomed especially where watersprouts in abundance indicate vigor. Such branches may have been killed by San José or oyster-shell scale through neglect to spray with proper materials.

RE-MAKING A NEGLECTED ORCHARD 211

After this dead wood and other unnecessary branches have been cut out and after the rough bark has been scraped off the trunks, collected in sheets placed to catch it so it may be burned to get rid of insect eggs and cocoons, the trees—provided they are dormant—should be given a thorough drenching with winter-strength lime-sulphur solution or a miscible oil. Never should such materials be used later than the earliest swelling of the buds or they will kill the newly

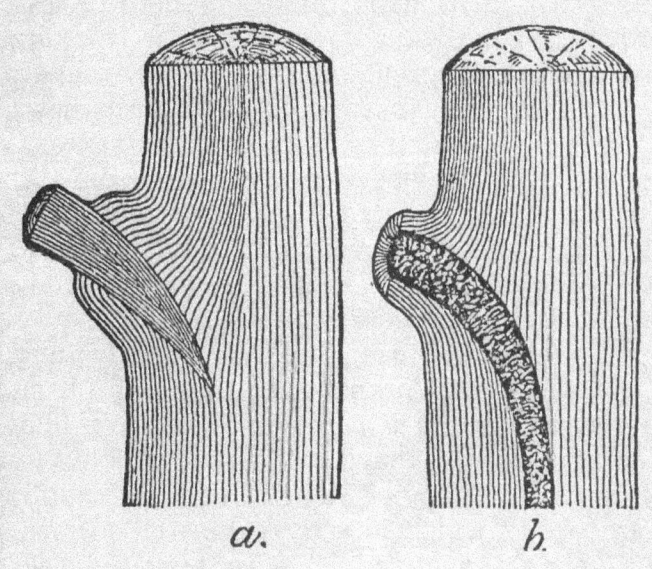

Fig. 50. How decay follows through the heartwood of a limb-stub to the heartwood of the main trunk.

forming foliage. This dormant spraying should cover every part of the tree—twigs, branches, limbs and trunk. It will therefore destroy not only scale and other insects but also the germs of many diseases.

As to the actual operation of pruning, one precaution is necessary: When a large branch must be removed always make a cut from beneath at a point a foot or more from its attachment to the larger limb or the trunk from which it springs. When the saw sticks (or "binds") remove it and make a second cut from above so as to meet the first. Before the cuts join, however, the branch will drop off. This will prevent tearing the bark and the wood from the supporting branch or trunk. Now the stub must be removed as close as possible to the flow of sap in the remaining branch or trunk. The closer this is, even though the wound may be larger, the quicker will be the healing.

In all renovation the dead wood should be cut out first. It is worse than useless because it evaporates water, harbors insects, impedes entrance of light and air, spreads disease and interferes with spraying and harvesting.

Sometimes the removal of dead wood is all that is necessary the first season. Usually, however, more or less interfering branches should be cut out. Often it is desirable to leave well placed watersprouts on the main branches and the trunk so as to fill in gaps and in time form new limbs. Therefore it is a good plan to leave most, or at least many, watersprouts during the first season so as to serve as "safety valves" for the excess plant food and vigor which must find vent after the pruning referred to is done. Unless this is done new supplies of watersprouts may be expected.

When new, if they appear where they are not wanted, they are easily removed during late May or early June, while they are still succulent. At that time they pull off and new ones do not appear at the same points, whereas if allowed to mature (become woody) they must be cut off, only to be replaced by new ones. A boy of 10 or 12 can do the pulling off.

In general, trees which already have well formed heads of medium height should be pruned only moderately so as to maintain their low-headedness and thus favor spraying and harvesting. Very tall trees, however, with few or no branches low down on the trunks or main limbs should have many small branches in the top thinned out so as to force the trees to develop watersprouts nearer to the ground. In two or three years new branches may be chosen among these sprouts and the undesired ones cut out. This method is usually better practise than the more common one of "dehorning" or cutting off large branches in the hope of developing new heads on the trees.

Until recently orchardists and other tree specialists have favored painting large wounds with various materials—shellac, white lead, creosote, tar and sundry proprietary preparations. Recent investigations, however, show that many of these are positively injurious, others of no use whatever and still others of doubtful utility. Small wounds usually heal in a single season and wounds 2" in diameter rarely require more than two years. For large ones, grafting-wax made rather soft appears to be as satisfactory as anything for keeping the edges from drying out and thus favoring the development of new tissue over the wounds.

If an orchard is located on a poor, light soil or on sod; if the trees are making slow growth, as indicated by short terminal twigs; and if the leaves are yellowish and small, probably heavy pruning and nitrogenous fertilizers will produce good results. Care must be taken, however, to discontinue these practises before they produce too vigorous growth or unfruitfulness. When diagnosis indicates that fertilizers are needed, the materials applied should be approximately as shown in the Appendix whether used singly or in combination. Doubtless nitrate of soda or some other quickly available nitrogenous fertilizer is most likely to give conspicuous benefit. To get most effect it should be well worked into the soil as far out as the limbs spread and just before the buds begin to burst. In peach orchards nitrogen has most consistently given satisfactory returns, though potash is often notably beneficial on light, sandy soils.

37

FRUIT TREE PRUNING

JE weniger wir zu schneiden haben am Baum, desto gesunder bleibt er und desto schöner entwickeln sich Früchte.
KARL KOOPMANN.

[Translation: The less a tree is cut, the sounder it will be, and the better its fruits.]

WHEN you read the above quotation or its translation you may have wondered why I should think it necessary to write a chapter on pruning! What would you think if you read it on the title page of my 420-page book, *Principles and Practise of Pruning?*

"The less a tree is cut" emphatically does not mean that

the more it is allowed to grow "as Nature intended" the better! Even the most casual observation shows that without assistance Nature develops branches that not only become useless but sooner or later prove a menace to the health and life of the tree!

It is often said that Nature makes no mistakes. Perhaps so —from her standpoint! But she is apparently just as much interested in producing a sour, hard, inedible, little crabapple as a luscious Mackintosh or a Fall Pippin! Perhaps she is also as well pleased with her methods of developing trees that will break down because of weak construction or the entrance of decay as if she built them on sound engineering principles and protected them from internal rot. If our human opinions differ from hers it may be because of our selfish bias! But if, as we are told, we are "lords of creation" why should we not "lord it" over her as much in the construction and care of trees as in the selection of varieties that please our palates?

In Nature, plants grow thickly; each crowds the others, so that superfluous individuals and branches are destroyed. Under cultivation the distances at which plants are set and the relatively greater abundance of available plant food often increase both number and growth of branches, so unless pruned many more would continue to live than would be "pruned off" by Nature's crude methods. Hence the necessity of our using improved methods of pruning to remove this superfluous wood, or, better, to prevent its formation and to favor the better placed limbs.

Methods of preventing damage are so simple and effective that anybody may apply them and as most of them cost nothing but a little time there is no reason why they should be neglected. They all depend upon a working knowledge of the underlying principles of plant life and growth. By understanding these even the veriest tyro may discover or indentify trouble in its incipiency and literally "nip it in the bud"—often years before it would become actually threatening to the well-being of the trees.

Prevention of damage should begin with the receipt of the trees from the nursery or with the seedlings that start to grow where Nature has sown the seeds. Before planting the former, every root as thick as a lead pencil and every scraped, broken, mangled or otherwise injured root should be shortened back to sound wood. Best cuts are made with a sharp pruning knife, though a hand shear is useful, provided both its blade and its holding jaws are keenly sharpened so as to make clean cuts—not chewed or crushed wounds. Quick healing follows smooth cuts.

When the tree is planted it is imperative to remove at least 50% of the top, of all except unbranched yearling trees. (Chapter 39 and later pages of this one.) These "whips" or "switches" should never be cut because they are the "leaders" or main trunks of the trees. Among commercial orchardists 75% of two-year and older nursery trees is pruned off immediately after planting. Though to the novice this treatment may seem "heroic" it is based upon the sound principle that a balance between top and root must be reëstablished or the tree will surely suffer, if not die. In any case where this pruning is not done the tree will be slow to recover.

The way to reëstablish this necessary balance is to reduce the amount of branch area. Professional tree planters and especially commercial fruit growers who follow this practise rarely lose as much as one tree in a hundred planted.

When making this reduction of top it is highly advisable to govern the cutting by the positions and sizes of the various branches. First, note which are the sturdiest, next the distances between these on the trunk. When they are 12" or more apart a stronger tree may be developed than where they are closer together and far stronger than when they are in opposite pairs or in bunches. Also a far better balanced tree may be developed if these three, four, or five strong, well separated branches point in as many different directions; so that, if viewed from above they would look like the spokes of a wheel with the trunk as the center. Never cut these until the last.

The first branches to cut off entirely should be the puniest and the poorest placed. The only puny branches that should ever be saved are such as are well placed and point in the desired directions, even though a thriftier branch near by but badly placed must be sacrificed. The puny branch will probably develop well when the competition with the other branches is reduced by pruning.

If at any time two branches are placed so close together and on the same side of the trunk that they would interfere with each other the one more poorly placed (usually the lower one) should be cut off completely, though sometimes it would be advisable to let it remain when its position is farther up or down from the next desired branch above or below. After puny and interfering branches have been removed so that only the strongest remain these latter may be shortened from 30% to 60% so as to help reëstablish the balance between top and root.

Whether or not the main stem or "leader" should be cut is a disputed question. For many years my conviction is that it should not be because, when this is done, new branches are almost certain to form in a cluster near the the top and thus lay the foundation of one of the commonest and most distressing tragedies of tree growing—the breakdown of the top due to the splitting of the Y-crotch or crotches formed by the branches at this point (Chapter 39).

The lowest branch of apple, pear and sweet cherry trees should be 30" to 36" from the ground; those of peach, sour cherry and plum, 18" to 21" because low headed trees are less likely to be injured by high winds, the fruit to be blown off and less difficult to reach at picking time, much of it being gathered direct from the ground or with only short ladders. Then, too, injury from sun scald of the trunks is reduced, so is the difficulty and cost of harvesting and spraying.

During the first few years judicious pruning with the knife will save not only much time and labor with saw and loppers in later years but will build stronger, more symmetrical specimens than by any other treatment. Still better, nipping undesired buds and soft inch-long shoots will save even knife work. The same principles apply to each frame limb as to the development of the main trunk—spacing the side branches far apart and pointing in different directions, except that on the under sides of the frame limbs no branch should be allowed to develop because it would be too much shaded by the main branch. Better the watchful eye than the active saw! It will see prospective undesirable developments and prevent the necessity of using the saw in later years.

Actually, while the trees are young—up to the fourth or fifth year with apple, pear and sweet cherry and to the third or fourth with sour cherry, peach and plum—about the only pruning necessary after the trees have had their preliminary treatment just after planting should be the removal of occasional branches that would sooner or later interfere with the ones desired and the more or less shortening of rampant branches that threaten to rob the others of food, light and air. The less pruning done during this period the better, because the removal of wood during the dormant season tends to the production of still more wood. Thus, severe pruning during winter may postpone fruit bearing, perhaps indefinitely, if persisted in annually.

Whenever a branch must be cut off make the wound as close as possible to the part that is to remain so there will be neither a stub nor even a shoulder. Closeness of the wound to the main flow of sap favors healing. When a stub is left decay is sure to follow. (Fig. 50.)

As fruiting age approaches be sure you know where to expect fruit. On apple, pear, cherry, plum and apricot most of the fruit is borne on "spurs" as the short twigs are called. Never cut them off simply because you like to see well manicured trees. One of my clients did this with the result that even such precocious varieties as Yellow Transparent and Duchess of Oldenburg did not bear a fruit until they were 12 years old, whereas they usually start when a third that age or even less!

Intelligent pruning for fruit presupposes knowledge of the appearance and position of the fruit buds. Blossom buds are rounder and plumper than branch buds. Apples and pears bear most of their blossoms at the tips of spurs in clusters surrounded by leaves also contained in the same "cluster buds." Occasionally flowers come on the sides of twigs produced the previous year. Because of the terminal position of the fruit on the spurs and because fruit production is an exhaustive process the direction of growth changes each year and fruit is borne on the spurs only each second year. With age the spurs become gnarled and crooked, but, if healthy may be as productive as younger spurs.

Cherries bear much of their fruit on spurs, but because the terminal bud is almost always a branch bud the spurs are relatively straight. Most of the other buds on the spur produce blossoms, though an occasional one may develop a branch spur. Blossom buds are also borne near the bases of annual growths of the previous year.

The plum and the apricot bear their blossom buds partly on spurs and partly on young growths, but in more varying proportions than with the cherry.

The peach is different. It produces some blossoms on wiry twigs on the larger branches, but these growths live for only a few years. By far the largest part of the blossoms are borne, one on each side of the pointed branch buds on growths of the previous season. They can be easily recognized, first because of their position and second because of their roundness. Sometimes the bud between the pairs just mentioned is a blossom bud.

Never prune off or break the spurs of any fruit tree unless there are too many or unless failing, because a spur removed is gone forever. On the other hand, always cut back peach shoots severely—often 50% to 75%. Unless you do the tree will extend farther and farther out each year and become more and more likely to break because of the increased leverage. Again, such annual pruning will concentrate the fruit bearing area in the reduced space and thus also reduce the amount of fruit thinning that must be done in midsummer.

The quince is again different. It bears its blossoms at the ends of new growths that spring from buds that have wintered over. Pruning for fruit, therefore, consists in keeping the tops of the bushes fairly open and reducing both the number of annual growths and shortening the remaining ones a third to a half.

Though many trees die prematurely from disease, neglect or insect attack, probably the commonest cause is breakdown. At first glance this may seem to be because of overweight of fruit, ice or snow or stress of wind, but examination will show that though the trunks may reveal interior decay the original cause is bad heading—the development of frame or main branches so close together they either pull against or shade each other until they break down or must be removed. In the former case the tree is ruined at once; in the latter decay is often admitted to the heartwood which is steadily eaten away until the interior is perhaps hollow. Hence the importance of training young trees so as both to avoid such disasters and to develop branches free from fault as to position with respect to the trunk and to each other.

The home orchards of my boyhood were developed according to the principle formulated by Downing; namely, "Every fruit tree grown in the open orchard or garden as a common standard, should be allowed to take its natural form, the whole efforts of the pruner going no further than to take out all weak and crowded branches." Many of the trees so developed by my grandfather and great uncles between 1850 and 1875 are still bearing, even though since the passing of these men, they have not been as well tended by later owners.

In contrast with such records tens of thousands of trees trained on the "vase-form" or "open center" plan, or because nurserymen cut the "leaders" have broken down at a quarter the age of these veterans. In such cases the branches having developed in clusters pull against each other and break down or when allowed to develop to bearing age before removal usually leave wounds through which decay enters the heartwood as already noted and thus precedes, in fact, assures, breakdown.

Removal of large limbs is always a menace to the well-being of trees. The only limbs larger than 1" in diameter that should be removed are dead ones, those so badly diseased that they cannot be saved, or those broken by accident. Even the removal of 1" branches may be avoided by proper training while the trees are younger than five years, and most of them during the first three years!

Because of the heavy losses of trees and differences of opinion as to what are sound principles of tree training various experiment stations have conducted investigations to determine correct ways to train young trees and have reported findings in bulletins. One such undertaken by the Illinois station is reported in a 125-page bulletin (No. 376) from which the following conclusions have been chosen and condensed. The italics are mine.

Growers attribute death to various causes, but do not realize the part that pruning plays directly or indirectly. Wounds are an important factor in death and the initiation of the unprofitable period. They are often attributable to the way in which the tree was headed. *Poor heads in trees now mature are due to the severe heading-back cut given when the young tree was a whip*. Efforts should be made to

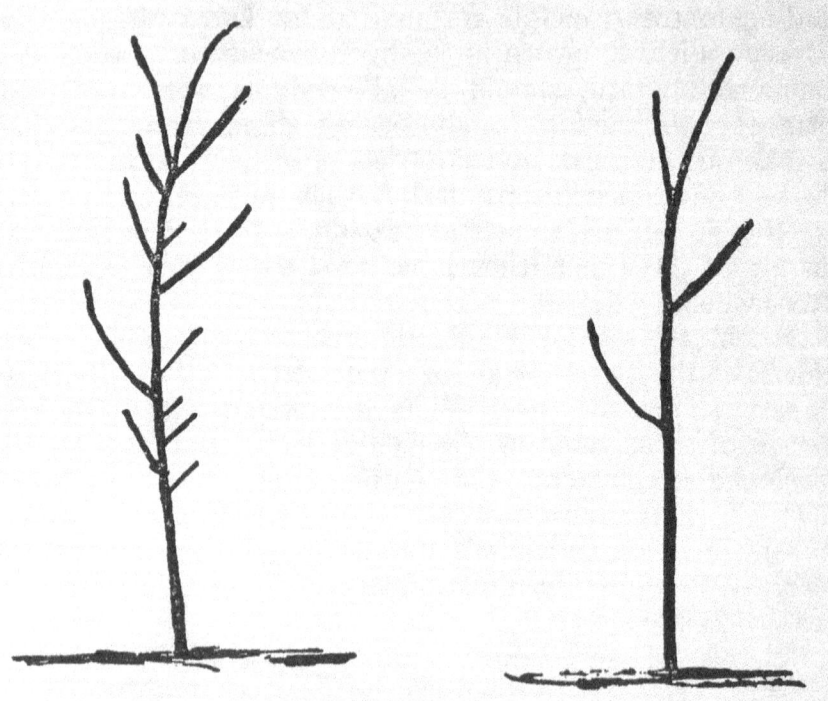

Fig. 51. Group method of disbudding yearling "whips" of orchard trees. Right, first season's growths; left, the same tree pruned, the best placed branches being saved to make the main framework. (From photos.)

produce frameworks in which equilibrium will be maintained as nearly as possible, especially late in the tree's life when wounds are likely to do most damage.

In forming the framework, very narrow angles and the excessive development of one main branch are to be avoided as these factors lead to splitting of the head. Vertical spacing of branches is desirable to avoid "smothering" of the leader. Problems of training are greatly simplified by starting the framework branches by disbudding to groups of buds, thus avoiding the dominance and the sharp forks that result from severe heading back. Uniformity is also secured among the main branches which are subordinate to the central leader. To train trees by disbudding the following steps are recommended by W. A. Ruth and V. A. Kelley, authors of the bulletin.

Of the three new methods of heading reported in the bulletin disbudding to groups of buds is considered the best and is therefore recommended in preference to the others. (Fig. 51.) The steps are based not only upon the data recorded, but also upon incidental observations made as the study progressed.

First season. Use vigorous one-year whips which have not been allowed to dry out before planting. [Fall buying and heeling-in may be necessary to assure plumpness as storage too often shrivels the trees. M. G. K.] Incline them slightly toward the prevailing wind. If they are tilted so far that the lower side becomes an underside, shoot development will be discouraged in that direction but will start to grow through the tree.

Just before growth starts disbud the whips with a sharp knife to groups of three or four consecutive buds, each group at the height where a frame-branch is wanted. Intervals between groups should be about 8" from center to center. [I prefer 12" or more because greater strength is thus assured. M. G. K.]

The whip should not be cut back. Let the tree grow undisturbed throughout the entire season.

Second season. At the beginning of the second growing season choose one branch at each height for the permanent framework. Select for uniformity of diameter and length (to secure balance), for proper direction (location in a spiral), and for an angle suitable to the variety [a wide angle makes a stronger limb connection with the trunk than a narrow one, which is likely to break down. M. G. K.]. The laterals left on drooping varieties like Jonathan and Winesap, should have a more upright direction than on upright varieties like Yellow Transparent and Delicious. In any case the angles should not be so close that bark will later be caught in the crotches. Proper angles will usually fall between 20° and 45° from the vertical.

Remove or head back lightly all vigorous laterals not to be left for the permanent framework. Their removal is sometimes the better treatment because it establishes the permanent framework branches at once and avoids the necessity of securing dominance gradually. It also avoids the difficulties which result from heading back. However, pruning vigorous trees heavily at this time induces such succulent growth in the laterals left for the permanent framework that they may be bent out of shape by wind and in the most upright varieties the angles between some of the framework branches and the trunk may become too acute.

Short horizontal laterals which will not compete with those selected for the framework, should be left to increase the diameter of the trunk as much as possible. It is not necessary or desirable to head back the laterals to be left permanently. If they are left alone, they will become branches coördinate with the central leader, an important step in the easy development of the modified central-leader tree.

Third season. At the beginning of the third growing season replace any poor laterals with better ones which may have developed from buds that remained dormant or from shoots that grew poorly during the first season. This may be necessary on poorly grown trees, on trees that have been mistreated before planting, or on trees that have been poorly planted. If necessary, higher laterals may be selected for the permanent framework at this time.

Remove any vigorous misplaced shoots. Let all other growth remain. It is seldom necessary to head back for balance, but occasional laterals may be removed for this purpose.

Laterals in the upper part of the tree should be thinned out if the tree tends to become top heavy. The central leader, however, should not be removed or headed back because it is to be used as the highest branch in the main framework. It is to be kept equal in size with those lower down, preferably by removing laterals.

Fourth and Following Seasons. Prune as little as possible. By this time the three to five main framework branches should have established themselves, and only an occasional vigorous new shoot should need removal. It may be necessary to remove a few branches to keep the tree balanced. Since the central leader is to be the highest main framework branch, coördinate in size with those lower down, its vigor should be reduced, if it tends to outgrow the lower branches, by removing some of its laterals.

38

GRAFTING FRUIT TREES

WHEN varieties are untrue to name there is nothing to do but to pull the trees out or top-work them. In many places it would be just as satisfactory from a commercial standpoint to pull the trees out and start over again as it is to top-work. Circumstances, however, alter cases, and if the owner does the work and looks after the trees he may succeed where an employee would fail.

SAMUEL FRASER,
In *American Fruits.*

ANYBODY may succeed with grafting; there is nothing mysterious or difficult about it. In fact, it is so simple that any boy can be at least 90% successful when he follows the simple directions.

Trees already bearing good fruit need not be grafted unless one desires them also to bear other varieties. It will not pay to graft very large, old trees because their trunks and main limbs are often unsound and may break in a few years. Unless they bear good fruit they should be cut down.

The trees to graft are those that bear worthless fruit but are sound and comparatively young. Far better make them useful than continue worthless. When such trees are to be grafted over do only part of the work in any one year. If it is done all at one time many of the grafts will fail to "take" and the tree may either develop innumerable watersprouts and suckers or it may die. Graft only a third, a quarter, a fifth or a smaller number of branches annually.

To prepare for grafting secure dormant, one year old twigs in late winter, preferably from the bearing parts of the tree of the desired varieties (not watersprouts and especially not suckers from the base of the trunk). Label them according to their varieties and store them in a cold, damp place so they may continue dormant and plump. The north side of a building and a cold cellar are good places. In either case bury them in moist ground or sand.

In cleft grafting, the style under discussion, the tools most needed are a fine-toothed, sharp saw, a grafting iron or grafting chisel, a very sharp knife and a mallet. (Fig. 52.) About 15" or 18" of an old pitch-fork or spade handle makes a good makeshift mallet. It is often convenient to have a pair of pruning shears and a pruning knife. A supply of grafting-wax completes the equipment. If you intend making only a few grafts you'd better buy a package of grafting-wax at a garden supply store; but if you plan to make a large number within even several years it will be cheaper to make your own supply. It's easy, as will be seen from the closing paragraphs of this chapter.

The best time to do this kind of grafting is just as the buds are swelling on the trees to be grafted. If the twigs to be used are kept dormant the work may be done even after the leaves begin to expand.

Branches ¾" to 2" in diameter usually give best results when grafted. These, when prepared, are called "stocks." The prepared twigs are called "scions." On the branches to be used as stocks choose places free from knots, buds and twigs. As a precaution against splitting, remove the branch 6" or 8" beyond the point where you intend to place the graft and then saw off part of the stub at the exact point.

Now split the stock 1" or 2" with the blade of the grafting chisel placed where the bark is smoothest. Then insert the wedge point of the tool in the middle of the cut to separate the sides enough to insert the scion which has been prepared as follows:

Cut the twigs so each scion contains three or four buds. Pare the lower ends to form long-pointed wedges (say, 1" long) with a bud just above the tops of the cuts. As they

GRAFTING FRUIT TREES 225

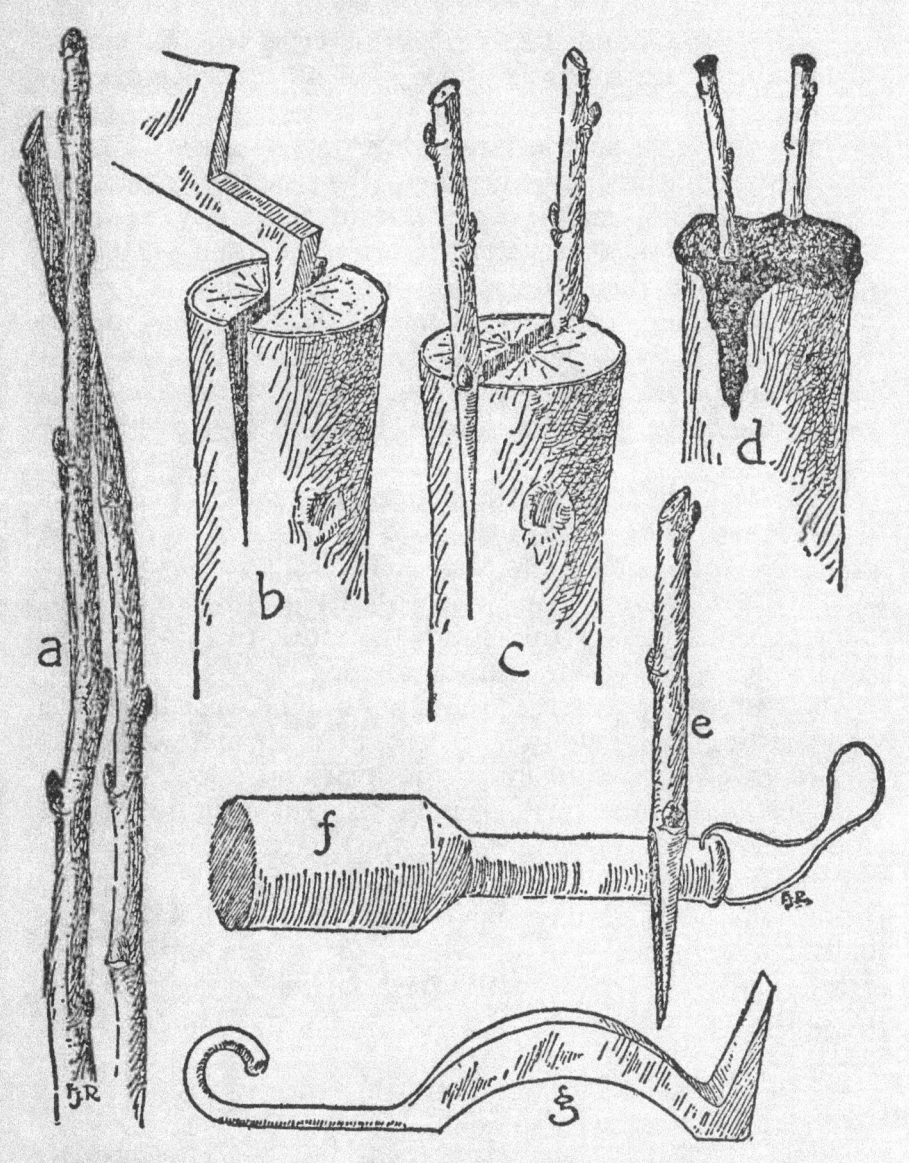

Fig. 52. Cleft grafting. a, scions; b, cleft prepared in stock; c, scions inserted in cleft; d, graft waxed; e, single scion showing wedge shape and position of lowest bud; f, mallet; g, improved grafting iron.

are made drop them in water or hold them in your mouth until ready to place them in the stocks, to prevent the cut edges from drying. In stocks less than 1″ in diameter one scion is sufficient; in larger ones two are desirable.

Success depends more upon the proper adjustment of the

scion to the stock and the thorough covering of all wounded surfaces with grafting-wax than upon all other points put together.

Place the scion so the lowest bud faces outward and is practically on a line with the top of the stock. Be particular to have the bark on the outer side of the scion cross the bark of the stock at a very slight angle—almost, but not quite, a straight line. The reason for this is that all growth in our common temperate climate trees occurs in the "cambium" layer of cells—the layer thinner than the thinnest tissue paper immediately beneath the inner bark but outside the outer layer of wood—between the bark and the wood!

Successful grafting depends upon the fusion of the cells of this layer in the scion with those of the same layer in the stock. In order to have the scion held firmly in place gently ease out the point of the wedge of the grafting tool, thus allowing the edges of the slit in the stock to close on the scion which must not be allowed to shift.

To finish the operation fill all cracks and cover all wounds with grafting-wax, mainly to keep out the air and water, but also to favor successful union of stock and scion.

Scions that "take" will usually start to grow in two or three weeks. The only attention the stocks will need during the first season is to prevent sprouts from growing on them, thus stealing food from the grafts and developing undesired branches. Should insects, for instance, plant lice, attack the new green growths they should be destroyed as promptly as possible as they will weaken if not kill the grafts.

When both scions grow (where there are two in a stock) they should be allowed to complete their first season without check because they thus tend quickly to cover the wound in the stock. In the spring of the second year cut off the poorer one near its base so as to force all food into the better shoot. If both are left a bad crotch is almost sure to form. Cover the wound with grafting-wax, and if the old wax has chipped off the old wound in the stock put fresh wax on it also.

When trees are left with grass, straw or litter near them

there is risk of losing them. Mice make nests in such material and when other food fails during winter they gnaw the bark, for several inches up and down, at and below the ground surface and often all the way around. Rabbits do similar damage but higher up on the trunks and without reference to anything lying around the trees. The result is death unless measures are taken to repair the damage.

Other injuries that often lead to death of trees are due to sub-zero weather which splits the bark; to body-blight, a form of fire-blight that destroys foliage, twigs and branches of apple, pear and quince; and to collar-rot, a disease of uncertain origin that attacks some varieties of apples (especially Grimes and Tompkins King) at and just below the ground level.

Mouse damage is easy to prevent in three ways: 1. Keep the ground bare for at least a yard around the bases of the trunks. 2. Mound up and pack earth 6" or 8" high around the trunks just before winter. 3. Encase the base of each trunk for 15" to 18" with ½" mesh galvanized hardware cloth (wire netting), making sure that the lower end of each is buried 3" in the earth. A combination of these methods will make assurance triply sure. You invite damage, if not disaster, when you do not take such precautions.

Rabbit injury may be prevented by swabbing the trunks and lower limbs if less than 3' from the ground at the approach of winter, using one ounce of red pepper thoroughly mixed with two or three pounds of lard or other soft grease. When applying this mixture use rubber gloves to prevent the severe smarting which the pepper will cause if worked into the skin.

Should trees become girdled do not doom them; they may yet be saved if the work of salvage is done promptly. In the spring, as soon as the snow has melted enough so the bases of the trunks can be seen, examine each tree to determine which ones have been injured. If the mouse or rabbit injuries are less than a couple of inches wide (up and down) and extend less than half way around the trunk, trim their edges smoothly with a sharp knife and cover the wounds with grafting-wax. The earlier this is done the better.

Should the wounds be wider and extend more than half

way around the trunks bridge-graft them. To do this choose watersprouts or other long, straight twigs of only one season's growth and of the same kind of plant as the girdled tree; i. e., apple sprouts for apple trees, pear for pear, and so on. Place the sprouts where they will keep cold and moist until used. The best time to use them is when the buds on the trees to be grafted are beginning to swell in spring. But

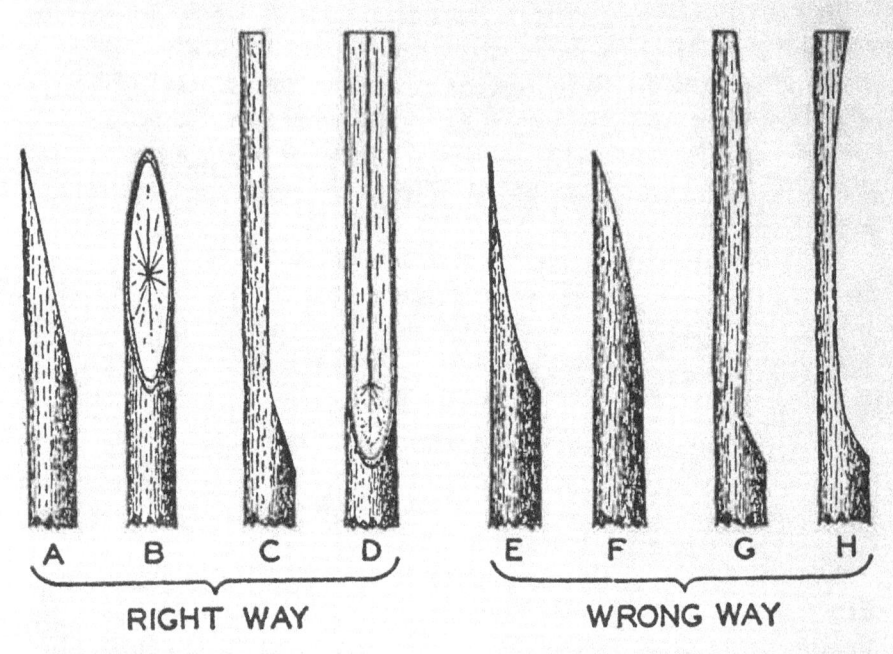

Fig. 53. Right and wrong ways to make scions for bridge grafting. A, B, proper taper cut for bowed insertion; C, D, cuts for inlay insertion; E, F, G, H, wrong ways of cutting scions. These make poor contacts between the cambium of scion and stock.

if the sprouts are kept dormant they may be used even as late as the development of the leaves on the trees to be saved.*

Trees of all ages may be bridge-grafted, most of them successfully, if the work is done shortly after they have been girdled. However, some types of injury cannot be profitably so treated. For instance, under most conditions it is advisable to replant trees younger than four years old. With one-year trees, if the girdling is above the graft-union,

* Part of what follows has been condensed from Circular 381 of the Illinois Experiment Station.

the trunk may be cut off just below the girdled area and one of the new shoots that push out used to make a new tree. Shoots from below the graft-union should not be saved, since they are from the seedling stock. Where the girdling is below or close to the graft-union, the trunk may be cut off, a scion of the desired variety grafted in the trunk and a shoot from this scion developed into a new tree.

It usually does not pay to bridge trees that have been completely or almost completely girdled for a year or more. They are generally in such a low state of vigor they will not respond.

Bridge-grafting over body-blighted areas will not prove profitable unless the blighted areas are thoroughly removed well back of the diseased areas and the exposed wood surfaces are sterilized and kept covered with grafting-wax for the first year.

Peach trees do not respond so satisfactorily to bridge-grafting as do other trees, so it is doubtful if much will be gained by working them.

As to the operation itself, anyone may be successful if he will follow these directions:

1. Remove all trash on the ground, hoeing away the earth if necessary to get at the lowest parts of the wounds. 2. Trim the edges of the wounds with a sharp knife to remove all dried and ragged bark and leave a smooth surface on the edges of the wound all around. 3. Disinfect all the wounded parts with corrosive sublimate solution (1 part in 1,000), Bordeaux mixture, lime sulfur solution or any other fungicide. 4. Cut the scions 2" to 4" longer than the vertical height of the wound, depending on the distance, up and down, to be bridged. Short distances require relatively shorter twigs than long distance. 5. With a sharp knife cut each end of each scion tapering 1" or more (Fig. 53), both cuts facing the same direction. 6. Vertically below each frame limb place at least one scion so as to bridge the gap at such points, making vertical cuts about 2" long and 2" to 3" apart through the bark of the wounded tree on the upper and lower sides of the wound.

7. The thickness of the bark will determine the type of cut to make to receive the scion. For young trees with thin

bark, cut a single slit about 1" long through the bark at each point where a scion end is to be placed. Make the slit in such a way that the scion can be inserted between the bark and the wood without lifting the bark between the scions (Fig. 54 A). With older trees where the bark is thick, make two slits, each 1" long (B), as far apart as the scion is wide. On old trees with very thick, tough bark, and on old roots, the removal of a piece of bark 3" to 4" long and

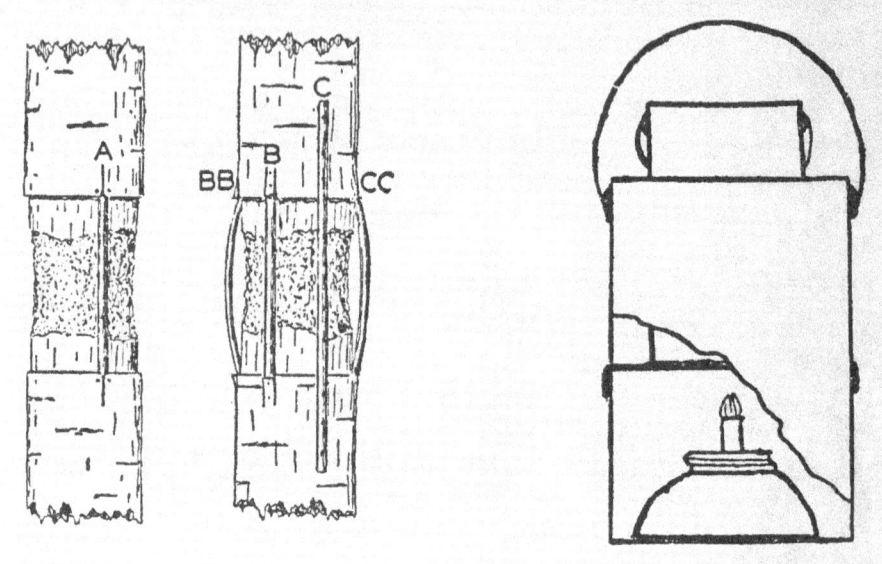

Fig. 54. Bridge grafting. Left, A, single slit; B, double slit; BB and CC, arching of scions. Right, Home-made grafting-wax heater made from pails.

as wide as the end of the scion will prove most satisfactory (C). When the scions are ready to be inserted, loosen the edges along the single split so the scion may be readily inserted between the bark and the wood (A). When using the double split, raise the tongue between the 2 cuts (B).

By following the above directions usually 100% success will result. The scions will grow in girth until, in time, they join to form a new, smooth trunk with no ridges to show where the repair or bridge was made.

A good grafting-wax may be made as follows: Weigh separately three pounds each of rosin and beeswax and two of hard beef tallow, the first in small lumps. Place the rosin in a clean pot that need never after be used for anything else(!), warm it over a gentle fire until it is all melted; then add the beeswax and tallow. After these have also melted stir until the mixture is uniform. Then remove from the fire and pour in a tub of cold water. As soon as cool enough to handle, knead and pull it until it looks like molasses taffy. To prevent the stuff from sticking to the skin grease the hands. The finished wax will keep indefinitely either in the pot, in sticks or in balls.

For a softer wax add a little more tallow; for a harder one, slightly increase the beeswax. In cold weather soft waxes are easier to apply than hard ones, but hard ones usually cling to the wounds better in hot weather.

During the past few years paraffin has been used instead of grafting-wax. The only objection to it is that it must be melted in order to apply it. When much grafting is to be done a railway switchman's lantern makes a convenient heater as it may be easily carried from place to place and either set down or hung by its handle.

To adapt it cut out the ventilator which is replaced by a metal cup to hold the paraffin. Even in cold weather the flame of the lamp will provide enough heat to keep the paraffin melted and on warm days it will not become too hot if the flame is turned low. Melted paraffin may be most easily applied with a small, flat, long haired painter's brush. Both scion and stock should be *completely* covered with the paraffin to check evaporation. When the buds start to grow they will easily push through. This would be good for grafting-wax also; provided the wax is melted, otherwise it cannot be done satisfactorily.

A good brush wax may be made by melting and thoroughly mixing five parts rosin, one part beeswax, and one-fourth part raw linseed oil. Boiling is not necessary. Another formula calls for six parts rosin, one part beeswax, and one part raw linseed oil.

"When brush wax is used, a container for keeping it warm is necessary. A convenient and inexpensive one may be made from a one-gallon and a one-half-gallon syrup bucket and two pieces of heavy wire. If the wax is melted before going to the field, a small alcohol lamp will keep it melted except on very cold days. If no such lamp is available, one may be made by cutting off about three-fourths of the spout of a common machine oil can and pulling a piece of woolen cloth up through the cut-off spout to serve as the wick. A small thimble may be used as a cap over the spout and wick to keep the wood alcohol from evaporating when the lamp is not in use." (Fig. 54.)

39

HOW TO AVOID NURSERY STOCK LOSSES

CHEAP trees are seldom, if ever, a bargain; the grower should insist on having first class trees and should be willing to pay for them.
PADDOCK AND WHIPPLE,
In *Fruit Growing in Arid Regions.*

RECENT investigation and experiment indicate that at least 90% of the disappointment and expense caused by improper management of nursery stock can be prevented in proportion as people learn what to avoid, what to choose when purchasing; how to plant and how to take care of nursery stock, especially while young. The essentials are so simple and so easily applied that anybody can manage them, whether or not he has ever planted a tree before. Better still, the most important items cost nothing, except a few minutes to do them. Having spent good money for trees and shrubs it is as foolish to neglect these essentials as to run an automobile without oil!

Most beginners assume that trees as received from the nursery are "full of wim, wigor and witality" and therefore "rarin' to go," or rather grow. Perhaps they were before being dug; but in spite of the most careful digging 50% to 75% or even more of the feeding rootlets are inevitably cut off and only part of the conduit roots are left.

Experiments with countless trees and bushes and many species have supported the business tree planters' contention that these remnants of conduit roots cannot do feeding-root duty and supply the trunks with the water necessary to develop leaves and new branches. Though the roots strive nobly to meet this demand, the branches attempt to carry on as if nothing had happened. The result is that, unless the top is reduced, the tree always suffers and probably dies.

To prevent this calamity it has been proved that a balance must be struck between top and root. (Chapter 37.)

Experiment and experience have also proved that freshly dug nursery stock properly handled and planted recovers more quickly and grows better than otherwise equally good stock that has been kept in storage. Whenever possible, therefore, freshly dug stock should be given preference. Except with autumn planting, modern nursery methods and popular demand almost preclude this practise with fruit trees and shrubs. The great bulk of such stock sold in spring is dug during the previous autumn and stored from three to six months before the buyer plants it! The exigencies of the nursery business have compelled the development of these storage methods. In spite of the best care the nurseries can give, a tree in storage not only loses vitality but the later it is planted the poorer the chance it has, especially in the hands of inexperienced planters, of overcoming the daily more adverse conditions of air and soil as spring approaches summer.

Still another handicap all nursery stock has to meet is the drying of the roots, branches and trunks (in the order named) when dug and exposed to sun and wind, when packed loosely, when shipped long distances, when unpacked and again left exposed to sun and wind, when planted in loose, dry soil and when left with unpruned tops as already explained.

Experiment and experience agree that fruit trees and shrubs of ordinary nursery sizes may be shipped safely without soil around their roots but with damp packing materials of various kinds. If it has been stored or has been long in transit it will be benefited by being plunged, root and branch if possible, in a pond, a stream, a barrel or a deep tub brim-full of water and so kept for a day or longer before being planted. It will thus "plump up" and have a far better chance of growing than if planted at once, or just as it arrives. Burying in sopping wet soil for three days to a week will give equally good results. I have saved almost brittle-dry stock in this way.

Analyses of countless cases have proved that attempts "to get fruit soon" are responsible for more failures and disappointments in amateur planting than is perhaps any other one thing. The fact that professional tree movers succeed in transplanting mature trees, perhaps in full leaf, tempts many inexperienced people to buy large stock. They do not realize that the tree movers' trees are specially handled by experts to insure success. Many of them are prepared for the ordeal of transplanting by perhaps years of tedious and costly root pruning, or previous transplantings; others are dug with exceeding care to save the largest possible amount of roots which are kept moist by wrappings of wet burlap. Finally, care is given at frequent intervals for at least a year, especially as to watering.

All this is very different from planting the ordinary "bearing age" trees which some nurseries offer for sale at advanced prices. Unless these are prepared as just explained the losses of roots in such cases are so great that the trees rarely recover, much less continue to develop into beautiful, fruitful specimens. Generally, therefore, smaller, less costly trees become established promptly and not only catch up with, but outstrip larger ones of the same kinds.

Modern machine methods of digging ordinary sizes of nursery trees are far quicker than old fashioned spade digging but they split, break, tear and scrape the conduit roots as well as cut off the feeding rootlets. The larger and older the trees, the greater the loss. Though English experiments seem to prove that roots so injured recover without doing damage to the trees through decay or permitting disease to enter, American tree planters agree that it is safer to cut off these injured parts so as to concentrate the healing energy and thus hasten both the recovery of the trees and the establishment of new feeding roots.

Commercial tree planters differ as to methods of planting, yet recent experiments have proved that the old fashioned plan of spreading the roots out in all directions is better than crowding them in a bunch as many commercial orchard planters do. For when they are spread out there is little or no danger of any root strangling another and thus impairing its usefulness, if not starving the tree!

Trees by the thousand are killed annually by placing manure or fertilizer in the holes dug for them. These materials come in contact with the roots of newly planted trees and when wet make such strong solutions that they literally burn the forming rootlets until the tree gives up attempting to develop them. The only safe way to apply these materials is to mix them with the surface layer of soil after tree planting so the solutions may become diluted before they reach the tender rootlets. When digging the holes throw the good top soil in one pile and the lower or poorer subsoil in another. It is safe and sane to place the best available soil in close contact with the roots and the poorer soil on the surface. The former encourages development of new roots; the latter discourages weed growth.

A study of hundreds of failures to make good stock grow shows that the stock was planted loosely. Practical tree planters always require that the earth be packed firmly around the roots as the plants are being set. Unless this is done too much air is left in the soil, moisture evaporates rapidly, new roots fail to develop, the old ones dry out and the tree dies. It is important to maintain 1" to 2" of loose soil above the hard packed earth in which the roots are embedded as this tends to check evaporation of water from the lower soil and to discourage weed growth.

After each tree is planted it is imperative to remove the label which the nursery has generally wired tightly to the trunk. If desired it may be attached to a branch by a loop of as large a diameter as the wire will permit. If the wire is left wound around the trunk or a branch it will girdle or "strangle" the stem at that point. The part above the constriction may blossom the following year but die soon after.

When the above essentials and precautions are practised even the beginner should succeed as well as the professional in making trees grow—practically 100%. Then by the application of some other principles he may build so strongly that until they approach senility they will carry heavy loads of fruit, ice or snow and stand the stress of high winds without breaking.

It is always advisable to buy high grade trees, not necessarily the largest but well grown ones. First choice should be those whose principal branches, if any, are far apart (12" or preferably more in the case of apple, sweet cherry and pear) and pointing in several directions. The inferior and poorly placed ones should be cut off. The strongest, best placed ones make most symmetrical trees. Among nursery trees as ordinarily grown such trees are rare because of the intensive methods of growing and because many nurserymen still cut the "leaders," or main trunks of their yearling trees, thus favoring cultivation between the nursery rows, but forcing the development of branches in bunches just below the cuts. Rather than buy such trees it is better to choose young, straight, unbranched trees of these species. With these unbranched "whips" it is easy to develop structurally strong trees as already explained. (Chapter 37.)

When such trees are bought in the fall they are usually chosen in the nursery row and dug individually by hand. They are therefore probably better specimens and in better condition than those dug by machine, stored over winter and hastily grabbed from a pile during the nursery's busy season. When planted in the fall they are likely to start growth long before spring planted trees; if planted in spring the work should be done as soon as possible after the frost is out of the ground so as to give the trees the best and earliest possible start. In those two cases no part of the "leader" should be cut off.

The symmetry and strength of such trees may still further be developed by treating each of the frame limbs as if it were a trunk—allowing only a few secondary, distantly spaced branches to develop on each until after the second or third year. The upward extension of the trunk may be treated as during the first year—nipping off all except three to five frame branches each year, preferably the smaller number. In due time as the tree will reach its "desired" height it will reduce its upward development and fill out. The strength and beauty of such specimens are well worth the small effort to secure them.

Sometimes a tree is received from a nursery with a Y-like construction of trunk and two evenly developed branches. (Fig. 55.) Unless such cases are treated one or both of these branches will break down sooner or later. Yet the

HOW TO AVOID NURSERY STOCK LOSSES 237

case is easy to treat in several ways. Perhaps the best is to cut off all but 4" to 6" of whichever branch seems to be the inferior and a year or so later to remove the stub close to the trunk. If cut back altogether the first year the wound will be proportionately larger so drying of the trunk at this point might be so great that the tree might die.

Fig. 55. Best way to treat Y-crotch in young tree. (From photo.)

Another way is to shorten the inferior branch 50% to 75%. It will thus develop as a branch upon the superior one which will become the main trunk.

In cases where both arms of the Y have been allowed to grow for several years the symmetry and the well being of the tree would be injured if one were then cut off. Yet the inevitable breakdown may be prevented in either of two easy ways. One is by braiding and tying two branches together, one from each arm of the Y extending toward the other. In a few years these two will become united to form a brace of living wood between the two arms. After the union (not before) the twigs beyond the brace should be cut off a little at a time each year until only the brace remains.

The other method applicable after the Y-arms have become large is to bore a ¾" hole in each arm at least 6' above the crotch, insert eye bolts, place large washers and nuts on the ends, join the eyes with a stout cable or a chain and draw up the bolts tightly. This work is best done while the trees are leafless as the arms are then closer together than when loaded with fruit or even with leaves.

Though the methods of management outlined will enable anyone to train young trees in the way they should grow, cases sometimes arise where neglect or accident almost compel the owner of established trees to do more or less pruning and simple tree surgery. For such people the following supplemental suggestions should be found helpful.

It is important to avoid attaching to trees anything that may restrict growth. A clothes line or wire wound around a growing branch or trunk may be almost buried by new growth in a single year. The next year it will begin to strangle the part and the third or fourth kill the upper parts.

Fence wires and the ends of fence rails nailed to trees have often been buried by rapidly growing wood. Benches placed so firmly between trees that there is no "give" often have their ends "swallowed" by the trunks.

When necessary to attach wires to trees the best way is to use bolts as already detailed.

Repeated investigation and experiment have proved that the time of year has far less to do with the healing process than has the position of the cuts—the closer and the more nearly parallel to the sap flow the quicker the healing. To date no investigator or experimenter has questioned the old dictum: The proper time to prune is when the tools are *sharp!*

40
VEGETABLE CROPS TO AVOID AND TO CHOOSE

GARDENING for money requires unceasing attention, close and thorough management, considerable hard labor, and often more or less exposure to the vicissitudes and inclemencies of the seasons. Nevertheless it is true that the majority of the profession make altogether too much work of it, especially by neglecting to make use of the newer improved implements of tillage.

T. GREINER,
In *How to Make the Garden Pay*.

ONE of the most striking and interesting things that a visit to a large city market will reveal especially in New York, Washington, Chicago, New Orleans or San Francisco, is the varied assortment of vegetables offered for sale. The fact that many of these are not staples but yet are offered in commercial quantities indicates, first, that they are in demand and, second, that at least some growers consider them profitable because they appear year after year and in about the same amounts. They are bought mainly by people of Italian, French, German, Chinese, Japanese, Negro and other foreign descent.

Should you wish to grow such crops you will probably have to search through a dozen, a score or more catalogues to find seed because few American seedsmen carry more than one or two each of the more common ones. The others would have to be bought from foreign seedsmen who specialize in what, to Americans, are oddities. In fact, you might have to write to the growers of such vegetables to discover where seeds may be obtained.

So far as growing them for market is concerned, at least where there is no resident foreign population, you might find no demand because people are so slow to try strange foods that you might lose money by raising them for sale. They should therefore be the first to omit from any list of vegetables to be grown for commercial purposes.

The next vegetables to discard should be those known as

"gamble crops"—the ones reputed to be difficult for one reason or another. To be sure, many growers find them profitable because they give the extra care these crops require. For instance, though mushrooms spring up spontaneously in lawns and pastures during favorable seasons in late summer and autumn they require "caves" or "cellars" in which to grow at other times and they are so subject to the attacks of insects and other enemies that many people fail with them.

If cauliflower, another gamble crop, were subject to only the insect and disease enemies of its close relative, cabbage, anyone could grow it, at least after a fashion. But as it demands cool weather, ample moisture in the soil and special attention to shade each head individually to make it blanch or whiten properly, it is a crop for only the man who will provide these favorable conditions and is willing to do the necessary fussing.

Watermelons, cantaloupes and cucumbers are gamble crops in many parts of the country because of "wilt," a bacterial or fungous disease which destroys the vines usually just before the fruit would normally ripen. Though it is possible to prevent such disaster you, as a novice grower, are almost sure to belittle the methods or to follow them in an imperfect way with the result that after failing one or more times you will condemn them as worthless.

For these reasons it is advisable to limit yourself to the more or less simple crops until you have learned how to manage each one successfully. Before deciding to grow any of even these, however, be sure to learn as much as you can about each one so as to know what to expect, especially the points brought out in the reading pages and the tabulated data in the appendices of this book; notably, the proper time to sow each kind, the amount of time it requires to reach edible maturity, its ability to withstand frost; for many of the disappointments, if not total failures and losses, may thus be avoided.

To illustrate: If you sow corn when parsnip seed should be sown the seed will rot because the soil then is too wet and cold for it, or if seedlings do appear they are almost sure to be killed by frost. Conversely, if you sow parsnip seed at corn planting time it will probably not sprout because the soil is then too dry for it, or if it does start to grow the seedlings will be burned up by the sun or will have too short a season in which to mature.

When your aim is to make money quickly, avoid the long season crops such as leeks, salsify and parsnips because they occupy the ground from early spring until late fall and because at best they usually sell in only small quantities and at comparatively low prices. Choose the quicker crops that are in positive, good demand; for instance, scallions (green onions from "sets"), radishes, spinach, and lettuce for spring sales; garden peas, round beets, and "horn" carrots for early summer; early cabbage, tomatoes, corn and peppers for midsummer and early fall; and late cabbage, summer sown round beets and carrots for late fall and early winter. Not only do these crops each occupy the soil for only a short time but they combine well with one another as companion or succession crops (Chapter 32) so the same areas may be made to produce several crops instead of only one in a single season.

Though it is usually desirable to grow sufficient quantities of staple vegetables such as late potatoes, late cabbage, ripe onions and turnips to supply the family table, it may not be profitable to raise these crops for sale because of competition with commercial truckers who have special equipment for handling them on a large scale. And yet this suggestion must not be applied too sweepingly because some of them may be locally profitable in spite of commercial competition, especially when an extra choice variety is grown and can be sold fresh, when the quality of the stock is exceptionally fine or when the product may be placed on sale when the market is bare, at least of a comparable quality. For instance, home grown, well ripened tomatoes that come in competition with southern ones, particularly when the local ones are exceptionally early, will always command a premium price.

One of my truck-growing neighbors always plants his tomatoes a month to six weeks later than do his competitors, most of whom at first jeered at him for his "folly!" But, just before frost he gathers the fruit, ripens it on deep straw in coldframes which he covers at night and in cold and wet weather and sells during October and November when the market has only short supplies and therefore pays high prices. He makes more money out of his tomatoes at such prices than do his neighbors when the market is well supplied—and does it with much less work!

For repeat sales it is desirable to grow only high quality varieties. There are such of nearly all kinds of vegetables. The extra early varieties, particularly of corn and garden peas, have only their earliness to commend them. They are profitable because of this and because people are so glad to "get a first taste" that they are willing to forgive even serious short-comings; but for mid-season and late the high quality kinds are what produce repeat sales and enhance the local reputation of the grower. Therefore, be on the constant lookout for new kinds superior to anything you have already grown. Test the novelties in a small way the first year they are offered for sale; in fact, if you will gain and never betray the confidence of the seedsmen you may have the privilege of testing such novelties a year or more before they are offered to the general public.

For sale, some varieties or some crops are better keepers than others. Hence when keeping does not oppose high quality these should be given preference for both home use and market. The seedsmen generally indicate which varieties are noted in these ways.

In some cases vegetables may be utilized or sold in several ways. These, therefore, offer more than one chance to make money out of them. For instance, set onions that cannot be sold as scallions may be allowed to mature for sale as dried ones and seedling onions may be thinned out and sold as scallions or "green boilers"; beans, especially white seeded and "kidney" varieties, may be sold as green "string beans" or dried seed; limas sell readily both as green and ripe beans; tomatoes have ready sale when ripe and in the fall green ones also sell fairly well for making pickles and relishes of various kinds.

When canning and pickling are added to the sources of income, tomatoes, peppers, onions and celery make many fine relish combinations which need only be sampled to be sold. Besides these surplus gherkins, "dill size" cucumbers, baby beets and carrots, pickling onions, rhubarb and asparagus all may be canned or pickled and sold at a profit, as many a farmer's wife can attest.

41

SEEDS AND SEEDING

THE cost of seed is ordinarily a trifling matter in comparison with the expense of the season's labor and the value of the crop.
L. H. BAILEY,
In *The Principles of Vegetable Gardening.*

ONE of the hardest lessons for a beginner to learn is that cheap seed is the most costly to buy! Why is it cheap? It may be—probably is—not true to name! It may be old—50 to 100% dead or at least weak! It may have been—probably has been—cheaply and therefore carelessly grown and poorly "rogued" (if at all) or otherwise carelessly handled. In no branch of farming is it so true that the penny wise, pound foolish policy is so often or so strikingly illustrated as in the buying of cheap seed.

As no one can grow a successful crop of anything from poor seed, the time and attention devoted to the plants will be largely, if not wholly wasted. The difference in first cost between cheap and "costly" seed is so slight that no one who has his best interests at stake will hesitate to pay it; for the so-called costly seed will, or should meet all the requirements of "good seed"; namely, viability (ability to "come to life") when conditions—moisture, oxygen and heat —are supplied; freedom from weed seed and debris, 100% true to name, disease-, and insect-free, and ability to produce a crop of uniformity and excellence.

Since none but the best seeds are good to buy as much information as possible should be gathered about the seedsman and each lot of seeds before the purchase of any. When there is any doubt as to which is the better of two strains, samples of each should be grown under identical soil conditions and treatment to decide, then to order that strain which gives best account of itself. To be sure of getting such a strain it is essential to buy samples under guarantee that the same strain, either named or numbered, can be supplied the following year and for a series of years. For, contrary to popular belief, no seedsman grows all his own seed; he contracts with specialist seed growers, each of whom may supply only one variety or strain of a kind; for instance, Chantenay and no other variety of carrot; Metropolitan and no other variety of sweet corn. Thus both the seedsman and his patrons may be sure of pure seed.

It is highly desirable to conduct strain tests because, as the experiment stations have proved with practically every grain and vegetable grown from seed, there are wide variations within the limits of a variety as to time of maturity, yield, uniformity and many other important features which make one strain worth more than another. Hence the sooner the best strain is located the better for the grower. To illustrate this point, that old stand-by, Detroit Dark Red, my own favorite variety of beet, now differs so widely in its strains that some might well be called different varieties!

Each reliable seedsman makes a specialty of some one or other variety or strain of vegetable; hence the further advisability of "shopping around," not for prices but for profits! This means that you may buy your cabbage seed from Jones, tomato from Brown, carrot from Smith and beans from Robinson.

Though the business plantings should always be of known standard varieties and strains, it is highly important to test the novelties in moderation the first year they are introduced; for thereby some genuine "find" may be located a year or even several years before the less alert growers find it and profits may be reaped before they wake up. On the other hand, no matter how reliable the seedsman, it is almost never safe to launch out extensively (and probably expensively!) with any of these novelties until after they have been tested on one's own place. Stick to the old, reliable until the new have proved themselves better or at least as good.

With standard varieties of some kinds of seeds it is advisable to buy more than will be needed during any one year; for thus trueness to desired type may be determined by means of sample planting the first year and the balance of the seed stored for future business planting if of proper quality.

Some seedsmen hold certain kinds of seeds for a year so as to test their trueness to type by growing them for a season before offering them for sale. They, and others, often sell various kinds of seed left from one season to the next, provided its germination tests indicate that it should give satisfactory stands of plants. Some of these latter seedsmen stamp the percentage of germination on ounce and larger packages of seed so the buyer may govern the rate of sowing—thin for high, thick for low percentage of germination.

The grower, however, usually has no way of knowing beforehand what to expect. So it is advisable for him to conduct a germination test for himself. This is easy to do by placing a counted number of seeds (25, 50 or 100) between sheets of white blotting paper in a plate, covering them with another plate inverted to check drying and keeping them moist, not wet, until the test is over. In a warm room the seeds if "strong" will sprout in a few days; if "weak" in perhaps many; if "dead" not at all. Each day they should be examined, the sprouted ones counted, recorded and thrown away. Rapidity and uniformity of sprouting indicate strength; slowness and irregularity, weakness. High percentage recommends thin sowing; low percentage, thick sowing. Weakness suggests that the plants may fail to grow in the open ground. Radish and other members of the mustard family usually sprout well within a week; carrots and other members of the parsley family may take three weeks.

During the past 40 or 50 years countless experiments have been conducted to determine the difference in germination, sturdiness of plant, earliness, yield, etc. between plants grown from large or heavy seed and from small and light seed. Without going into detail let it suffice that large seed is superior to small of the same variety and strain. So it is advisable to buy greater quantites than necessary of most seeds, to sift out the small ones and either to sow them separately from the large ones or to throw them away.

The superior value of large seed over small is perhaps most strikingly illustrated by Country Gentleman sweet corn, one of the favorite varieties for canning. Bulk or general run seed produces such uneven stands and such variation in maturity of the ears that the Indiana Canners Association asked the state experiment station to develop better strains. The following are striking features and comments published by I. C. Hoffman in the *Journal of Agricultural Research* (Vol. XXXI, No. 11).

Large and small kernels were chosen from individual ears and from bulk seed, planted and grown under identical conditions with the results that the large seed germinated better, produced larger plants, more 2-eared stalks, larger ears and *the ears were ready for gathering an average of 5 days earlier* than ears borne by plants produced by small seeds. They also produced fewer barren and unproductive stalks.

When ears on the large seeded plants were ready for canning those on the small seeded plants were "so tender and watery that the corn was unfit for use." These results were attained not only in experiments but were also proved upon an acreage scale by growers coöperating with canning factories.

The reason for such results is ascribed to the fact that large seeds contain more reserve food material than do small ones. Therefore, it is recommended that seed be graded before sowing, the large and heavy kernels being separated from the small seeded ones and each lot sown by itself in order to approximate uniform stands.

Just before I started to write today an amateur gardener told me that he believed "the best way to get reliable seed is to grow it yourself." He is not so sure now! When one is merely interested in getting a radish or some other kind of vegetable, home seed saving may answer fairly well. But when it is important to have a specific type of product, to have it uniform as to type, earliness, color, size, and quality, in fact, to have it seem as if poured from a caldron into a mold—"well, that's something else again!"

Seed growing and saving are such highly specialized forms of business and demand such intimate knowledge of the plant in question, the needs of the business grower and the demands of the public that the grower of vegetables for sale may well leave it to the specialists who devote their lives to it. There are too many risks to run. Yet after one has had experience in growing vegetables he may find it to his advantage to produce some kinds of seed; but this is a branch of work which, to get best results demands at least a working knowledge of plant breeding—a subject beyond the scope of this book.

Though * germination depends upon moisture, air, viable (i. e., *able-to-live*) seed and temperature we generally think of only the last two factors. No one expects dead seed to grow, but many people fail to grasp the importance of favorable temperature and of proper management of seeds and seedlings.

Some seeds (oats and shepherd's purse—a weed) will sprout on melting ice; others (portulaca and pussley—a weed) do so only when the soil is hot. How foolish then to sow the former seed when the ground is dry and hot and the latter under reverse conditions! The former, not finding sufficient moisture, will usually remain dormant or if it does sprout the seedlings may shrivel and die; the latter in wet, cold soil will either rot or wait till the temperature rises and the moisture lessens. To prevent such experiences follow the seedsman's directions.

The importance of sowing and planting at proper times is also emphasized by the fact that some plant species are killed by light frosts, others are not; and likewise that plants of even tender species when "hardened off" can stand light frosts without harm.

Seasonal and local conditions so greatly influence temperature that no rule-of-thumb as to sowing and planting is safe to follow blindly, especially that one which declares that the sowing and planting season advances northward 100 miles and up mountain sides at 500′ a week. For instance, Denver, Colorado, is 5,270′ above sea level; Philadelphia, Pennsylvania (in practically the same latitude), is only about a tenth of this above high tide; yet because Colorado prevailing winter and spring winds are from the warm South

* Much of what follows is quoted from the author's book *Modern Guide to Successful Gardening*.

and those of Philadelphia from the cold North the temperatures at Denver are often 20° or more higher than those of Philadelphia during February, March and April.

Again, localities vary as to temperature; one is "early" another "late" as explained in Chapter 8.

Such influences, coupled with character and condition of soil may make one or two weeks' difference in the advisable date of sowing seed or setting plants of any specified kind. So likewise may the date the last killing spring frost normally occurs.

Commercial gardeners nowadays govern their spring sowing and planting largely by the weather maps and still more by the daily forecasts of the United States Weather Bureau. Isotherms, or lines of equal temperature, serve as guides to safe seedage and planting. For instance, the isotherm of 45° is the northern safe limit of transplanting hardy plants (cabbage and lettuce) from coldframes to the open ground; and that of 60° for tender plants (tomato and peppers). To have plants ready on the dates when these temperatures normally arrive the seed is sown under glass 5 or 6 weeks earlier.

Another popular, satisfactory way to govern seedage and plant setting is to follow natural signs, sowing hardiest seeds outdoors only after buds of the earliest shrubs and trees of the neighborhood swell—spicebush, forsythia, peach, red maple; those somewhat less hardy when june-berry, Japanese quince and plum are in bloom; the semi-tender kinds when apple, cherry, pear and sugar maple are flowering; and the tender ones when mock-orange, grape, raspberry and horse-chestnut are in blossom.

Ground temperature more than that of the air governs seed sprouting and plant growth. In spring only 3" or 4" below the surface the soil temperature may be 40° though that of air is 70° or 80°. So anything that will raise this temperature will favor germination, plant development and earliness. Good drainage and darkening the soil after the seedbed has been prepared for seeding both help to warm soil. Enough black swamp muck, to darken the surface will help in the latter case; it absorbs sun heat and thus raises soil temperature. It will also become humus after being turned under.

SEEDS AND SEEDING

Properly managed hotbeds and coldframes produce superior seedlings and flats (shallow boxes) are better than flower pots or rows in hotbed or coldframe. For minute seeds shallow cigar boxes are fair substitutes for the pottery "seed pans" popular with plantsmen.

You may buy standard cypress flats or make similar ones yourself. As wet earth is heavy, avoid sizes larger than 12" x 18" x 3¾" deep, preferably only 3" inside. Standard sizes fit in hotbeds or coldframes without waste space. End pieces should be of ¾"; bottoms and sides of ½" material. In the bottom have five ½" drainage holes arranged as on a 5-spot playing card.

To prepare a flat for seed sowing place pieces of broken flower pots, crockery or glassware over the drainage holes, next a ½" layer of pulverized, thoroughly decayed manure, sphagnum moss or granulated peat to absorb and retain water but let excess drain off. On this spread sifted soil until the flat is level full while loose, but a ½" below the edge when somewhat pressed down. To press it use a smooth board about 4" wide, almost as long as the inside width of the flat and with an easily grasped handle.

As the roots are perhaps the most important parts of plants develop them as well as possible before transplanting. Therefore in the flats use a rather poor soil mixture—one about half fine sand, a quarter humus, and the other quarter only moderately rich soil, well combined and sifted free of lumps. This will make the roots forage and tend to develop them well but make small tops. When transplanted to open ground such plants take hold well and develop far better than those with more top and less root.

You may scatter seed direct from the seed packet in the rows; but may get it too thick and produce inferior (because crowded) plants. A safer plan is to pour it into the palm of one hand, pick up "pinches" with thumb and finger of the other and gradually work it out as the hand moves along the row. A still better way is to use a seed-sowing convenience offered by garden supply stores.

The seed may be sown either in drills (rows) 4" to 6" apart or broadcast. The former is better for medium-sized and large seeds (cabbage, beet); the latter for small to minute ones (celery, thyme). Drop the individual seeds farther apart in the rows than recommended by seedsmen to reduce the work and waste of weeding and thinning and to increase sturdiness. Merely press small seeds into the surface and cover the larger sizes with only once or twice their diameters by scattering finely sifted soil or sand over them and pressing it somewhat.

To avoid errors in plant names follow the greenhouse and nursery rule to place a label at the front end of the left hand row and the next where the next variety starts. Thus you will always read from front to back and from left to right in each flat. If you follow the rule in the outdoor nursery rows you will know exactly what plants you are growing and will thus avoid mistakes due to lack of system.

When broadcasting small lots of seed divide your flats into two, three or four areas by pressing the edge of the firming-board hard enough in the soil to make a slight depression from side to side and thus mark the bounds of each "seedbed." (Fig. 56.) Then scatter the seed of one variety thinly in each bed and press it down. Finally, label each area. On every label, starting at the square end write the date of sowing, then, writing toward the point, add the variety name, using 2 or more lines if necessary.

The best way to water a newly sown or planted flat is to place it in lukewarm water only an inch or so deep, let it stay until the surface shows wet spots then remove it, tilt it slightly to drain away excess water and when drainage is complete, place in a hotbed, a coldframe or on a greenhouse bench, cover it with a pane of glass to check evaporation and a newspaper to keep out light. When the seeds begin to germinate remove the paper and when the seedlings show well above the surface take off the glass. Repeat the watering when necessary as described or by using a fine rose sprinkler, taking care with the latter to avoid washing the seedlings out of the soil.

To avoid mixing, separate varieties of one species by at least one row (preferably two or three) of some different species so the differences in type of growth will show which is which. For instance, seedlings of cabbage, cauliflower, broccoli and other varieties of the mustard (botanical) family look much alike but differ widely from those of onion, carrot, beet, lettuce and tomato which belong to 5 other botanical families.

Instead of using soil for the surface ⅛" or ¼" use finely sifted sand that has been cooled after baking to destroy the spores of damping-off fungi to prevent killing the seedlings. As this trouble is worst when the soil is wet and the air stagnant it may appear in spite of the sand. When the glass shows

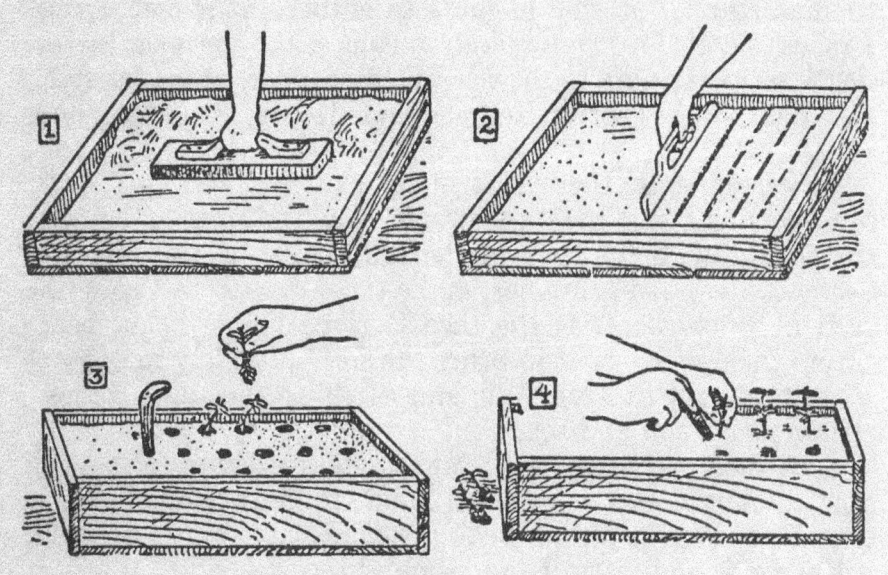

Fig. 56. 1, Tamping sifted soil in flat; 2, marking rows for seed sowing; 3, pricking holes in soil with dibble; 4, pressing soil around seedlings.

water drops on the under side remove it until the surface soil becomes somewhat dry, or if no glass is used avoid watering for a day or two, loosen the surface soil with a small stiff wire and dust the whole flat with finely powdered sulfur. On the other hand never let a flat dry out because though the plants might survive they might be checked and stunted.

Nursery beds outdoors are often used to start plants of hardy species (parsley, cabbage) to transplant later. Early beds, for this purpose, usually not more than 3' wide, are best placed on the south side of a wall to benefit by reflected sun-heat and thus force rapid growth. By sowing the seed thinly either broadcast or in rows, pricking out may be almost wholly avoided, only crowded plants being thinned out and placed in vacant spots. In order to prevent baking of the surface soil scatter finely sifted peat moss, or sphagnum moss over the beds or use burlap screens.

If you use a properly adjusted drill to sow the larger sized vegetable and flower seeds in the open garden the machine will do the work correctly; but if you sow by hand you must govern the depth by the size of the seed and the condition of the soil. Some gardeners *say:* "Cover the seed only 3 or 4 times its diameter." But who is going to bother about measuring? I never have! This rule merely means make the seed furrow depth proportionate to the size of the seed—shallow for small and light ones (carrot, onion) and deeper for larger ones (bean, corn).

In spring while the soil is moist (but not wet) make the outdoor drills shallow and merely firm the seeds a little in the ground with the back of the rake. After it becomes dry, even powdery as in summer, make them deeper and pack the seed in much more. If the seed is large (corn, bean, beet) tramp the earth over it to bring the soil intimately in contact with it, get rid of excess air and establish moisture connection between soil and seed.

Seedsmen and many gardening writers recommend sowing seed thickly to make sure of a good stand of plants. This is wise if the seed is old or weak, certainly not, in my opinion, if it is fresh and lusty. I practise and recommend thin—even sparse—seeding, not to make the seed "go further" but to reduce competition for food and water, to make the seedlings sturdier and to reduce thinning of plants.

Another point: Some seedsmen and writers advise relatively broad rows with seed scattered widely in them. I don't because this increases finger and thumb weeding. When I don't use a drill I always try to sow the seed in as narrow a row as I can by hand so as to reduce thinning and weeding.

42

TRANSPLANTING

It is a question whether the time at which tender plants shall go into the ground is a matter of prudence or of courage. . . . Transplanting is not to be recommended where plants must be purchased, or all one's stock is put into the ground at once.

IDA D. BENNETT,
In *The Vegetable Garden.*

UNTIL about 1890 many gardening rules-of-thumb taught by the Old World apprentice system were followed blindly in America, but from that time until now these have been challenged and tested, mainly by scientists in our agricultural colleges, experiment stations and the Department of Agriculture to determine whether they are founded on scientific principles, whether they are necessary, and whether for economy's sake they should be modified or discarded. Where only a few plants are to be transplanted the matter may be of small concern, but when thousands are to be handled, the elimination of "superfluous and frequently injurious practises" translates itself into economies that may mean the difference between profit and loss.

Among reasons that suggest the advisability of transplanting, the following seem to be reasonable, at least for American conditions. 1. Saving of costly seed and therefore money needed to plant specific areas—only an ounce or even less of some species being necessary to produce enough plants to set out an acre, wheras if sown direct in the field the quantity would run into pounds. 2. Avoidance of thinning and consequent waste of seedlings, time and money if seed were sown in the open field. 3. Prevention of the risks of open ground seeding. 4. Assurance of a good stand of transplanted plants. 5. Starting plants under glass in early spring, "pricking-out" about a month later and transplanting still later to the open field assures earliness of the crop. 6. Concentrating weed, disease and insect control and other care of the seedlings and thus reducing costs. 7. Assurance of ideal soil and other conditions in the seed bed until plants are sturdy enough for field exposure. 8. Utilization of the field by a crop previous to transplanting the one that is to follow it.

These reasons, which seem all to be well founded, indicate that transplanting is necessary, even though it may be a necessary evil! But many of the ipse dixits of Old World gardeners are so questionable that investigators have been conducting experiments to learn the truth. One of these, W. E. Loomis, in a 60-page bulletin ("Memoir 87") of the Cornell Experiment Station, reports a two years' investigation and series of practical and scientific experiments to determine the immediate and secondary effects.

As a preliminary to the investigation progressive growers were asked: In comparison with non-transplanting: 1. Does transplanting cause more heavy growth? Answers were, 15, yes; 11, no. 2. Does it make better root systems? Yes, 25; no, "1 (?)." 3. Does it hasten maturity? Yes, 13; no, 10. 4. Does it produce heavier yields? Yes, 5; no, 18.

These opinions, writes Loomis, show considerable divergence from earlier recommendations, when we were told to "transplant freely—nearly all vegetables are better for it"; to "transplant to induce productiveness"; and that "frequent transplanting is of great benefit but should not be done merely to give the plants more room."

To epitomize his discussion, the experiments indicate that though strongly recommended by older gardeners and earlier writers many growers doubt the advisability of more transplantings than required in economical crop production. The tendency is away from transplanting as a cultural practise toward transplanting for economy. Early lettuce, cabbage and tomatoes are transplanted to grow them out of the normal season for a given locality. Tomatoes, sweet potatoes and similar vegetables may also be grown in higher latitudes if the plants are started with artificial heat.

Artificial protection usually requires transplanting, because it is most convenient to have the plants concentrated in a small area while the protection is being given. The same is true when special care in cultivation, watering, protection from insects and so forth are required by seedling plants. A third factor, which accounts for most of the transplanting in greenhouses and some done in intensive cultivation, is the saving of space and of expensive seeds. For these reasons transplanting will be practised as long as profitable crops can be produced by the method. On the other hand, it is an expensive operation and not to be performed unnecessarily.

The immediate effect of transplanting is to slow down or stop plant growth for a period which apparently varies directly with the amount and duration of the reduction of the water supply. When a large proportion of the active root system is retained and adequate moisture supplied, there may be little effect from transplanting. Conversely, with more susceptible plants and more rigorous treatment the check may be severe and permanent.

Recovery from transplanting is affected by environmental and internal conditions, but no factor has been found which does not appear to be based, in its final effect, on a change in the water supply of the plant. The amount of suberization [cork-ifying of plant tissues] of the older roots, the proportion of root system normally retained in transplanting, the rate of new root formation, adaptations of top to prevent water loss or increase resistance to death by sudden wilting, as well as soil moisture, temperature and other factors which may be concerned in the recovery of a plant from transplanting, all affect growth through the water supply.

After transplanting, if the plant, through abnormal accumulation of carbohydrates or other products, develops an excessive number of main root branches and a consequently reduced vigor and spread, or if it becomes stunted, either through structural changes such as lignification [conversion into woody tissues], or through permanent changes in metabolism [chemical changes in the plant], these are fundamentally the effects of a reduced water supply. Transplanting, however, is not the only operation or condition affecting water supply, so it is not surprising that there are variations in the response to this treatment.

Transpiration may apparently be significantly reduced by previous hardening treatments only when the rate of loss is low and reduction is not an important matter. Under conditions favoring rapid transpiration, water supply becomes the limiting factor, and differences between hardened and tender plants tend to disappear. The water supply concerned is the supply moved to the leaves. A reduction of the moisture content of tender leaves insufficient to cause noticeable wilting may more markedly affect the transpiration rate than all the hardening which can be given the plant.

There is, however, an important relationship between hardening and transplanting in the resistance of the hardened plants to death by drying. In those plants not capable of being hardened the intracellular moisture necessary to maintain life is less strongly held, and with the great reduction in water supply following transplanting the older leaves or, in extreme cases, the entire plant may be killed by drying.

The roots are the most important factor involved in resistance to, or recovery from, transplanting. Apparently three factors are concerned: The proportion of root system retained in transplanting—dependent upon the size of the plant and the character of the root branching; the effectiveness of the roots in absorbing water during the first few days after transplanting—an effect presumably correlated with the amount of suberization in the older roots; and the rate of new root formation—dependent upon the kind and age of plant and possibly the amount of accumulated food.

The immediate effect of transplanting is a reduction in the water supply, and the immediate and long-time results are dependent upon the severity and duration of such reduction. Therefore large plants will be more seriously injured than small ones by transplanting because the proportion of roots will be less. It follows also that those plants having root systems of such a form as to be largely lost in transplanting or whose older roots are heavily suberized so they must depend upon the outer portions for their moisture supply, will be more seriously injured than plants having a more branched and less heavily suberized root system.

No well-defined relation between the rate of water loss from the top and the ease of transplanting has been established. As an average, however, those plants easily transplanted are capable of being more thoroughly hardened. The value of such hardening probably lies more in the increased retention by the cells of the moisture necessary to maintain life than in effect upon transpiration. Hardening may also build up a food reserve to be used later in replacing the mutilated root system. Both factors are probably active in certain plants.

The rate of new root formation is the most important consideration in the reëstablishment of transplanted plants. Other factors, such as absorptive capacity of the transplanted root system, resistance of the top to death by wilting, rate of growth, and susceptibility of the plant to stunting, are important only as they tend to bridge the gap between one root system and the formation of next or as their action is allowed to continue because of a slow rate of replacement.

Transplanting may be an important factor in saving valuable greenhouse or garden space, in allowing for better care of slowly developing seedlings, or in saving time or seed. While there are isolated instances of crops which appear to have made a thriftier growth after an early transplanting, these instances will seldom bear statistical analysis, so the conclusion that transplanting is a harmful but frequently necessary operation seems to be required. Conversely, the field data do not justify a large expense to avoid early transplantings, since the only shift which has consistently either earliness or total yield in cabbage and tomatoes is the final one to the field. The use of pots at this stage has produced earlier crops but has not had an important effect on total yield.

In the case of commonly transplanted crops handled when not too large and under favorable conditions, the effect of transplanting seems to be proportional to the check in the growth of the transplanted plants. When growing plants in the greenhouse this difference may be overcome by earlier sowing and is frequently lost, even in experimental work, by slight differences in watering or other conditions. It would not seem, therefore, to be important. Heavy root pruning of larger plants, transplanting under unfavorable conditions, or moving crops whose rate of root replacement is slow may seriously retard maturity and, with crops slow in root replacement, may result in permanent stunting.

To supplement the above findings and principles let me add a few suggestions of a practical nature, such as deal with the actual operation.* Every experienced man knows them, so the sooner you as a beginner learn them the better; for you should keep them well in mind whenever you have work of this kind to do.

Newly turned, deeply worked soil favors transplanting because the roots quickly become established in it. Good plants—sturdy, stocky ones—are surer to succeed than spindly ones, which latter are likely to collapse. When the soil is dry, cool, cloudy days favor success, so do late afternoon and early evening. When possible, transplant just before a rain. Never transplant just after because the soil, unless sandy, will puddle and bake. Wait until the surface has lost its excess moisture. Always press the earth firmly around the roots but leave the immediate surface loose to check evaporation.

Certain species of plants are difficult to transplant successfully (melon, squash)—those which the seed packet directions say: "Sow where the plants are to remain." However, some of these may be successfully transplanted if their seeds are sown in 6" squares of inverted sod, in strawberry baskets or paper pots 4 to 6 weeks before time to set in the garden. When transplanting do not disturb the roots. Paper pots are not removed because they soon decay; but strawberry boxes must be cut at their corners and the pieces carefully slid from below. After planting water such plants until they become established.

At least 3 hours (preferably longer) before you plan to remove seedlings or transplant potted plants soak the soil thoroughly so the plants will be filled with water and so earth will cling to their roots. Whenever possible do transplanting outdoors on a cloudy day, before a rain, toward evening, or after sundown and, unless the soil is moist, water the plants as soon as set.

After the seedlings in the flats have developed 2 to 4 leaves prick them out 1" or 2" apart each way with their roots extended downward full length, never curled up, in another flat and keep them fairly cool so they may develop stockily and sturdily.

* These are mostly quoted from the author's *Modern Guide to Successful Gardening.*

TRANSPLANTING

For very small seedlings a large crochet needle or a double pointed stick will lift and place the plants more speedily than fingers. Seedlings 1" or 2" high are best pricked out with a dibble about as thick as one's middle finger but more pointed. When the plants are 4" to 6" high a flat trowel is better than a dibble because when skillfully used there is less danger of leaving an air-space around the roots. Always press the soil from the bottom upward and by a sidewise push to make it firm around all the roots. Avoid surface

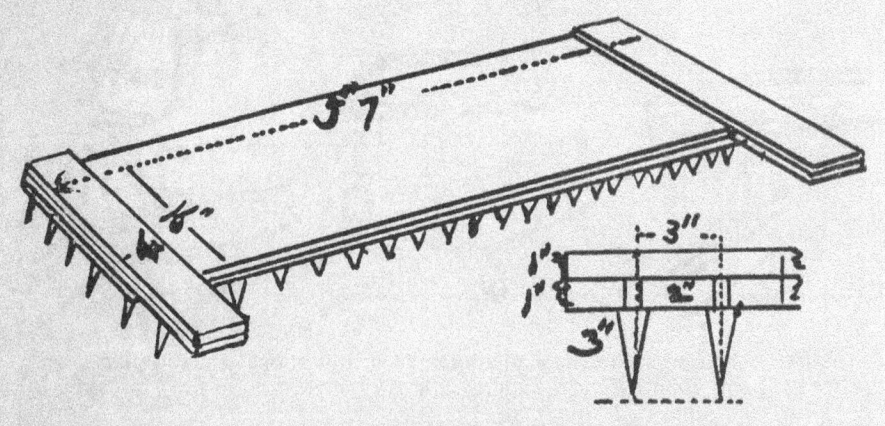

Fig. 57. "Spotting board" to locate positions for seedlings to be pricked out in flats.

thrusts downward as these are likely to make air pockets.

When the leaves again touch prick out each alternate plant into other flats or outdoors if conditions are favorable but place the flats in a coldframe otherwise.

The day before transplanting to the open ground thoroughly soak the soil so the plants will be full of water. To remove them first tilt the flat and lift it a few inches, then strike its bottom edge of one side on the ground hard enough to shift the soil and plants a little so as to break connection of the soil with the sides and bottom. Then lay it flat on the ground and remove a few plants at one corner. After these are out the others may be easily lifted with goodly lumps of soil attached to their roots.

Puddling (dipping the roots in thin mud) is usually not a good practise with small plants because it bunches the roots instead of keeping them separate. When plants are lifted with little or no earth attached to them they should have their tops wetted and should be protected from sun and wind with wet burlap. Clipping about half the tops is a good practise with large leaved plants, especially those lifted without much earth attached. It is also a good plan to set spindly plants as deep as their seed leaves, particularly when the soil is rather dry. This places the roots in moister, cooler

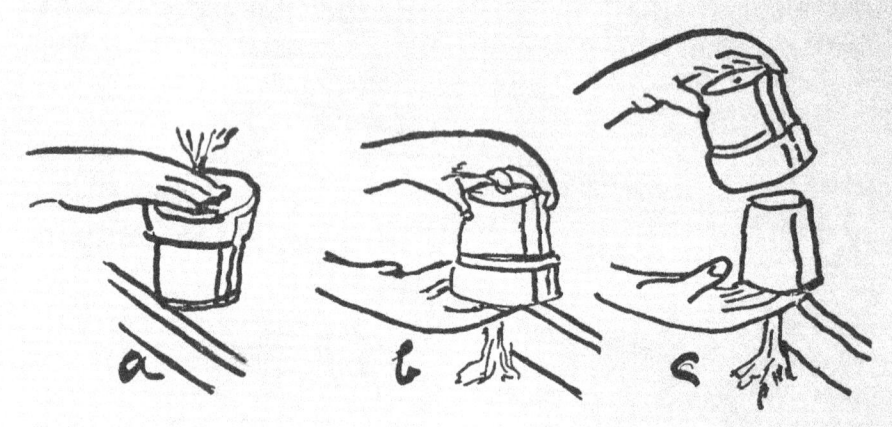

Fig. 58. Right way to take a plant out of a flower pot. a, hand placed on soil with plant stem between fingers; b, pot inverted and rapped on bench; c, ball of soil separated from pot.

soil than does shallow setting and tends to keep them from falling over.

In a small way and in dry soil plants may be shaded from the sun for a day or so after being transplanted by thrusting a shingle slantingly into the ground on the south side of each. But when the soil is moist and other conditions favorable this is not necessary. What is more important is to press the earth firmly around the roots and to leave the surface earth loose to serve as a mulch.

For large scale transplanting various power machines are available. They do excellent work. On a smaller scale I have found a metal plant setter more satisfactory than either trowel or dibble in stony land. It drops, sets and waters each plant as the operator strolls slowly across the field. In well prepared stone-free soil from 5,000 to 8,000 plants may be set in a day.

43
PLANTS FOR SALE

MARKETING is the culmination of any production enterprise. Growing and selling depend absolutely upon one another. The most skillful production is in vain if the marketing is not well done. On the other hand, the best methods of marketing cannot save an enterprise if yields are too low for economy or if the quality is not sufficiently high to command ready sale at favorable prices.

PAUL WORK,
In Tomato Production.

HAVE you ever noticed the spindly, pale, discouraged-looking plants offered for sale each spring by drug, grocery, hardware, and "general" stores? Of course, everybody has! Have you ever wondered how anybody with even a smattering of garden knowledge could be tempted to buy such futile stuff? If so, has it ever occurred to you that in the very towns where these plants are displayed are good opportunities to make profitable sales of *really well grown plants of good varieties* for transplanting?

Those italicized words hold the secret of success in making first sales, repeat sales and developing a profitable business. Here's an instance: One year when I was planning to grow several acres of late cabbage I paid $10. for a pound of seed of the very best strain of Danish Ballhead I could locate, though I could have bought general stock at $6. or even $5. a pound.

The nursery bed was made where the sun reached it all day in sandy loam, not rich but well supplied with humus, and the seed was sown very thinly, the drill being tested on a floor to drop only two or three seeds to the inch in the rows because I wanted to have as sturdy plants as possible and to avoid the work of thinning and the loss of valuable plants as would otherwise be the case.

Just before the plants were ready for transplanting I inserted an advertisement in the local paper to the effect that I had well grown, stocky plants of this specially fine strain for sale, invited buyers to call and see them and quoted a price no higher than for ordinary stock. Next day a man called and bought enough plants to pay for the ad! The day following another came and bought all I was willing to part with! Half a dozen others arrived later—too late! Among them I could have sold 10 times the quantity I had grown. It would have been more to my interest that first year to have sold such good plants in smaller quantities to these or a larger number of buyers because the stock would have advertised me better and would have led both to repeat sales and new business in subsequent years.

Instances like this prove that even when one does not have a greenhouse, a hotbed or a coldframe (as I then did not have) plants may be profitably grown if they are stocky, of good color and of choice varieties or strains. Still further, it is easy to sell such plants even when the price asked is a little higher than that of ordinary stock. The small difference in cost of best over medium or low grade seed is far more than offset by sales of stock due to the reputation of the variety and the plants themselves.

When one has coldframes the scope is increased considerably. Hotbeds widen it still further and a greenhouse, even though a small one, used in conjunction with these two accessories gives a field limited only by its size, the available hands and the means of disposing of the plants while in prime condition.

The following instance * will indicate how well a greenhouse, hotbeds and coldframes may be made to pay in a small town, especially when there is no similar establishment in the vicinity. On a property of which I had charge for a couple of years were greenhouses and coldframes which had not been paying their way because the man who operated them had not been giving them his undivided interest and who had been dismissed in consequence. Yet within 15 months a competent man whom I engaged made them pay all expenses, including his salary, and yield a worth-while profit—the first in four or five years. As some of the ways this was done may help you I shall outline them.

Preliminary to starting operations the new man recon-

---

* Much of the following paragraphs is condensed from the author's articles in the *Florists' Exchange*.

noitered the town; first, to get an idea of the possible area of gardens and the quantity of vegetables and flower plants that might be needed during the first season; second, to estimate how many house plants, bulbs in flower, and forced vegetables might be sold; third, to make a similar estimate of the cut flowers that the community might use; and fourth, by inquiry to learn about how many balls, dinners and other gatherings would likely occur and at about what dates. With these data in hand we made production and sales plans.

Without going into details let it suffice that these plans included a series of "special sales attractions"; for instance, cineraria, cyclamen, narcissus, tulip, gladiolus and geranium. Besides these we bought a few plants of striking interest, not for sale but solely to pique curiosity and serve as "bait" to lure the unwary public to the greenhouses *when we had something else to sell* and which our visitors would buy on sight because of its high quality!

Whenever one of these remarkable plants reached its showy stage we would put a news item written in editorial, not advertising, style in the local paper. It would tell something about the plant and suggest that the people go to see it.

All attracted visitors, many of whom bought house plants, bulbs in flower, cut flowers, or forced vegetables all of which, when they arrived, we made it a point to be working with as if to fill outside orders. The high quality of the plants and the freshness of the vegetables contrasted with store stuff and therefore sold themselves on sight.

We also advertised in the local paper, but two or three days before any "ad" was to appear we mailed postal cards inviting our best patrons to come and choose the best stock before the general public arrived. Such plants we always sold at somewhat higher prices than those to be advertised and patrons were always pleased because they felt that they really got the full value of their money.

Our sales were usually advertised to continue one week, the date of starting and closing always being stated. Toward the close when there were odds and ends we sometimes announced a bargain sale of "remnants" or of "surplus stock," but whenever possible we scheduled such sales for the same periods as the advanced sales to select patrons were to occur. Our reason was that we wanted to let the "bargain hunters" see what really good stock is like and thus, if possible, teach themselves to appreciate higher qualities and values.

Among other ways by which money may be made by growing plants for sale are by propagating special varieties, strains or stocks of plants from seed or stock which patrons supply. Agreements in such cases, always in writing, might be at set prices for the delivery of specified numbers or percentages of plants so grown to certain stages of development, the grower to own the balance. This plan has worked especially well among people who come to the country from the city only in spring or early summer and who want to have plants of specified kinds ready for them to start their gardens.

Where there are local garden clubs or where a large enough number of amateurs can be brought together, it is a good plan to invite them to visit the establishment on a specific day—always when it is looking its best and when timely topics of special interest at the time are to be presented and discussed. On such occasions the owner should outline ways by which he may serve the gardening community, especially in the propagation of stock for their gardens as already outlined.

For such reasons as those sketched in this chapter, "glass" may be made one of the most profitable departments of the small farm, especially in localities where no such equipment is already established.

44
SOMETHING TO SELL EVERY DAY

THERE is no safer place of existence than the moderate sized farm. It is not often practical to employ a large amount of machinery and a large area of land in attempting to turn agriculture into manufacture of some single great staple. But the family that makes the farm an old fashioned home with diversified crops, fruits and domestic animals sufficient to meet the household needs will still find agriculture one of the most satisfying forms of existence.

CALVIN COOLIDGE.

To BE most successful the small scale farmer may well adopt as his slogan the old advice: "Don't hurry and don't worry, but keep moving and *make every stroke count*," especially the last!

Ways to live up to this dictum are: 1. To plan the whole general scheme well in advance of any action; 2, to decide which are the major features or crops and get them established first; 3, to fill in details in logical order; 4, to review the experiences, especially the mistakes, of each season and modify the plans when necessary or advisable for the following season so as to emphasize the one and avoid the other; 5, to make important changes only after deliberation.

The general plan should take into consideration what branches of farming are best adapted to the land—those which will probably produce the best yields; what ones will be in sure and great enough demand to pay well; what ones are least risky, either because of hardiness, ruggedness, fewness or insignificance of enemies; what ones may be disposed of in the greatest variety of ways; and so on. After making this decision, if fruits or perennial vegetables are to be grown their planting should be done the first year of ownership, even though no money return can be expected for several years. The cost of stock and development may seem excessive on the start but after the investment gets under way, being permanent, it should pay well and for many years with only the annual cost of up-keep. To offset this expense until the permanent crops pay, annual crops may be grown between the trees and bushes, and certain kinds even between the plants of perennial crops as outlined in Chapter 32. These though planted for profit should always be such as never interfere, but through their careful culture actually aid the permanent investment.

However, even under ideal conditions, with hotbeds and coldframes to start, and with proper storage to extend the season it (Chapter 50) will rarely be possible to have "fresh" vegetables to sell during the whole 12 months, usually not more than 7 or 8, and with fruits not so many. So other branches of farming, yes, and manufacturing, may well be added to fill in the gaps. Among the best are dressed poultry, eggs, honey, plants for transplanting, flowers (Chapters 22, 23, 43), canned fruits, and vegetables, soups, jams, jellies, pickles, and fruit syrups.

The principal advantages of adding the preserving branches to the list of departments are first, that unsold, surplus and cull fruit and some kinds of vegetables may be converted from waste and loss to saving and profit; second, that sales may be made every day in the year. Though all the fruits may be worked over in one to several of these ways (and others), many of the vegetables are either not adapted to such preservation or are too "cheap" to allow a fair margin of profit. For instance, out must go the salads and the potherbs such as spinach.

The ones best suited to canning and for which there is greatest demand are tomatoes (canned, preserved, catsup, juice, cocktail, pickles), "baby" beets and carrots, little onions and cauliflower (pickles), and rhubarb (canned, conserve, juice and wine). With the exception of the last, all these, also celery, pepper, leek, garlic and flavoring herbs such as parsley, sage, summer savory and thyme in multifarious combinations may be used in pickles, relishes, chutneys, sauces and soups.

Lest anybody doubt the demand for such comestibles let me say that during the past five years three women of my acquaintance have sold every container of marmalade, jam, jelly and soup they have offered for sale through the woman's club of their village or otherwise and have had repeat orders, even a year in advance—when new crops would come in! Had their supplies of raw material been larger and their equipment better adapted to manufacture they could have marketed everything they could put up.

Another woman, suddenly left without support, started in, but soon outgrew her kitchen, migrated to an outbuilding, outgrew that, built a small factory, enlarged it, contracted for freshly gathered fruit *by the ton* direct from growers of desired varieties and at the time of my last visit, though the season was scarcely half over, had her store rooms packed with more than $100,000. (!) worth of preserved, canned, jellied, jammed and other classes of goods! Most of her products are sold through "fancy" grocery stores, though much also goes direct to regular, personal customers.

If the goods are only good enough—as hers were and are —they need only be tasted to assure a steady demand at reasonable prices. But the quality must always be kept up to standard.

Besides all these are specialties such as lily bulbs, everbearing strawberries, pussy willows, Japanese lantern plant, ornamental gourds, madeira and cinnamon vine tubers, straw flowers, culinary herbs, holly grown from cuttings, bittersweet vine, winterberry (*Ilex verticillata*), tigridia, snowberry, montbretia and many others.

45

STRAWBERRIES

THE three essentials that should never be overlooked before going into the culture of strawberries are, first, the best variety suited to the soil, strong, vigorous, pure bred plants; second, a well drained and prepared soil; and third, thorough and frequent cultivation.

HUGH FINDLAY,
In *Practical Gardening.*

NO FRUIT is easier to grow, quicker to yield a crop, surer of a demand, or more likely to be profitable than the strawberry.* Whether you have only a backyard or whether

* This chapter has been edited by Mr. W. Lee Allen of Salisbury, Maryland, General Manager of The W. F. Allen Company, which for nearly fifty years has specialized in the testing of strawberry varieties and the production of plants.

your farm is measured by the square mile your strawberry "patch" should be at least large enough to supply all the fruit your family needs; and if your funds are so limited that you must make every cent work, the strawberry should be your first choice as a money fruit crop, because even at the height of the picking season the supply is seldom equal to the demand and because prices, no matter how low, are usually surer to be in excess of cost than are those of any other fruit. To make strawberries pay, however, it is essential that the plants be well grown, yield sufficiently and the fruit be properly handled.

Naturally you will want to know what to expect from a specific area. According to the United States Department of Agriculture the average American yield is about 1,800 quarts to the acre; but there's no sense in having such small crops. Commercial growers, on a field scale, get 5,000 to 10,000 quarts to the acre and specialists, growing in small areas, have attained yields that would make the acre rate 20,000 quarts! Of course, such high yields are attained only by intensive methods and exceptional attention to details. I mention them merely to indicate the possibilities and to show the importance of good care.

As costs and prices differ widely it is safer to calculate on the total time required to grow a crop, for then local costs and prices may be used to estimate local costs. The 5-year averages of six Kentucky strawberry farms were somewhat more than 2,400 quarts, 100 horse-hours and 300 man-hours an acre. It may be said that though many instances of profits of $1,000 or even more an acre have been realized and though $500. may be easily attained by experienced growers in favorable years, you must remember that frosts, drouths and other adverse factors may cause complete failure any year—perhaps for two or more years in succession—so if you can average $100. an acre year in year out you should be pleased because you will be earning several times as much as growers of most other crops. Higher averages may be expected from areas of ¼ to 2 or 3 acres than from larger ones.

One of the chief factors in securing high yields and therefore profits is the variety. During the present century improvement has been so startling that, of more than 400 varieties I tested up to and about 1,900, only four are included among the 46 catalogued by a nursery which for nearly 50 years has specialized in the production of strawberry plants, and which it grows by the million. The policy of this nursery has long been to discard all varieties that it has proved to be inferior to the ones it catalogues and to point out the failings or weak points as well as the merits of each one it lists.

Some varieties called pistillate or imperfect will not set fruit when planted alone. The reason is that their stamens are either aborted or missing altogether so cannot produce the pollen upon which fruit formation depends. To make them fruitful, perfect kinds must be planted near-by—not less than one row of perfect to five of imperfect. In my opinion there are so many excellent kinds to choose among and alternating the rows so often causes trouble at harvest time that I would confine my choice to perfect kinds. Catalogues indicate perfect ones by "P" or "Per" and the imperfect as "Imp" or "Pis."

The best variety I have grown is Premier. Concerning it this nursery says in part: "It has given more general satisfaction and has been a better money maker over a wide territory than any other variety ever introduced. For home garden, local market, or for shipping moderate distances it has outclassed them all."

Since 1915 when Premier was introduced at least a score of really excellent varieties have been selected from perhaps a thousand offered. Of these two of the most notable are Fairfax and Dorsett which originated with the Department of Agriculture. As yet I have not had opportunity to fruit them but such is my confidence in the judgment of the nursery referred to that I quote: "Where they have been tried, Dorsett and Fairfax outclass Premier as berries for the home garden, local market and for shipping, just as completely as Premier outclassed the others when it was first introduced. We believe they will maintain this superiority over most of the territory where Premier has been so good."

I would not quote the above statements if I could not indorse them. Nevertheless, I strongly advise everybody who grows these varieties for home use or sale to grow other kinds for comparison because differences of constitution, of soil or other factors may or may not be favorable and because the aim should always be to have the best that will do well under local conditions. For these two outlets you should also include at least one of the everbearing varieties to supply dessert fruit in late summer and autumn.

Among varieties that have done well for me are Big Joe (alias, Joe Johnson, New Hope and Joe), Glen Mary, Marshall, Aroma, Chesapeake (Lateberry), William Belt, Gandy, and Orem (Frostking). These cover a season from earliest to latest—3 to 5 weeks in early summer. Others highly and widely recommended are Blakemore, Bellmar, and Southland (for the South)—all developed by the Department of Agriculture.

Everbearing varieties have two bearing seasons each year, one in spring, the other in late summer and fall. The crop is usually too light to be of commercial value, but the second crop makes a welcome addition to the home bill of fare and often finds ready sale to personal customers and at roadside stands; but unless you have one or the other of these outlets it would be wise to feel your way before planting them extensively.

Of all the everbearing varieties I have grown Champion (Progressive) is of best quality but the berries are smaller and much less abundant than those of Mastodon, concerning which the nursery already quoted says it "is good enough in all respects to stand out among everbearers just as Premier has done among spring bearing kinds."

Last year, under exceptionally unfavorable conditions (gravelly soil and unavoidable shade much of the day) my plants grown by the hill system averaged about a quart each for the season. Under more favorable conditions I feel sure that this average could be easily exceeded.

When plants must be bought to start a strawberry "bed" it is advisable to get them from a nursery which specializes in strawberry plants rather than to buy from a neighbor or even from a "general" nursery. After a start has been made many northern growers dig plants from their own beds to make new ones; others buy from the nursery. In the South renewal of stock from the northern nurseries at least every two years is necessary because the rest period of the southern winter is not long enough to maintain the vitality of the stock.

In my opinion digging one's own plants is an undesirable practise because, 1, digging and cleaning the plants demands skilful work, is costly and likely to be done improperly by inexperienced hands; 2, the consumption of valuable time due to unhandy methods of digging and unskilled trimming; 3, probable delay of digging due to unfavorable weather or soil condition; 4, likelihood of getting plants inferior to those bought from a specialist nursery, because diseased or infested with insects; 5, the loss of berries such plants would bear the same year; 6, probable injury to the plants near the ones dug but left in the bed.

On the other hand, the advantages of buying from the specialist nursery are the reverse of those just ennumerated and also include the certainties, 1, of getting high grade plants trimmed properly and ready for planting; 2, of having them arrive early—dependent only upon early placement of the order; 3, of having plants with straightened roots, thus favoring speed and correct planting. Such plants would probably give better performance because usually grown on lighter soil than that of the bed to which transplanted. In short, purchased plants will probably cost less and produce larger crops than those dug from one's own bed. However, no higher price should be paid for so-called "pedigreed" plants because these have been proved to be no more productive than "unpedigreed" ones.

You need not hesitate to give specialist strawberry nurseries your orders. Their packing methods are so scientific that they can ship by express from coast to coast and yet—barring unreasonable delay in transportation—guarantee every plant to be in prime condition on arrival. If the plants are dried out and wilted soak them until roots and leaves are thoroughly revived and plumped up, several hours if necessary.*

As the strawberry is a perishable fruit it must be picked

* From this point forward many of the statements have been condensed or adapted from Circular 64 of the West Virginia Experiment Station.

as soon as ripe and disposed of without delay or rough treatment. Planting should, therefore be near a market large enough to absorb the crop, provide plenty of cheap picking labor and avoid a long haul over rough roads. Conversely, it should not be on too high priced land since interest may run away with profits.

Statements concerning site (Chapter 8) apply with special force to the strawberry. Make sure of good air and water drainage, a southern exposure for earliness, a northern one for lateness and protection of the blossoms from spring frosts and heaving of the plants by alternate freezing and thawing during winter.

Any medium fertile, well drained soil with high water-holding capacity is suitable; but since the crop demands considerable attention it is better policy to give it the best land available. Poor land is sure to be disappointing. An ideal soil is a sandy loam underlaid with clay because this gives a warm surface easily worked and a subsoil retentive of moisture and fertility. Sandy soils are warm, suited to plant production, and early fruit, but likely to lack fertility and moisture-holding capacity. Addition of organic matter may correct the latter fault. Clay soils are colder therefore better adapted for late crop production. Their worst fault is that they may bake during hot weather. It seems slightly acid soils are preferable to neutral ones. Perennial weeds such as quack-grass should be eradicated before the plants are set, else they will cause trouble if not destroy the patch. Fields devoted to hoed crops are likely to have few weeds and grass and to be economical to cultivate. However, a rank growth of grass and weeds when plowed under while still lush and green and before seed matures is valuable for humus.

Never should strawberries follow a grass sod of several years' standing because cutworms, wire worms, but especially white grubs (larvæ of May beetles), will be present to destroy the roots. A pure clover sod is safe as it is not infested with white grubs. As 1,000 quarts of berries require about 1,600 pounds of water a good crop will take 3 to 5 tons of water from the soil for the fruit alone, and several times as much to grow the plants. Hence preliminary treatment should aim to augment the water-holding capacity of the soil.

When possible, an inter-tilled crop should precede strawberries, preferably for two years at least, to kill weeds, starve out the insects mentioned, put the soil in good physical condition and betray wet places that need drainage. A heavy dressing of 10 to 30 tons of manure the year before strawberry planting is the best method of increasing the humus content of the soil and of adding fertility. It may be applied to the preceding crop or in the fall.

Green manures such as rye and vetch (together) sown after the cultivated crop will also help build up the humus content (Chapter 29) especially when manure is not available. When cover crops are not used, fall plowing 7" to 8" deep the year before plant setting is desirable, provided the furrows are left rough until spring, then disked, harrowed and fertilized as early as possible. Most soils will be benefited by a dressing of 250 to 300 pounds of superphosphate and small doses of potash muriate (50 to 100 pounds) to the acre harrowed in at this time.

When lime is necessary for the cover crop it should be applied at least a year before strawberry planting because this crop seems to resent its presence if too prominent. Better be safe and make it three years.

If the ground is not ready when the plants arrive heel them in temporarily in a well drained, shaded spot and in well prepared soil. To do this make a V-shaped trench about 6" deep, spread out the plants so the roots of each are in close contact with the earth then firm the soil well around them but not over the crowns. Other rows may be made 2" or 3" away from and parallel to the first one. Be sure to keep each variety by itself and properly labeled.

Strawberries are best planted in early spring. Conditions are then most favorable and the most nearly perfect stand may be secured. When necessary to plant in the fall the plants should be well mulched to prevent or reduce losses by heaving.

By marking the ground before plant setting the rows may be kept straight and subsequent care favored. A hand-drawn marker is best for a small patch; a horse-, or tractor-drawn one for a large area. Marking in only one direction is necessary for most systems but in two at right angles for the hill plan because the plants may be set in checks so as to permit wheelhoe tillage being given in two directions.

The most important factor in obtaining a good stand is to prevent the plants suffering for lack of moisture from the time they are dug until again established in the soil; the next most important is to set each plant at the correct depth—no shallower and no deeper than the crown (Fig. 59). When these precautions are taken plant losses should be small. Poor stands are likely when plants are set in late May after the rainy season has passed.

The essentials of setting are the same whether the work is done by hand or by machine (Chapter 42). Plants should be dropped only a little ahead of the planter when planting is done by hand. The roots should be spread out fan-shaped so each will intimately contact the soil which must be firmed thoroughly around them to prevent air spaces. This is especially necessary in late planting and when the soil is dry. Too deep planting often suffocates and rots the plants; too shallow allows the roots to become dry. Both finally kill the plants. (Fig. 59.)

Planting distances depend upon the system of growing, though soil fertility and, in many cases, plant-making characteristics of the variety are also often considered. Kinds which produce many runners should be set farther apart than those that make few. The former are best adapted to the matted row systems; the latter to the hedge row and the hill method.

The matted row system is considered best for commercial production, plants being set 18" to 24" apart in rows 42" to 48" apart. For sparse plant producers the distances may be respectively 15" and 36" or even less. The various distances may be increased in highly fertile soils and reduced in poorer ones.

During the first season plants in the matted row systems are allowed to produce many runners in rows 15" to 20" wide. Wider rows than the latter are harder to pick. Though most commercial growers do not thin out the plants, best results are ordinarily obtained when plants range from 4 to

6 to the square foot in the filled-in row. The sooner a full stand of plants can be developed the better will be the yield the following year.

Larger berries but smaller yields are usual with hedge row and hill systems. However, these make picking and cultivation easier, cost of plants less, removal of some or all runners necessary, but prices for "fancy" berries often offset the extra work and smaller yield.

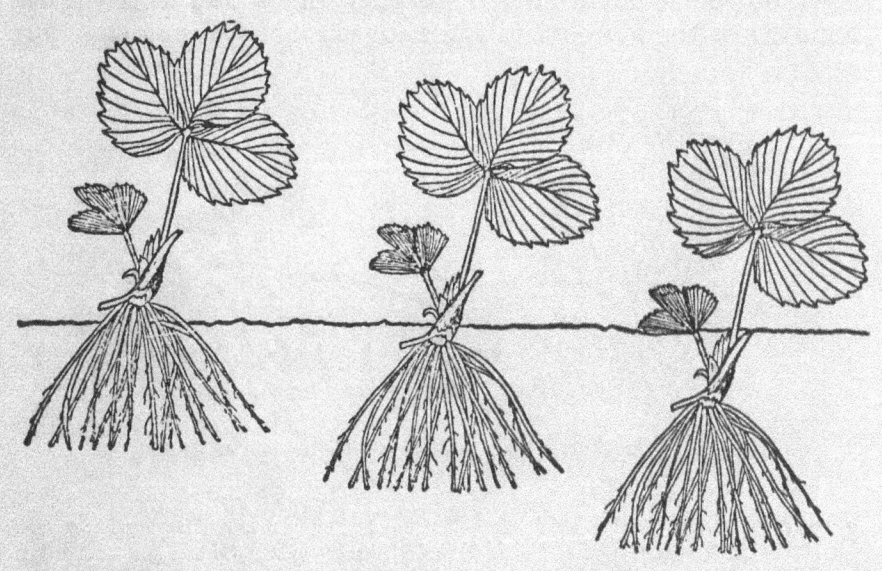

Fig. 59. Strawberry plant setting. Left, crown placed too high; plant will dry out. Right, crown too low; plant will rot. Center, correct; crown at ground level.

In the single hedge row system plants are set 15" to 18" asunder in rows 24" to 30" apart. Each plant is allowed to form two runners which are placed in the row, one on each side of the parent plant. The final result is a single straight row with plants 5" or 6" asunder. In the double hedge row a slightly greater distance is allowed between rows and each plant allowed to make six or eight plants, two of which are placed in the row as before, the others spaced on each side of the parent plant so the final result is three rows 5" or 6" apart and about the same distances between plants.

Plants in the hill system are set 12" to 24" asunder in

rows 30" to 48" apart, but preferably in checks 18" to 24" apart so as to favor wheelhoe cross cultivation. All runners are cut off as soon as seen so as to increase the size and strength of the original plants. This increased size does not usually make up in yield for the smaller total number of plants in the bed, but the fruits average much larger and thus command fancy prices.

During the first season care of the plants includes tillage, flower removal in all systems and either training or removal of the runners in the hedgerow or hill systems and

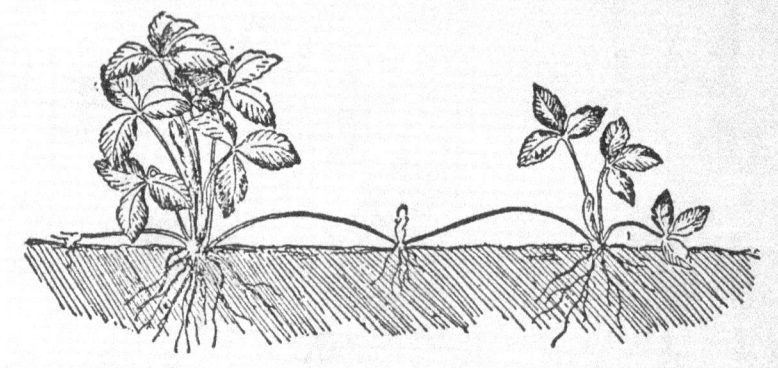

Fig. 60. Strawberry plant showing well developed runner.

thinning plants of free growing varieties in matted rows. Cultivation should begin immediately after planting and be repeated at weekly or 10-day intervals until fall. It should always be in the same direction with the matted row and hedge row systems, but in different ones with the hill system. The object of one direction is to keep the runners in the rows instead of spreading them too widely. Placing plants where wanted and hand hoeing to keep down weeds close to the plants will be necessary. When the first hoeing is being done covered crowns should be uncovered. Later when runners appear these should be anchored with clods or pebbles where wanted. They soon take root. Excess runners may be cut with a sharpened 8" disc run between the rows at the desired distance from the row centers.

When allowed, plants will produce a few flowers and fruits the first season, but at the cost of vigor and runner production. To have enough plants for the next season all

STRAWBERRIES

flower stems should be pinched off, except that those on everbearing varieties should be allowed to remain after about July 1 so as to have fruit later in the season.

About a month after the plants are set 125 to 150 pounds of nitrate of soda to the acre should be top dressed to help runner growth during the first season and a second dose in mid August to increase and strengthen fruit bud formation which begins in early September. A third application of the same amount is sometimes given in the spring of the fruiting season, but may not be needed. If the bed is to be renewed, 250 to 300 pounds of a complete fertilizer high in nitrogen should be applied after the crop has been gathered.

In my opinion the best materials for mulching are buckwheat straw between the rows and buckwheat hulls around the plants, the former spread 3" or 4" deep (at the rate of about three tons to the acre), the latter about 2" deep around the plants, both being applied after the ground has frozen hard enough to bear a team and a loaded wagon. These materials are usually free from weed seeds, especially of perennials, the straw becomes brittle during winter, soon breaks up by tramping pickers' feet and is easily plowed or dug under after berry harvest. The hulls work well among the plants which they not only protect but keep cleaner than do most other mulches. In addition they have an appreciable fertilizer value when they decay.

Shredded corn stover when obtainable at low cost is almost as good as buckwheat hulls. Both may be spread with a large scoop shovel as evenly and as deeply as desired. Marsh or salt hay, though harder to apply, may be used for several years if removed at the close of the picking season, thoroughly dried, and stored under cover or out of contact with the soil. Oat, wheat and rye straw are less desirable, harder to spread and to plow under and are almost sure to contain weed seed. The most undesirable materials are hay and litter from the horse stable because these are sure to introduce weeds and grass.

The time to remove mulch varies according to whether late or early berries are wanted. To get late ones the mulch should be left on until the leaves show slight blanching, no longer. The loose material is then raked off the plants and tramped down between the rows; the small stuff left among the plants. New foliage will grow through the thin covering left on the rows.

Irrigating strawberries will insure the crop in dry seasons and increase it in others, but the cost of installation is usually warranted only when other crops are to be grown on the same land so as to apportion the expense. In a permanent business it should more than pay for itself over a period of years.

Harvesting lasts from two to five weeks, depending on the season and the succession of varieties. It starts about a month after the blossoms appear. The fruit is usually picked in quart boxes placed in four-, six- or eight-box carriers. It is marketed in 24-, 32- or 48-box crates. Many growers withhold ½¢ a box to be paid to the pickers who stay the entire season. Otherwise some of them might leave when the berries become scarce. Enough pickers should be employed to pick over the field once in two days. Picking should not start until the dew has dried off the plants nor continue later than noon. Berries should not be picked while wet.

A competent supervisor should see that the following precautions are observed: 1. Fruit should be picked only when properly ripe. For home use and roadside market it should be mature and well colored; for near-by retail markets it should be hard ripe and almost fully colored. 2. The patch must be picked clean at each picking. Over-ripe and misshapen berries must be picked off and discarded. 3. Each picker must stick to and finish his assigned row. 4. The fruit should be picked with a ¼″ or ½″ of stem, and only a few berries should be held in the hand at a time. 5. Boxes must be well filled. 6. Plants must not be knelt on. 7. The fruit must be kept out of the sun after picking, preferably in a shelter shed near the patch, the berries being taken there to be packed as soon as the carrier is full.

The best method of marketing the crop will depend upon the location. In most localities selling in a local market will be best. This applies especially to roadside marketing. Pan-grading the fruit and arranging the top layer attractively often build up a fancy trade.

Though most commercial growers prefer to plow under

STRAWBERRIES

the plants after only one crop has been gathered and immediately to plant a late vegetable crop such as cabbage or a cover crop such as crimson clover, many others crop their matted row beds a second time. The best way to do this is to renew the beds once. If insects and diseases have been

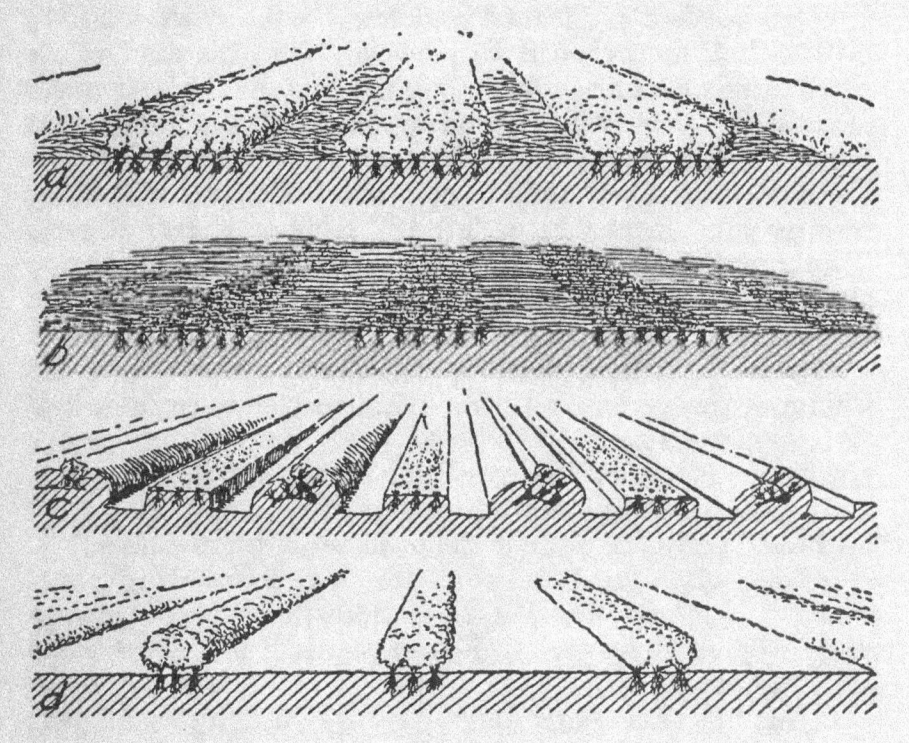

Fig. 61. Strawberry bed renewal. a, Matted rows and mulch before treatment; b, tops cut and burned with part of mulch; c, strip plowed at each side of matted row; d, new foliage on renovated plants; runners will develop later.

very bad, renewing even the first time is not advisable. The best time to renew is immediately after the last picking. The usual method consists of the following steps:

Most of the mulch is raked off the field and either made into artificial manure or compost. If infested with insects it should be burned. The foliage is next mowed with the cutter bar of the hay mower set high enough to avoid cutting the crowns. The chief purpose of mowing is to control foliage diseases. When the mowed leaves are dry and when a fairly strong breeze is blowing in a safe direction the dry material is fired in several places on the windward side so as to burn rapidly across the patch. If the plants are too dry, or the mowed leaves too wet, the crowns may be injured. Burning over destroys insects and diseases; where these are not present it need not be done.

After the bed is cleaned and burned the rows must be narrowed. One method is to plow one (say the east or the north) side of each row including the center. The best width of row to leave will depend upon local conditions and season. If the season up to the time of treatment has been dry and seems likely thus to continue few plants will likely form so the row should be left 12" wide or wider; if more favorable, it may be as narrow as 6". Character of variety also must be considered: kinds that produce many runners may be more closely narrowed than those that produce few.

A second method, most applicable where the rows are wide and the variety a liberal plant-former is to plow two furrows toward each other between the rows, a second pair down the centers of the rows and then cultivate these furrows down smoothly and finely. Both these methods get rid of the old plants and favor the production of new ones.

A third plan merely narrows the rows by plowing a furrow on each side of the rows and the spaces between (Fig. 61). As it leaves all the old plants it is less desirable than the other two.

Plants in the narrowed row are usually somewhat crowded so some must be removed to give the next year's fruiting plants a chance to grow large and vigorous. Runners begin to form soon after berry harvest.

Cultivation should begin as soon as the rows are narrowed and continue until fall. A top dressing of 250 pounds each of nitrate of soda and super-phosphate an acre is advisable just before the first cultivation.

Strawberry insects and diseases are usually not troublesome where clean tillage, rotation, correct renovation and the planting of healthy stock are followed. Setting plants in soil which has been in sod the previous season, and especially for several years, should always be avoided because the white grubs are almost certain to do serious damage. Where facilities are available, one or two combination sprays of bordeaux-lead arsenate before blossoming is insurance against leaf spot and chewing insects such as leaf rollers. Root aphis seldom give trouble if clean plants are obtained and set in land not occupied by strawberries for several years. Weevil may be controlled by setting chiefly pistillate varieties since the larvæ feed on the pollen.

46

GRAPES

By using the different classes of varieties of grapes for the different latitudes, soils and purposes, there is scarcely a farm between the Great Lakes and the Gulf but can successfully grow grapes. Of only one or two other fruits,—the strawberry and blackberry—can as much be said.
T. V. MUNSON,
In *Foundations of American Grape Culture.*

GRAPES will thrive on every kind of soil that will grow other cultivated crops and be more likely to yield better under adverse conditions than any other fruit. They require less space than even the smallest standard fruit tree; in fact, the area usually allowed in business orchards for a mature peach tree (20' x 20') is greater than necessary for six vines as grown in eastern commercial vineyards. The flowers open late in spring so are more likely than those of other fruits to escape frosts.

Trained as described further on vines may begin to bear the year after planting; when managed according to usual commercial practise a year longer is usual. I have repeatedly gathered as much as six pounds of fruit from strong vines planted only 17 months before and 25 to 30 pounds the year after. With ordinary care vines are likely to yield abundantly for 25 years at least. On one farm that I owned is a Concord vine 45 years old that averaged more than half a bushel of fruit a year, though it had been so neglected that when I bought the place I cut out a small wagon-load of dead and worthless wood and the former owner said the vine would die! But it bore that year and every year while I owned the place.

Red, "white" and blue varieties cover a great range of flavors from mid August to October without storage and when properly stored to March or even April. Some varieties are especially delicious as dessert, some adapted to jelly, jam or conserve, others for juice or wine. Various European varieties grown in California make raisins. No fruit plant is easier to train or so adaptable to all sorts of situations and styles of support. No matter how little your present experience you may count upon being successful after learning two fundamental principles and may grow choicer grapes than are usually sold in the markets and stores.

Having decided upon varieties suited to your locality order the vines as early in the year as possible so as to get the kinds you want. First class one year vines are generally better than those two years old and far better than still older ones which not only cost more but are slower to bear fruit because of the loss of roots. On arrival handle them as suggested for other nursery stock. (Chapter 39.)

Give the vines the best soil and the sunniest site at your disposal. When the soil is poor or the site trying (as against a building) dig a hole as wide as a washtub and somewhat deeper for each vine, throw a peck or more of bones in the bottom and fill the hole with good soil before planting. The bones will supply plant food as they decay.

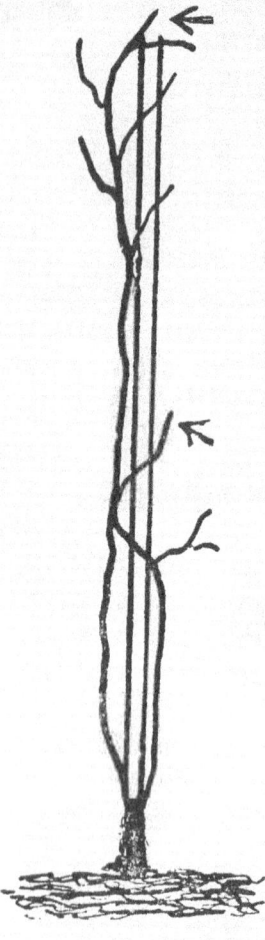

Fig. 62. Two-trunk grape vine grown tied to stake and "tip-pinched" at arrows.

After the plants are planted cut off all puny shoots and shorten the strongest stem of each to only two or three joints with a plump bud at each. This will concentrate the plant food so the new shoots will be sturdier than if all or a larger number of buds were left to grow.

As soon as the strongest shoot gets woody at its base (in early summer) shorten the weak ones to only one joint and one leaf. Until then it may be easily broken off; hence the advisability of leaving two or three shoots temporarily. By allowing each of the temporary ones to retain one joint and one leaf after cutting them back, the main stem will not be wounded and drying or dying will be avoided. In this method of treatment lies the secret of gathering fruit the year after planting—but on only strong vines.

Fruit is borne on the green shoots and on no other or older part. The shoots develop from buds formed in the angles of the leaf stems. Where the growing season is short these buds are developed during the previous summer. They are conspicuous during winter at the joints on one year old canes. Occasionally similar buds may appear on older wood but these may be disregarded except when it is desired to have new branches or trunks develop to take the place of old, decrepit or diseased ones where these buds are located. Sometimes a second or even a third set of shoots may develop and even bear fruit in a single season but the fruit will not ripen unless the season is long enough.

Most pruning and training depend upon the use of buds formed the previous year. From these arise green shoots which consist of joints, leaves, tendrils and flower clusters from which last the fruit is developed. The tendrils and the flower (or fruit) clusters present gradations between themselves thus proving that one is a modification of the other.

Each green shoot has one chance—*only one*—to bear fruit. If it fails it never has another; the buds it bears take up this function and produce fruit on *their* green shoots.

Because only green shoots bear the fruit, the fruit-bearing parts of untrained vines get farther and farther away from the roots each year. The result is that each successive year the vine wastes unnecessary amounts of energy and plant food in pumping sap from roots to leaves and elaborated food back from leaves to roots so that in nature and neglected vines little or no fruit is produced. This is where man steps in and compels the vine to produce liberally by pruning off parts that have survived their usefulness. He reduces the distance between root and branch and forces the vines to focus their attention upon fruit production. The removal of excess parts also conserves plant food so that the clusters of fruit are larger and the berries both bigger and of improved quality.

When these two principles are understood and acted upon, grape vines will produce abundant high quality fruit every year no matter how their parts are placed on supports. However, when trained upon posts the fruit is often injured by being whipped about by wind. So this method should usually be avoided.

Two kinds of canes are worthless—puny shoots often found on young, weak, neglected and very old vines; and "bull" canes, the thick, long-jointed, burly fellows that often grow 20′ in a single season. The canes that produce best are those of medium diameter—about the thickness of an ordinary lead pencil.

On arbors and summer houses more work is required to prune the vines than by any other method of training. Unless special care is taken to develop and train branches low down on the sides such structures will become bare below and the roofs too densely overgrown. Thus most of the fruit can be reached only with a ladder and pruning will be a tedious and difficult task.

On such supports the best way is to lead one main trunk direct to the peak and each winter cut back the canes along this trunk to short stubs (called spurs) of only two joints. The shoots developed from these spur buds will require tying where their foliage is needed to supply shade for the interior.

When trained against walls the vines should be on trellises, not fastened directly to the walls. The trellises should be such that they may be lifted free of the walls and laid on the ground when the walls need painting, which is best done during the dormant season.

The European varieties grown on the Pacific Coast are of much more stocky growth than the native American kinds and their hybrids. They are, therefore, more often cut back to stubs or spurs than trained with long arms. For this reason

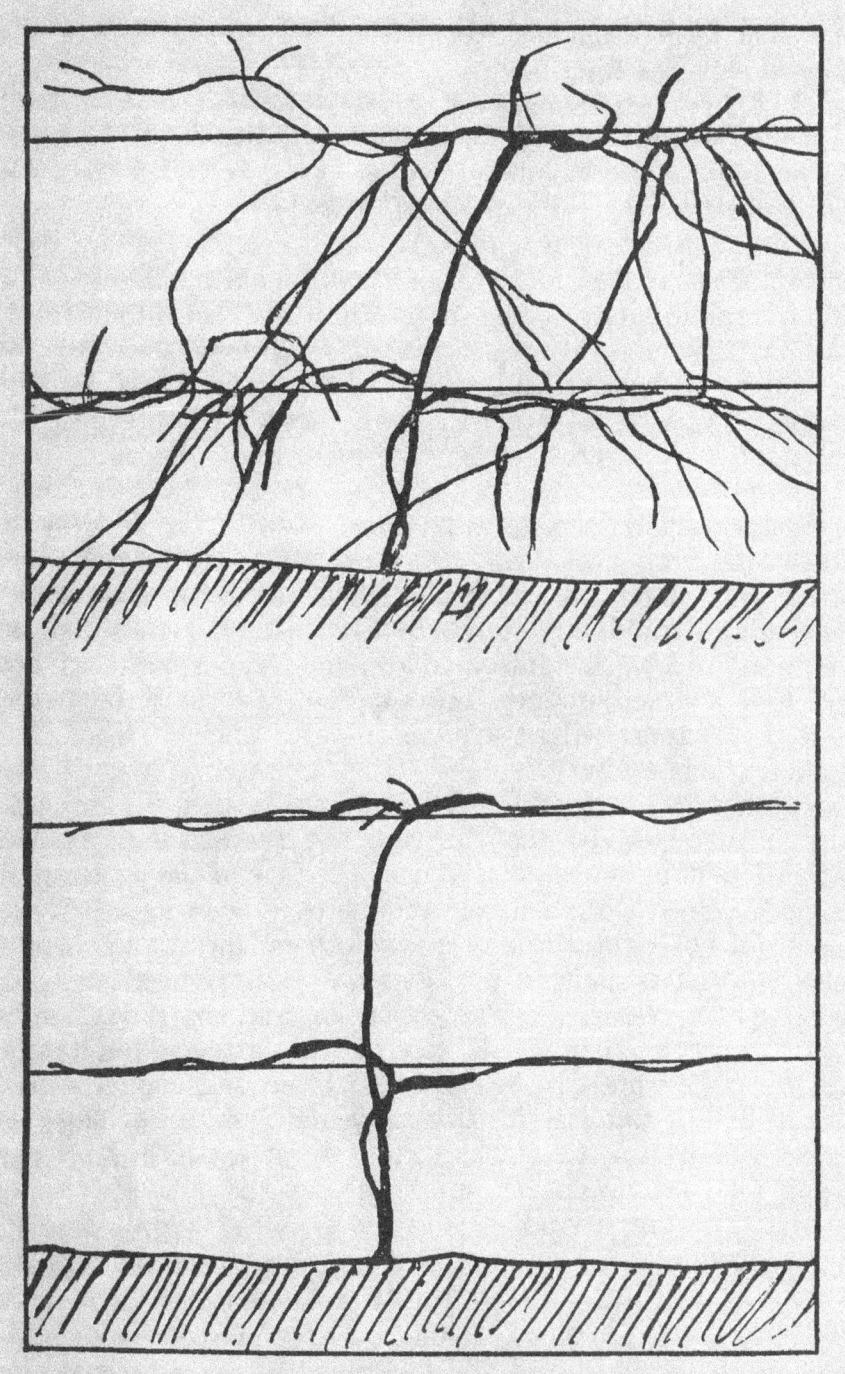

Fig. 63. Before and after pruning grape vine trained on four-arm Kniffin system. (From photos.)

they are better adapted than American varieties to walls and other restricted spaces.

After having tried various styles of trellises and training I have discarded wood because it is short lived and the framework loaded with vines is too likely to be blown down by wind or to break when loaded with fruit.

Perhaps the Kniffin (Fig. 63) is the most popular system of training used in the eastern half of the United States and for American varieties. In its modifications it meets economic considerations and differs from the Munson method mainly in having its two wires one above the other instead of at the ends of cross-arms. The principles of pruning are the same.

As in ordinary farm fencing heavy end posts of rot-resisting wood, such as locust, cedar or white oak, 8' long or somewhat longer are set well below the local frost line, thoroughly braced and tamped to withstand the strain of wind and fruit. The intermediate, lighter posts should be 7½' long and set similarly but not braced. Usually they are spaced 30' apart with two vines planted each 5' from the posts and one midway.

Number 10 or 11 galvanized wires are generally used, the former preferred because heavier and therefore more durable. The upper one is placed near the tops of the posts; the lower 2' below. Both are securely fastened to one of the end posts but not tightly to any of the others, the staples being merely to keep them in position, not to prevent the slight movement necessarily due to expansion and contraction due to temperature changes. If too tightly fastened the posts will be pulled out of place. At the loose end enough wire should be left to allow for this movement and a device provided to take up the slack during warm weather. Various types are sold by garden supply stores and nurseries that specialize in grape vines.

One form of the Kniffin system requires two upright main trunks which start at or near the ground, one reaching to the lower, the other to the upper wire. Another form uses only one trunk but has a pair of branches ("arms") extending in opposite directions on each wire. There is little if any difference in yield.

GRAPES

In the first case two strong shoots (instead of the one described under the Munson system) are developed from the newly planted vine during the first season if possible. As they grow they must be tied loosely to stakes in order to

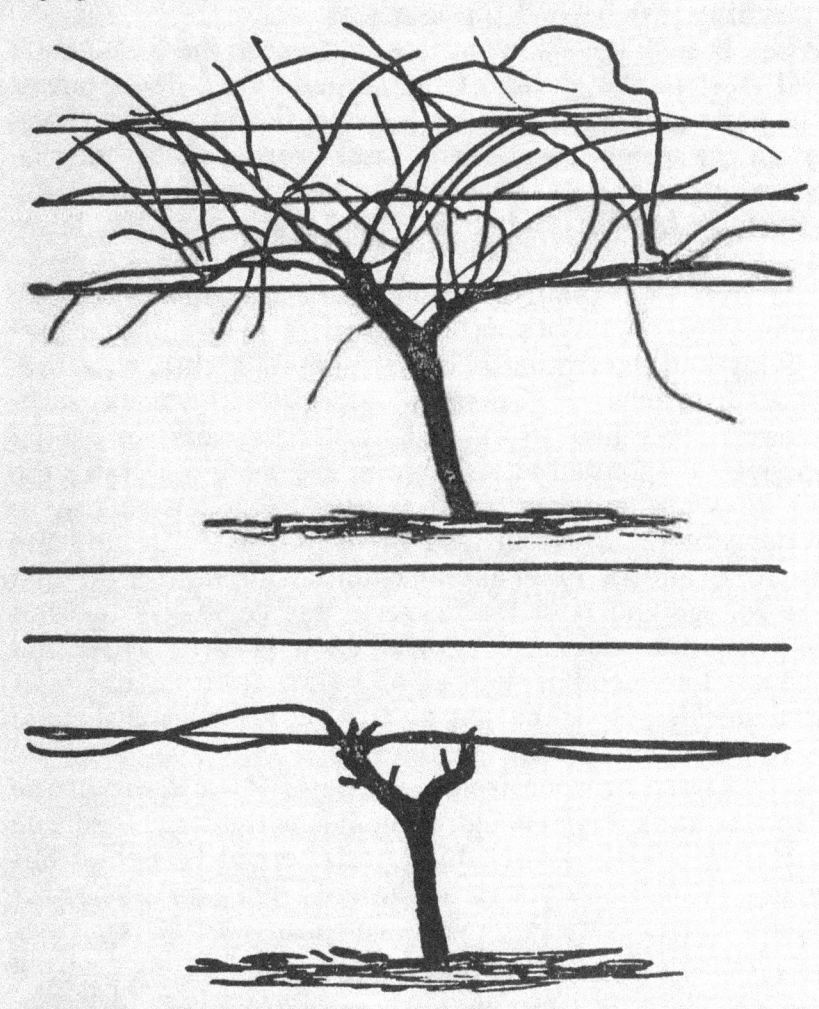

Fig. 64. Munson system of grape training. Above, vine unpruned; below, vine pruned and arms fastened to lowest wire. (From photos.)

make straight trunks. Preferably when one reaches the height of the lower wire its tip is pinched and the other is similarly treated when it reaches the upper wire.

Generally it is more convenient to erect the trellis during the second spring or the first fall after planting and to use only 6' or 7' stakes the first year. Usually pinching encourages the development of branches of which the uppermost two on each trunk should be trained in opposite directions on the wires. All other shoots should be removed, preferably by pinching just beyond the first leaf.

When branch development does not occur the first season it will start in the spring of the second. Then development of the pairs and the treatment just mentioned will be necessary. In the former case several fruit-bearing shoots may be expected from one or more of the arms; in the latter the shoots may bear fruit but not be branched at the lower joints.

The style of training I prefer is the Munson or canopy system. (Fig. 64.) With it: 1. The work of pruning, spraying, tying and harvesting is breast-high or higher, thus preventing backache. 2. Currants, gooseberries, black raspberries or other low growing plants that do best in partial shade may be planted alternately in the same rows with the vines, thus saving space. 3. It is easy to pass from row to row beneath the vines instead of having to go around the ends. 4. It affords excellent distribution of heat, light and air to foliage and fruit. 5. There is less resistance to wind than when the vines are trained on a vertical plane and hence less likelihood of their being blown down. 6. The fruit is well shielded from the sun by foliage. 7. The trellises are easy to make and, 8, their cost is low.

To make the Munson trellis, use posts 5" or larger at the top for the ends, long enough to be below the local frost line and 5' or 6' above ground. Line posts should be lighter but as long and as deeply buried as the ends. They are placed far enough apart to place three vines between; i. e., 30' for vines to be planted 10' apart, the usual distance for strong growing vines. The vines nearest the posts are placed half their distance from the posts—5' in the above case.

Across the top of each end post is nailed a crosspiece of 2" x 4" scantling 2' long and on the tops of the line posts 1" x 4" pieces. Previous to nailing, a notch ½" deep is cut 1" from each end and on one side. When nailing the crosspieces on the posts these notches are turned upward for the upper trellis wires to rest in them without stapling. The lowest wire is stapled 6" below the top not too tightly to all the posts except one of the end posts. The crosspieces need not be put up until the spring of the third year as the vines will not reach them until then.

After the posts have been placed, run a number 11 galvanized wire (copper is better but more costly) from end to end 6" below the tops of the posts, fasten securely at one end and provide a tightening device at the other.

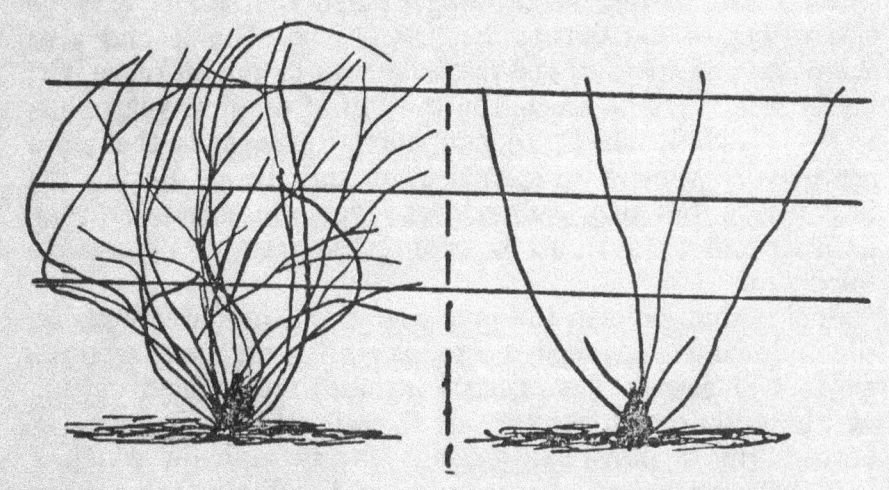

Fig. 65. Fan system of grape training. Left, vine before pruning; right, as pruned. (From photos.)

The vines having grown one season on poles are pruned in winter to only one cane. If this reaches above the wire (after the wire is in place) cut only the excess and tie the cane to the wire. If not long enough to reach the wire, cut it back a half to two-thirds; if very weak, cut to only two or three joints, each with a strong bud. In the last case treat the new shoots as during the first year.

Kniffin and Munson winter pruning are the same—a renewal system. Each winter the two year old and older wood (except the trunks) is cut and replaced by ripened canes of the previous season's growth. Each system makes necessary only one cut of each arm near the head of the trunk. Before cutting, however, the renewal arms must be chosen and freed from attachment to the other canes so that when the cut arms are pulled off the trellis these renewal arms shall not be broken or otherwise injured. These new arms should start as near the trunk as possible. Often as the vines grow older it may be necessary to get rid of accumulated wood where the arms have started. In such cases a cane may be cut back to one or two joints. From the buds on this spur new shoots will develop. The best one of these should be chosen to become the new arm in the following spring. The other must be cut off.

Whatever system of training chosen the shoots may be tied to bean poles during the first season. The second year the posts and wires of the trellis should be put in place. Or, better, they may be erected in the fall of the first year when work is usually slack. In commercial vineyards the vines are usually allowed to sprawl upon the ground during the first season, but they cannot thus receive the attention they deserve and as a result bearing seldom starts before the third year.

Shoots that develop the first year from the lower buds on the main cane (now called a trunk) often bear one to three clusters of grapes. Commercial growers recommend cutting off these shoots to strengthen the vine but I have never noticed the slightest weakening of any vine on which I have allowed them to mature. For I believe that a vine ambitious to set fruit knows more about its resources than any human does; so why should I interfere with its plans, especially when these lead to a gift of choice fruit!

Summer pruning, though formerly advised, has proved inadvisable, so don't do any.

During the winter of the second year the Kniffin trellis must be completed by stretching the upper wire from end to end. To complete the Munson trellis two wires are similarly stretched in the notches on the crosspieces, fastened at one end and provided with a tightener at the other. Then every stub and cane except the ones wanted for arms must be cut close to the trunk and the arms shortened to eight or ten buds. The shoots that spring from these buds should yield two to four or even five clusters, so it is easy to calculate at pruning time how much fruit to count on the following fall. As the shoots develop they are allowed to form a "canopy" in the Munson system.

In the winter of the third and subsequent years the arms attached to the lowest wire and all inferior canes are cut off and only the strong renewal canes left. These selected canes must be carefully shortened as already described and fastened to the wires.

47

BUSH AND CANE FRUITS

BUSH-FRUITS occupy a place of their own in the pomological field. They do not meet the universal demand that does the strawberry, and cannot be grown on such an extensive scale under most conditions. . . . Each kind is a favorite of some persons, and all are relished by most people.

FRED W. CARD,
In *Bush-Fruits.*

BUSH and cane berries offer better opportunities to a larger number of people than do any other lines of fruit growing. Everybody is hungry for them. They furnish a great range of flavors and a wide variety of uses—dessert, canning, jam, jelly, juice, wine, etc.

By proper selection of varieties and the application of simple methods of culture their season may be made to extend from June until October. No other fruits except the strawberry begin to bear so young—the second season after planting. Some bear more or less even the first year and reach full bearing the third or fourth. Under good care all continue productive for 5 to 10 years or even longer. They need comparatively small space, are of simplest culture and are therefore specially adapted to limited areas. Still more important, they are the fruits that should be chosen first to supply the home table and to sell locally because they may be allowed to reach the perfection of full ripeness before being gathered. The ease of their culture often encourages those who start with them to branch out and grow grapes and tree fruits, so that in due time the family will get most, if not all its fruit, from the home place and during most if not all the year.

Finally, when once the start with berries has been made the grower need expend no money when he desires to enlarge his plantation or make a new one. By the same token, when the plants have become established he may give away or sell plants to the neighbors, at no disadvantage to himself or the plantation. No other fruits offer opportunities that compare with these.

When a choice of site for a berry patch is possible, give preference to a northern exposure and an elevation above surrounding ground. The former is less affected in summer by the direct rays of the sun and the latter is less likely to be frosty than lower ground while the plants are in blossom. But if one has little choice of site, then select the spot that will be most convenient, never fearing that feeding and care will offset possible disadvantages of site and location. Gentle slopes are generally better than level and low lands, so should be preferred.

With reasonably good care berries do well on practically all soils, especially deep, strong, moist but well drained ones. When possible avoid the lightest sands and the heaviest clays, though dewberries thrive well on the former and currants and gooseberries on the latter, especially when the surface is kept loose by frequent shallow tillage or by mulching. But don't let a somewhat adverse soil prevent planting, for it is surprising what cultivation, fertilizing and mulching will do. Where a range of soils may be had red raspberries and blackberries should be placed on the lighter soils and the black caps on the heavier. But since varieties differ somewhat in their requirements those that normally grow rank should be placed on the poorer soil and the weaker ones on the richer land to offset and correct these habits.

Ability to retain moisture during dry times is the main point to secure in a soil for berries. This may be improved in any soil by maintaining the supply of vegetable matter through manuring, mulching and cover cropping. (Chapter 32.) Never use rye or vetch in the berry patch, unless strictly confined to the center half between the rows. The former is dangerous because it starts early in spring and grows rapidly, uses great quantities of moisture and if left too long becomes woody. In this condition it is slow to decay. The objection to vetch is that it is almost sure to clamber upon the berry plants.

Where the growth of the berry plants is lusty, nitrogen is not needed so no legume should be sown nor any nitrogenous fertilizer applied. In such cases it is better to use buckwheat, barley, oats, rape or some other crop sure to be killed by winter. Where a winter legume such as crimson clover or hairy vetch is not desired but nitrogen is needed soy beans are good to sow during June. Though killed by frost they add abundance of nitrogen to the soil.

Red raspberries won't stand soil acidity. They are benefited by dressings of wood ashes or applications of hydrated lime once in four to six or seven years. Black cap raspberries are almost the opposite; they prefer slightly acid soils. Blackberries, currants and gooseberries are much the same, though each may be helped by lime if the soil is very sour.

Though most berry planting is done in spring as soon as the ground can be worked, currants and gooseberries are best planted in the fall after their leaves have dropped because these plants start to develop new roots very early. Probably 95% of the failures in spring setting them is due to planting late. Blackberries and red raspberries may be planted in either fall or spring but black caps and dewberries should not be planted in the fall because being "crown" plants and shallow rooted they will likely be heaved out of the ground by alternate thawing and freezing during winter.

Where horse or garden tractor cultivation is to be done in only one direction the popular distance for setting plants of all but blackberries is 3' x 6'; where cross cultivation is to be done, 4' x 4' or 5' x 5'. The latter is decidedly the better distance because it reduces hand work. Blackberries being large growing are generally set 5' x 7'. Currants, gooseberries, dewberries and black cap raspberries, which do not propagate by means of suckers, make "stools" or "bushes" and "stay put." Allow 5' or more between bushes.

Red raspberries and blackberries, which have the suckering habit, are usually allowed to develop narrow "hedge rows" in which only the stoutest canes are kept after the spring thinning, these preferably not closer than 8" for the raspberries and 12" for the blackberries. When only three or four stems are allowed to remain in the black cap and dewberry stools, larger and finer fruit will usually be produced than if more are left when the spring thinning is done.

When planting black cap raspberries and dewberries always leave the bud of each plant uncovered, preferably in the bottom of a slight hollow. The other bush berries may be planted more deeply as they can stand rougher usage. In every case spread the roots out somewhat and thoroughly firm the soil about each plant, but leave the immediate surface loose. Give clean, shallow cultivation each season until the end of June, or to within a week of fruit harvest. Then sow a cover crop as already suggested.

Avoid digging or plowing close to the plants in spring because blackberries and red raspberries thus treated tend to produce forests of suckers and the roots of currants and gooseberries are sure to be more or less seriously injured by soil stirring at that season. If ever necessary to dig or plow around the former, the best season is shortly after fruiting as this results in few suckers; around the latter, mid fall is better because the plants are then slowing down for the winter and will overcome any injuries before spring. When suckers appear between the rows or other place not wanted *pull them up* when about a foot high. If this is done before they become woody new ones are not likely to form at the same places. If they are allowed to become mature or woody they cannot be pulled up, so must be cut. In such cases other and often many more suckers will form at the same places and close by.

Shallow rooted, low growing vegetables, such as dwarf peas, bush beans, cabbage and spinach may be grown to advantage among the berry plants the first year and perhaps the second; but no crops such as potatoes and parsnips which demand deep stirring of the soil, nor corn, which casts too much shade, should ever be so grown.

None of the bush berries need pruning the first summer. In the spring of the second year, before growth starts, the bramble berries may have their main stems and branches

BUSH AND CANE FRUITS

shortened somewhat. Usually nothing is needed by currants and gooseberries unless their stems are spindly, when they may be shortened to thicken them up. They will need no pruning until the following spring when rampant growth may be cut back. (Fig. 66.)

Whether or not the bramble fruits are benefited by pinching is a disputed point. Some growers, depending on their preferences, pinch off the tip of each young stem as it

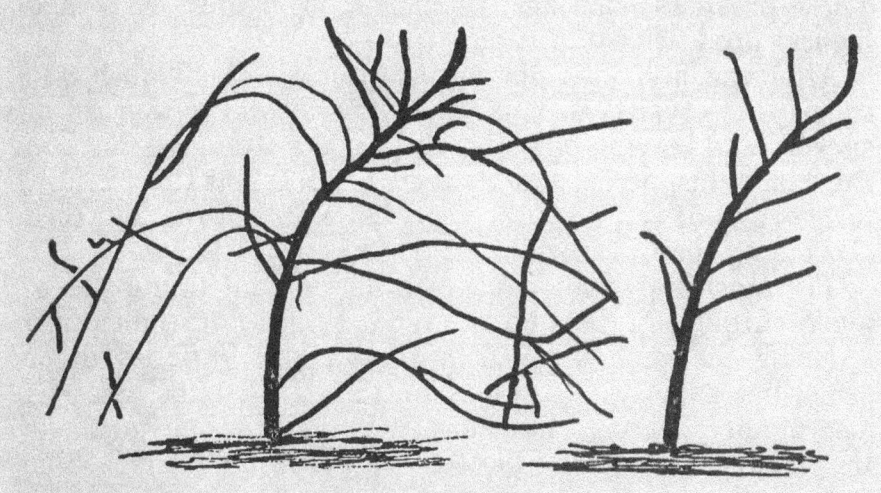

Fig. 66. Diagram of raspberry cane, unpruned and pruned. (From photos.)

reaches 18" to 30". Shoots thus checked elongate somewhat but develop stout, sturdy branched, stocky stems which need no staking or trellising to keep them off the ground. But it is claimed they do not bear quite so much fruit as the stems that develop naturally. Experiments to test this matter have given conflicting results. To do this pinching only the finger and thumb are needed—no knife or other tool.

Cutting out the bramble fruit canes after the fruit has been gathered in July or August is better than leaving them until winter. These canes will die then anyway, they obstruct light and air, become breeders of disease and insects, and therefore are a menace to the young canes. Moreover they are easier to cut in summer because softer than in the following winter when they are dry.

With currants and gooseberries the case is different. But

though the stems of these plants may live for perhaps many years they bear poorer and less fruit each year after they are four years old; so during mid summer, after fruit harvest, most growers cut out those stems that have borne three times. At the same time they get rid of the puny young shoots that come from the bases of the bushes and leave only the two or three stoutest new shoots as renewals. The bushes thus consist of eight to ten stems each after pruning. These practises maintain the bushes in greatest vigor and highest productivity of choice fruit.

After the first year the spring pruning of all bush and cane fruits consists in removing puny shoots, getting rid of suckers and leaving only the two to five strongest canes in the black raspberry and dewberry stools, making narrow hedge rows of red raspberry and blackberry rows and then shortening the branches of all these bramble fruits.

The most important point of bramble fruit pruning is to avoid cutting branches on the canes back too far, and thus reducing the chances of fruit production. Varieties differ somewhat as to the positions of their fruit buds. Seasons also modify bearing. Therefore it is advisable to postpone the shortening until the blossom buds can be recognized as they develop. Then there can be no mistake, for one can see how much bearing wood is being left. The larger the number of blossoms that remain, the smaller the size of the fruits. For specially large fruits the number of flower buds must be reduced. This concentrates the available food in the remaining fruits. It also tends to make the new canes stronger, favors vigor and greater productiveness the following year.

Currants and gooseberries being hardy require no winter protection, though a mulch of buckwheat straw, cornstalks or other loose material is helpful when the plantation has been fall plowed. In many sections the bramble fruits are more or less tender, some varieties being subject to frost, others hardy, but both affected by the character of the previous summer and fall. Mulching is a help, but in some sections protection of another sort is given. A favorite way is to bend and fasten down the canes after the leaves have dropped in the fall and to cover them with earth. When this

BUSH AND CANE FRUITS

is done the earth must be removed in spring before growth starts, otherwise the canes will suffer.

Damage from insects and diseases may be prevented to a large extent by timely attention. The currant worm which also attacks gooseberry bushes is easily destroyed by spraying the lower parts of the bushes with arsenate of lead shortly after the first leaves form. The eggs of the first brood are laid on the under sides of the leaves near the ground. For aphis nicotine sulphate or pyrethrum extract is satisfactory, provided it is applied before the leaves become much crumpled, as they do when these insects attack them. Always it is important to spray from below upward because the eggs of the former and the insects themselves of the latter are located on the under sides of the leaves. An angle nozzle is a great help in doing this work.

Gooseberry mildew may be prevented by spraying with Bordeaux mixture. Begin when the berries are as large as little peas and repeat at intervals of two weeks. The stains may be removed by dipping the berries in vinegar then rinsing in water and drying them in the wind.

Though some diseases of bramble fruits are apparently incurable no one should fear them or the insects that attack the bramble fruits, because timely attention will prevent their spread. These diseases are seldom seriously troublesome and with the few exceptions may be kept in check by spraying with Bordeaux mixture applied the first time while the blossom buds are small and the second soon after the petals have fallen or before the berries have formed. Orange rust attacks only weak plants but may be kept from spreading by digging up and immediately burning affected plants whenever discovered. These are easily recognized because the leaves become stunted and orange colored on the under sides.

The insects that attack the bramble fruits are mostly borers whose presence may be detected by the wilting of the affected parts. These should be cut off and burned at once. Occasionally the leaves are eaten by caterpillars or slugs. Arsenate of lead will kill them or hydrated lime dusted on while the plants are dewy may be used for the latter.

How many plants of each kind to order is a matter for

individual calculation because people differ so much in their likes. The easiest way to judge is to reckon from known yields. Well managed red currant bushes have yielded more than 12 pounds; black ones more than 10 and gooseberry bushes 15. In commercial plantings raspberry bushes average a little more than a pound to the plant. Double this is not uncommon. Blackberries planted 4' x 6' apart often average 2 pounds to the plant, sometimes three times as much. For calculation these maximums had better be cut down a half or two-thirds to get probable yields.

The first selection of varieties should be among those known to be hardy and successful in the neighborhood, always giving preference to those of highest quality. But where such information is not obtainable locally the kinds that do best in the largest region should be chosen. It is advisable to test at least three kinds until the one best suited to the local conditions has been found. A dozen plants of each variety of bramble and three or four of each kind of gooseberry and currant are enough to make satisfactory tests. For lists of varieties see Appendix.

In the North and the East the loganberry, a cross between the red raspberry and the blackberry, is not a safe venture because it lacks hardiness. On the Pacific Coast it is a wonderful commercial success. With occasional exceptions it has been found too tender from New England to the Rocky Mountains, but in the middle South has given promise of at least partial success. As it is a wonderful and delicious fruit and is as easy to grow as the dewberry it deserves testing wherever this fruit can be grown.

Bush and cane fruits may be grown for five to ten years or even longer in the orchard between the trees. (Chapter 49.) However the plants nearest the trees will become too shaded and the soil they occupy too robbed of water and plant food as the trees grow large, so will have to be removed, a few each year after the sixth, perhaps the fifth. Such plants need not be destroyed. If dug carefully and with plenty of roots they may be planted elsewhere to form a new berry patch.

In the orchard the plants should be set in rows to correspond with those of the trees in each direction so that during the first year and perhaps the second cultivation may be given in both directions. Thus with trees set at 25' apart and the berry plants at 5' there will be four rows of berries between the trees each way. When the plants in the two middle rows begin to fail they should all be removed and the whole area between the trees given clean cultivation and cover cropping for the benefit of the trees.

48

SMALL FARM FRUIT GARDENS

AN increase in the number of fruits demands an increase in the number of flower-buds, and as these are generally formed in the preceding season, the horticulturist must extend his operations to that period. Everyday experience teaches us that the period of sexual reproduction does not, as a rule, set in until the vegetative growth (production of leaves) begins to diminish or to cease entirely.

PAUL SORAUER,
In *Physiology of Plants*.

WHETHER or not the farm orchard and fruit garden, as ordinarily planted and mismanaged, is an asset or a liability to the "average" farm and the farmer is a debated question upon which the Missouri Experiment Station has recently published a bulletin (No. 307), not based upon speculative ideas but upon actual records made in a three-quarter acre orchard planted for this very purpose and maintained until 13 years old.

From the data presented it seems that efforts were made to imitate the hit or miss methods of the "average" farmer and to prove them slip-shod by the actual records. As this case is unique (at least, so far as I know) a rehearsal of its main points should serve as a warning to whoever is about to plant a fruit garden. I condense the main statements in the following paragraphs and, in brackets, add a few comments of my own.

Standard nursery stock was bought. It included 12 apple, 6 cherry, 3 plum, 4 peach, 4 pear, 8 nut, 13 currant, 12 gooseberry, 50 raspberry, 25 dewberry, 25 blackberry, and 12 grape. After planting, the land was cultivated for 3 years, though from then forward it was kept in grass to prevent further soil washing. During twelfth year some potatoes were planted. Spraying, pruning and other attention was given, not as professional fruit growers would do them but about as they would be done by the "average" farmer.

Cost items included nursery stock, harvesting, spraying labor and materials, cultivation, pruning and training, mowing and mulching, planting and staking, fertilizers and application, but not "overhead" such as rent of land, taxes and deterioration of tools.

Though the text does not specifically state that the planting was done piecemeal during several seasons the tabulated data show that strawberries were not planted until the fifth year, but from the sixth to the eleventh inclusive they bore annually. Raspberries bore sparingly from the fourth to the tenth; dewberries from the third to the tenth; blackberries from the third to the thirteenth; grapes from the fourth to the thirteenth; plums began during the fifth and except for one failure continued till the thirteenth; cherries began a year later and also missed one season until the thirteenth; peaches bore only three times, and only one of these times liberally; apples and pears did best of all—began to bear the fifth year and bore steadily increasing amounts of fruit annually.

The cost of plants, planting, developing and maintaining during the 13 years was $148.15, an average of $11.43 a year; total returns for this period were $341.66, a gross annual average of $26.28 or a net of $14.83 for the three-quarter acre, or at the rate of $19.76 an acre—less overhead. This $19.76 is approximately 13.33% on the total cost of the orchard ($148.15) or 6% on $326 for one year. Profits on the principal tree fruits should increase annually for another 10 to 25 or more years but cease long before that on all the small fruits except grapes. [With better planning, fertilizing, inter-cropping and cover cropping, returns should have been several times the amounts made.]

The bulletin calls specific attention to the "unfortunate" location of this orchard—on a slope which washed so badly that cultivation was abandoned and, about the third year, the area seeded to grass [a generally bad practise] except where strawberries and potatoes were planted. Mulching did not prove profitable; the fruit plants suffered from weed competition and reduced the yields of berries and grapes. It was found unprofitable to plant Persian walnut, southern pecan, almond and Japanese "heart-nut" because the wood or the buds are destroyed by low temperatures.

In spite of the record the bulletin declares: "Many may think that the home orchard will require an undue amount of work. This is not true for the home orchard of one acre or less in extent; in fact, the care amounts to so little that no farmer, if his work is properly managed, should be handicapped or delayed in handling other farm enterprises."

[Whether or not the evident mistakes were made purposely is not clear from the bulletin. Certain it is, however, that they were made. For instance, planting on a slope so steep that cultivation can not be given; placing the trees so close together (36′ for the large growing trees, 18′ for the small) that before maturity they will crowd and grow spindly and tall and therefore difficult and costly to spray and harvest; placing the berry plants in such ways that cultivation can be given in only one direction instead of two; failing to use the ground between the trees for berry crops or vegetables or both while the trees are young and thus paying the cost of rearing them to bearing age or for several years longer; planting trees not known to be hardy or profitable; applying no fertilizer until the fourth year, none again until the tenth and a total of only $3.90 during the whole 13 years—including cost of application!

[Had a plan such as described in the following paragraphs been adopted returns would have started with vegetables and everbearing strawberries the first season, with "regular" strawberries and grapes the second, with bush fruits the third, with peaches and sour cherries the fourth, and with other tree fruits from that time forward according to the characteristics of the varieties and the type of care given the trees. Many a fruit grower following similar lines has brought his tree fruits into profitable bearing with "a clean slate" as to cost; indeed, has had money in the bank from crops grown between the developing trees!]

With improved methods of cultivation, better knowledge of pruning, harvesting and storing, etc., it is so much easier to grow choicer fruit than one can buy, that each year more people are planting home orchards to supply their needs. Many of the most satisfactory small orchards are developed along the following lines.

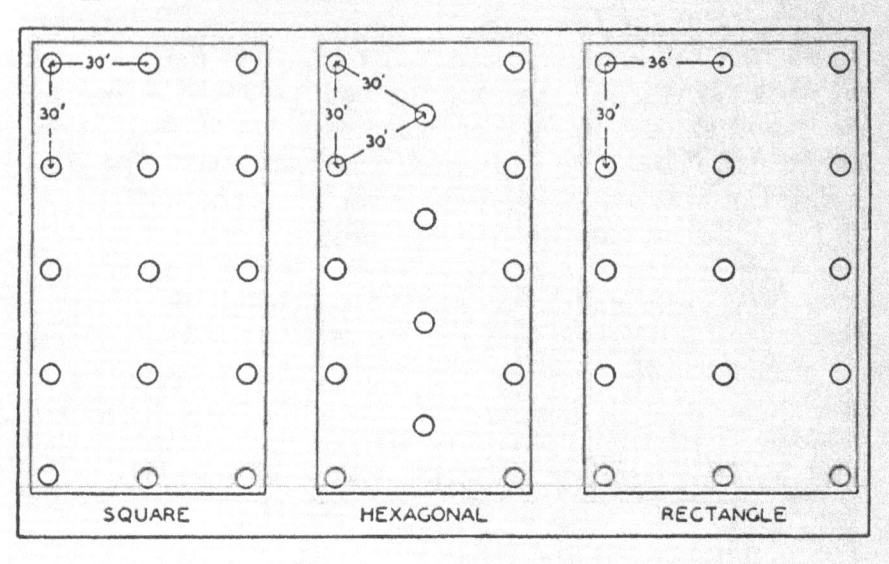

Fig. 67. Methods of placing orchard trees. These distances are much too close for apples, sweet cherries, pecans and walnuts. (See Appendix.)

When the available space is unlimited it is advisable to set out the various fruits in separate areas, just as commercial fruit growers do, planting apples in one section, pears in another, and so on. This facilitates operations in handling the various crops. The only conspicuous exception to this method is the "filler" plan of planting peach and apple trees alternately. (Chapter 49.) Usually, however, this is not an advisable plan.

When the space is restricted but still liberal enough to plant an "orchard" this plan may be modified by dividing the fruits into four or preferably five groups; namely, 1, tree fruits; 2, grapes; 3, strawberries; 4, cane fruits (raspberries and blackberries); 5, bush fruits (currants and

SMALL FARM FRUIT GARDENS

gooseberries), the last two groups planted adjacent to each other.

When only, say, half an acre can be devoted to fruit and where the greatest assortment of kinds is desired the best plan is to divide the space so each of the above goups will *ultimately* have its allotted space, though while the orchard is developing the space between the trees may be filled temporarily by the other fruits. Grapes, however, should never be planted in an orchard or near trees. They do best on trellises. If these were placed among tree rows they would interfere with cultivation. The vines are also likely to give trouble by climbing into the trees, especially if neglected—even for a single year.

One favorite way and a modification to get the greatest possible assortment of fruits in a limited space, such as half an acre or even less, is described below; but first the undesired kinds should be eliminated—poor quality kinds, purely culinary varieties and staples which can be bought in the markets and stores.

Next to omit, or at least plant sparingly, are varieties that ripen when the normal supply of other fruit is likely to be so abundant that there might be waste—during August and September when grapes, blackberries, peaches, plums and early apples and pears crowd each other for consumption. Each of these should be represented but none too liberally because they are highly perishable.

On the other hand, gaps must be avoided in the sequence of supplies. The most likely to occur is after September when there are no really choice peaches or plums, when raspberries and blackberries are rarities and currants and gooseberries never seen. During this period, however, some of the most delicious varieties of apples, pears and grapes are in season and everbearing strawberries are welcome luxuries. Very few of these are obtainable and these few at high prices in the markets or the stores, but with only ordinary care they may be had for four to eight or ten weeks hence are well worth having for home use.

The three most popular systems of orchard lay-out are the square, the hexagon (or equilateral triangle) and the rectangle. (Fig. 67.) Because of its simplicity of setting the square is most popular for small farm orchards, though the rectangle is a close competitor. The hexagon is used for large commercial orchards because 15% more trees may be planted in a given area and because, the distance between trees being the same, cultivation may be given in three directions instead of only two. It is not desirable for the small orchard.

Having decided upon the varieties, the next step is the plan of arrangement. For convenience of discussion suppose that the available area is 100' by 200'. This is slightly less than half an acre. By vizualizing the trees as if full grown and calculating distances on this basis mistakes as to crowding may be avoided. The distances given in the Appendix are mostly adaptable to business plantations. If they are adhered to strictly in the small orchard they will make proper tillage difficult or impossible and all appearance of symmetry and convenience will be lost.

The best way to avoid such objections is to adopt a unit distance that will provide both adequate space between the trees and permit the temporary growing of small fruits, but not grapes, between the trees. The plants should be placed so they may be cultivated by either horse or garden tractor, preferably (while young, at least) in two directions—for the first year or perhaps two.

For the small orchard 50' may look like an extravagant distance between apple and sweet cherry trees, but with small growing trees between there will be no crowding even when the trees are full grown. This would not be the case if a distance less than 40' between the large growing trees were adopted. Even at this latter distance the filler trees would either have to be removed while they were still bearing well or they would be almost sure to interfere with and perhaps actually injure the large kinds. Though the average commercial life of the peach tree is only about 10 years it may be extended to double or triple that time, even in commercial plantings. Why not give it the care that will assure this longevity?

As the orchard is to be a permanent investment it should be placed on the best land and given the best of attention. Preferably the land should be prepared at least one season ahead of the planting by breaking up the sod, getting rid of stumps, large stones, and by growing a well fertilized, well tilled crop of corn, early potatoes or other profitable vegetable on it, partly for the crop itself but mainly to get the soil in the best possible condition for tree planting. Really the seeming loss of a year will be a gain because the trees will thrive much better and make greater growth in well prepared soil than in raw sod freshly turned over.

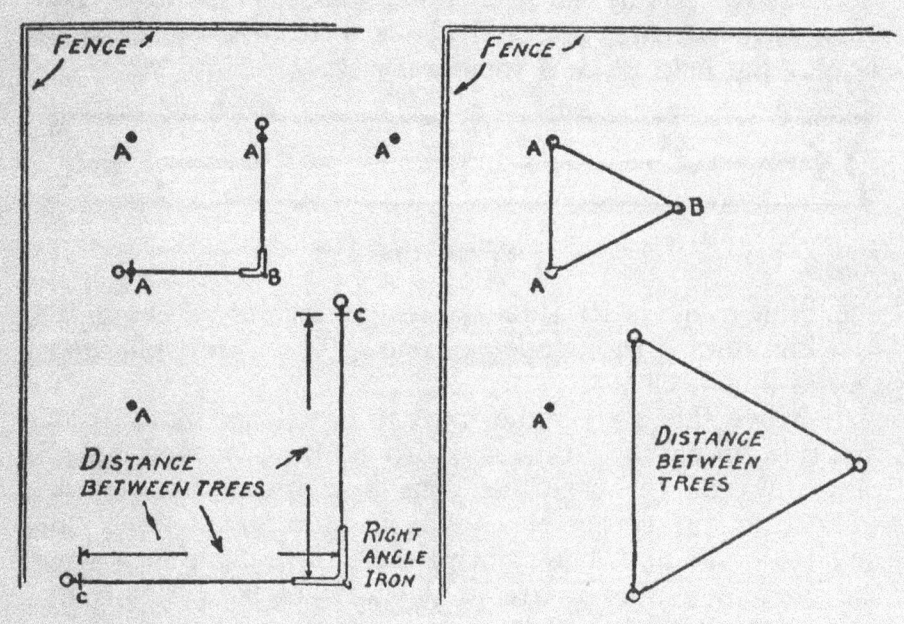

Fig. 69. Methods of laying out orchards by chains or wires.

Whatever crop is chosen for this preliminary work it should be one that must be harvested not later than mid September. Immediately following the harvest of this crop the ground should be plowed, harrowed and sown with a mixture of rye and winter vetch as a cover crop. If the sowing must be delayed until after the middle of September Canada field peas may be sown either alone or with rye. Vetch when sown as late as this will not make a worth while stand. (Chapter 29.)

The easiest way to get rows straight and the trees uniformly spaced is as follows: 1. Establish a straight "base" line across the longest dimension of the field at half the distance the trees are to stand apart—12½′ in the present case. This space is for the turning area for team or tractor.

2. Place a stake the same distance—12½′—from the end of this line to show the position of the first or corner tree.

3. Use this stake as a starting point and place stakes at 25′ intervals along this base line.

4. Start again at the first stake, measure off 8′ toward the second stake and place a temporary stake there.

5. Again start at the first stake, guess at making a right angle with the first line and about 6′ toward the opposite side of the field place a temporary stake.

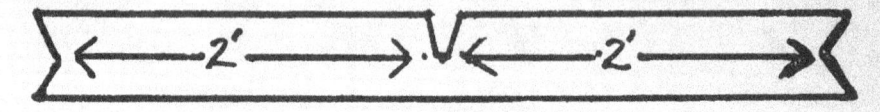

Fig. 70. Tree setting board. (See text.)

6. Now measure 10′ from the temporary (8′) stake on the first line toward the temporary one at 6′ on the (tentative) second line.

7. Move this (6′) stake until it is exactly 6′ from the first tree (or corner) stake and also 10′ from the first temporary stake on the base line. The line drawn through this stake from the corner stake will be at right angles to the first (or base) line. The principle here involved is that when the sides of a triangle are in the ratio of 3 to 4 to 5 the angle opposite 5 is a right angle.

8. With this angle established, extend the new line to the opposite side of the field.

9. For planting small fields these two lines will be sufficient; for large ones most planters make lines on the other two sides (or at various places across the area in each direction) and "sight" from side to side and from end to end to locate the stakes for the trees.

10. Measure off and place stakes at each 25′ interval on the second line.

11. Use two light (surveyor's) chains or stout wires each 25′ long, placing the end of one at the second stake in each of the two base lines and extend each wire parallel with the other side. The ends will meet where the next stake is to be

set in the second row of trees. The four stakes will be at the corners of a square.

12. Move the end of one wire to the third stake in the base line and the end of the other wire to the last stake (the one just placed) and extend the wires as before to place the next stake. Continue this method until all stakes are placed. When carefully done the rows will be straight and the trees evenly spaced. A man and two boys may work this plan better than one or even two men.

To insure getting the trees placed properly in position a planting board is a help, almost a necessity. The most popular style (Fig. 70) is about 5' long, 5" or 6" wide, ¾" thick, with a V-shaped notch 3" wide cut at each end and another at the middle on one side. To use it place the middle notch snugly against the tree stake. Then place a temporary stake in each end notch. Next remove the tree stake and the board and dig the hole where the tree stake stood. Then fit the board to the temporary stakes in the earth, place the tree trunk in the centre notch and fill in.

One style of planting board (Fig. 70) which obviates the necessity of using temporary stakes consists of two pieces, each about 30" long, hinged together, preferably with two strap hinges to make them rigid. One piece has a notch in its free end; the other has two 6" or 8" legs fitted snugly in it. To use it the board is extended full length with the notch snugly against the tree stake and the legs pressed full depth in the ground. The notched end is then swung backward, the tree stake removed, the hole dug, the notched end returned to open position, the tree trunk placed in the notch and the earth spaded in.

In all tree planting it is advisable to make the holes large enough to get both feet in so as to pack the earth firmly about the roots. Also it is best, when digging the holes, to throw the good (top) soil in one pile and the poorer (subsoil) in another so that when filling the hole the good soil may be thrown in next to the roots which will thus have the best possible chance to make a good start. The subsoil being placed on top will also reduce possible trouble with weeds. Under no conditions place any manure or chemical fertilizer in the tree holes because these will burn the roots.

The only safe fertilizers to use thus are the non-chemical ones such as ground bone, dried blood and tankage.

Another caution: Not until after the trees are planted do any pruning of the tops. Should this be done previously, twigs or buds that should remain may accidentally be broken without chance of making substitutions.

During the growing season keep the land cleanly cultivated for a diameter of at least 5' around each tree. If the whole area is cultivated (which is the best practise) cease tillage between July 1 and 15 and sow rye and winter vetch, but preferably after the last cultivation. Clean cultivation until midsummer will insure conditions that favor best development and will also reduce the danger of damage by borers which are always more destructive where weeds and grass grow around the trunks and thus form shelters. More peach trees are killed by borers than by all other causes put together.

49

SELECTION OF TREE FRUITS

IF THE general farmer will become an expert orchardist, he will find that year by year his ten acres of fruit will give him a larger profit than forty acres of grain land; but to get this result he must be faithful to his trees.

JOHN WILLIAMS STREETER,
In *The Fat of the Land.*

INCOME from fruits, especially tree fruits (most of all from the apple) is so largely dependent upon the variety or varieties grown that it is essential to success to ponder the main points before placing a nursery order. (Chapter 39.)

Recent surveys show that whereas Michigan and New York orchards contain nearly 250 varieties each of apples, Washington State has less than 75. This is because in the former states orchards were planted mainly to supply relatively local markets, whereas Washington growers aimed for distant ones, even the Atlantic Coast and Europe. Taking the country over, estimates place only 25 varieties in the prominently commercial class, of which 15 make up more than 70% and of which more than half (37% of the total) consists of five leaders—Delicious, Jonathan, Stayman Winesap, Winesap, and McIntosh.

More striking even than this: the survey brought out the fact that though in former years many varieties of high production and good color but of low quality were largely planted, recent plantings indicate that such kinds—Ben Davis, York Imperial, Gano and Arkansas, as instances—are being omitted from nursery orders and higher quality varieties substituted.

Before the advent of the automobile when shipments to market were made almost exclusively by rail and water, individual commercial orchardists restricted their lists of varieties to less than half a dozen—usually only three or four —because they shipped in carlots or cargoes. Now that auto-trucks are so largely used the tendency is to plant six to ten successionally ripening varieties so as to reduce the rush at harvest and thus employ smaller numbers of men over longer periods.

Except in orchards where special marketing advantages exist, summer and early autumn varieties of apples and pears are shunned by commercial growers because of their perishable nature. Winter varieties are far more widely planted because they can be handled to greatest advantage, especially for distant markets and where the grower has his own cold storage house, as many of the larger ones have. The result is city markets are poorly supplied with summer and early fall apples and pears.

So here is one of the most important points for the small grower to consider when choosing varieties for the local market, especially the roadside market. As the small farm cannot produce winter fruit in competition with the large growers these varieties should be the first to pass by. For the roadside market they should also usually be omitted because the season of such marketing generally continues only during the growing season, if that long—April to November—so winter fruit would have to be sold through some other outlet.

Conversely, summer and fall ripening varieties are ideal for local and roadside markets, first because of their usually short supplies and second because of their generally perishable character—they cannot be kept or shipped as successfully as winter fruit.

Two other advantages summer and early fall apples have over late fall and winter kinds are first that they generally begin to bear fruit when much younger—often in less than five years; second, early kinds usually bear moderate to even large crops *every* year, whereas most winter kinds have their "full" and "off" years, unless specially trained and their fruit well thinned. The assigned reason for this latter characteristic is that as the fruit ripens early the trees have opportunity after harvest to store up plant food and develop flower buds for fruit production the following year; whereas after winter fruit has been gathered there are only a few days or at most weeks before winter. Thus under normal conditions and treatment summer and autumn varieties not only begin to pay sooner than do the winter kinds but they assure an increasing annual income until they reach full bearing when they often yield larger profits than do winter kinds of the same age.

No matter whether the varieties chosen be early or late ripening kinds, too much emphasis cannot be placed upon high quality, especially for one's own roadside market or for "personal customers." In spite of the steady enhancement of quality in commercial orchards the markets are still full of inferior fruit. So when a customer discovers a reliable source of high quality fruit not only his custom but that of his friends is assured.

Though many of our choicest quality apples have unattractive color (for instance, Swaar, Pomme Grise, and Roxbury Russet) the public is so slow to learn and to recognize them that such varieties should be planted solely for home use and to share the surplus, if any, with a few discriminating friends and patrons. The general public "eats with its eyes"; therefore "any color so long as it's red" is what will open the public purse. If pleased with the quality there will be no difficulty in making further sales. So for profit it is wise to plant only red skinned varieties.

SELECTION OF TREE FRUITS

Before deciding on the list of varieties to plant it is advisable still further to know intimately the characteristics of each one so as to avoid trouble or loss later on. Some varieties such as Tompkins King and Grimes Golden are subject to collar rot, a disease that attacks the trunks at the surface of the ground and kills the trees. It may be avoided by "double working," that is, grafting the Grimes on a variety known to be resistant or immune to this disease. Some nurseries list such trees, always however at prices above those of ordinarily propagated trees. Alexander and Wolf River (baking varieties) are subject to more or less decay of the fruit while still on the trees, so they require extra spraying. Chenango Strawberry, a delicious dessert variety, is also subject to this disease. Baldwin seems specially subject to attacks of bitter rot and railroad worm, both of which are difficult to control. Canada Red shrivels badly in storage, Hubbardston also to some extent. Lady and Pomme Grise (the former beautifully colored, the latter not), are so small that unless one has a special market they will not pay. Esopus Spitzenburg, though delicious, beautiful and fragrant, often bears too few fruits to be profitable. McIntosh and Yellow Transparent are subject to fire blight. And so on! The peculiarities of each variety should be carefully weighed before a decision is reached whether to discard or to include it. You cannot know too much about a variety before ordering it!

For local market, personal customers, mail and express orders, and roadside marketing one of the most important points to assure in an orchard planting is selection of varieties that ripen successionally from earliest to latest—at least the latest that one desires to have the season extend, not necessarily to include winter varieties. By having such a sequence a customer pleased at the beginning of the season will continue to buy until the close, whereas if there are serious breaks his trade will be much more difficult to reestablish when a new supply begins to ripen.

So far as supplying home needs is concerned, every farm, large or small, should have its quota of tree fruits, enough trees of each species to supply, but not over-supply the family during the entire year with all the fruit it can use either fresh or preserved. Calculation as to the amount likely to be needed should be made on the basis of trees in full bearing, even though while young they will not bear enough to supply the demand for fresh fruit, to say nothing of canning. Otherwise the probability is that there would be over-supply and waste.

The varieties of each species should ripen successionally so as to avoid having a greater quantity of any one kind that can be conveniently handled. Thus, as maximums, one

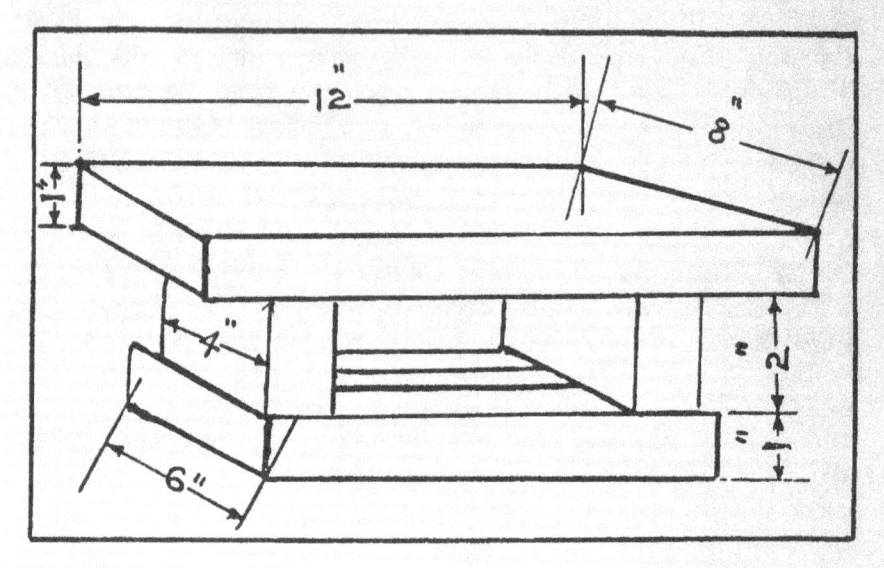

Fig. 71. Field mice poison station made from old 1" boards. Poisoned grain is placed in ½" depression beneath hood.

sweet cherry tree of each of three varieties—early, midseason and late—when full grown should produce fruit for even a large family. Sour cherry trees, being much smaller, two each of three varieties should be ample. Five or six varieties of peaches, two trees each of the earliest for slicing and three or four of the later ones for canning should give abundant fruit for eight to ten weeks used fresh and as canned throughout the year.

Early pears, like early peaches, are useful only for eating raw; they are too watery or soft for canning and too perishable to warrant planting more than one tree of a variety for home use. A total of eight or ten trees to cover the season

SELECTION OF TREE FRUITS

from August to December and for canning (in September and October) will be all that any family will be likely to need. Summer and early autumn apples are also perishable, therefore only one or at most two trees of a variety should usually be planted. Later fall varieties are less perishable, and among them are some of the best for canning, so two or even three trees may not be too many, provided space can be spared for them. Where space is too precious winter varieties may be omitted because cooking apples may be bought in the neighborhood or the markets until strawberries are ripe the following summer.

One or two quince trees will usually bear enough fruit for any family. As they are smaller than any orchard trees they should be planted separately. Though apricot trees are hardy, their blossoms are often killed by spring frosts. They should, therefore, be planted for ornament and in the hope that they may escape damage. Two or three trees of different varieties—early, mid-season and late—should be enough for a family. Two or three nectarines might be added more for curiosity than anything else.

Of course, if there is any surplus from any of these trees it should be easy to sell, provided its quality is sufficiently good to attract buyers.

Just what, if any, tree fruits one should grow on a small place in the hope of making them profitable is a question to be weighed seriously before deciding. As it is folly to plant trees closer together than they should stand when full grown (See the Appendix), the space they would occupy may perhaps be made to pay larger returns during the same period of years, but normally less after the trees reach full bearing.

Peach trees often begin to bear when three years old and though their commercial life is only about ten years, they may be made to continue in bearing for two or three times as long. Sour cherry and plum trees sometimes start to bear when four years old and unless killed by black knot or other diseases continue for 20 or 30 years. Sweet cherry may start when four years old but continue for 40, 50 or more years. Some varieties of apple, mostly early ones, begin to bear when three or four years old, others not until 10 or even 15, though much time may be saved by proper management. Often they bear well for 50 to 75 years or even longer.

Before discarding the idea of growing tree fruits for profit on the small place consider this point: While the relatively slow-maturing sweet cherry and apple trees are developing the spaces between them may be occupied with "filler" trees of smaller kinds which mature quickly, bear liberally, and under ordinary conditions are shorter lived. The peach is the favorite filler in apple orchards for these reasons. However, objections are often raised concerning it. Among them are the facts, first, that the trees require spraying or dusting with different (or weaker) mixtures than do apple and thus demand unreasonable expenditure of time and attention; second, the filler trees may be so profitable that they may be allowed to remain so long that the permanent cherry or apple trees may be injured by crowding, especially when they are set closer together than the recommended distances. The only way to prevent such a calamity (for so it may become) is to delegate their destruction to some heartless hireling and absent oneself from the murder of the innocents!

Even when permanent and filler trees are planted together it is not necessary or even advisable to let them have exclusive possession of the ground while they are small. Temporary, inter-tilled crops may be grown between the trees for several years. Under no temptation should such crops be hay, grain or any perennial vegetable like asparagus. These demand much moisture and plant food (the former especially), encourage neglect of the trees and make losses by borers much more likely than when the ground is cultivated cleanly.

Among the best inter-tilled crops are strawberries (cropped one year and then plowed under), tomatoes, cucumbers, cantaloupes, watermelons, early (not late) potatoes; in fact, any crop that is shallow rooted, does not require deep stirring of the soil after late summer, and preferably is gathered before the first autumn frost or soon after so a cover crop may be left in possession of the land for the balance of the season. By using such crops many a man has paid all the costs of his orchard from planting to profitable bearing age, and in some cases even made a profit before the trees began to bear. It must always be borne in mind, however, that the permanent crop—the orchard—is the real investment. This must not be neglected for any incidental profit that might be derived from any temporary crop. As the trees increase in spread the available area between them will necessarily become narrower until, eventually, they will shade the ground too much to make the growing of temporary crops feasible.

Unless the orchard area is very narrow in proportion to its length it is a good plan to have the rows of temporary crops extend lengthwise one year and crosswise in the following season. Such alternation of direction will break up and aërate the soil between the trees in each direction and thus not only better control perennial weeds but prevent the soil from becoming unduly compacted.

The 25-year-old apple orchard soil fertility experiments at the Pennsylvania Experiment Station support the opinions of fruit growers in the conclusions chosen and condensed from Bulletin 294 as follows: [Italics are mine. M. G. K.]

1. Orchard soil fertility is more than its plant food content. It involves the nature, depth, topography, previous treatment, use of fertilizers and manures, amount and nature of cultivation and covers or sods grown. Fertilizers are only part of the problem of soil fertility.

2. In this orchard, any treatment that has influenced the trees at all has done so in the following order: first, cover crops; perhaps several years later, leaf color; shortly after, branch growth and circumference increase; and last of all, yield.

3. The reason for this sequence of results is that the treatments—whether chemical fertilizers, manure or cover crops—have influenced yields chiefly by changing the organic content of the soil; that is, *those treatments which have resulted in larger cover crops have ultimately resulted in more fruit.*

4. The organic content of the soil has been a considerable factor in determining the amount of water in the soil.

Those treatments which built up organic content have enabled the soil to soak up rainfall rather than to lose it by surface run-off. A larger water supply, in turn, has produced more cover crops.

5. Good treatments have nearly offset the initial disadvantage of poor soil; but it is more economical to plant an orchard on good soil, than to attempt the improvement of a poor one.

6. A short, non-legume sod rotation is an efficient means of building up a depleted orchard soil. After a sod of any kind becomes thick, tree growth is checked and yields decline. Orchard sods should be turned under, or partially broken, frequently.

7. Moisture conditions are often more favorable in sod orchards than in cultivated ones. Run-off is checked by the sod and less water is used by a sod in midsummer, after it has been mowed, than by a heavy cover crop.

8. Under a non-legume sod, the soil nitrate supply becomes low in late May or early June, necessitating early applications of nitrogenous fertilizers. Annual, per tree, applications of 10 pounds of nitrate of soda or its equivalent of ammonia or other forms, have proved profitable. Superphosphate, in light applications, has increased sod and cover crop growth.

9. Trees that received annual tillage and July seeding of cover crops have not done so well as those under sod culture rotations. If the cover crops are seeded in early June the difference may not be marked.

50
STORAGE OF FRUITS AND VEGETABLES

THE roadside market presents an exceptional opportunity for small capital in a day when agriculture is becoming highly specialized. It takes a lot of money to start in almost any kind of farming. The roadside marketer may begin with a few acres and plant intensive crops, such as vegetables, berries and flowers, which bring cash returns without long waits. An inexpensive stand may be made as genuinely attractive as an expensive structure. The operator's personality will be his most valuable asset.

GILBERT S. WATTS,
In *Roadside Marketing.*

IN one house my father owned a dumb waiter between the kitchen and the cellar saved so many steps between these two rooms and was so convenient in other ways that I show a still better one! Ours used only the moderately cool cellar air for chilling and keeping food; the one illustrated [*] (Fig. 3) uses ice, thus securing much lower temperatures.

The refrigerator is built in the basement under the kitchen or the pantry and the dumb waiter is so arranged that the cage is lowered into the cooling chamber beside the ice-box. Both the cooling chamber and the ice bunker are fitted with a door opening into the cellar, so these chambers may be accessible for cleaning, filling the ice-box, if desired, loading and unloading the dumb waiter.

The cooling chamber may be 2' square, the cage of the waiter 18" square, and the bunker 2' x 2' x 4'. This bunker will hold 600 to 700 pounds of ice, thus reducing the necessity for frequent filling and the rate of melting, because of the large quantity of ice in the box.

Cheaper construction than shown might be used, but the meltage would be greater and the temperature in the cooling chamber could not be kept so low or so uniform. The size might be reduced if the family were small.

Successful storage of fruits and vegetables depends on maintenance of favorable temperature, ventilation and hu-

[*] Bulletin 306, Ontario Department of Agriculture.

midity. So far as storage is concerned vegetables are of two classes, warm and cool. The former (sweet potatoes, pumpkins and squashes) need a temperature of 50°, so some means of heating should be provided; cool vegetables (potatoes, cabbage, carrots, beets and turnips) do best when the temperature is between 34° and 38°.

The humidity requirements of the two classes also differ; warm vegetables must be stored in dry air; cool ones in moist. Usually, the soil of the storage floor supplies enough moisture provided it is not covered with concrete or stone masonry; but if the floor is not of earth it should be sprinkled often.

Ventilation is essential. Though this is less important for cool vegetables than for warm ones, except during the first four to six weeks of their storage, it is essential for the warm vegetables to prevent sweating and rotting. Sweating may also be prevented by proper heating.

Fruits need temperatures equivalent to those of cool vegetables. Ventilation is still more essential because fruits readily absorb odors. They should never be stored near strong smelling vegetables. Even when stored alone good ventilation and clean conditions are essential to the retention of high quality.

Where permanent facilities are not available, mound storage is the least costly, simplest method of storing cool vegetables. A well drained place 4' to 6' in diameter is leveled off and two trenches 6" deep dug at right angles across it for ventilation and drainage. These are covered with stiff, galvanized hardware cloth (½" mesh wire netting) and from their intersection a woven wire flue erected to serve as a chimney. The earth floor is next covered with fallen leaves, straw or other loose material 4" to 6" deep and the vegetables conically piled around the chimney. Thus the whole interior will be well ventilated.

When the pile is about 30" high it is covered with 4" to 6" of the same kind of litter and then with 6" to 10" of soil, except for 4" to 6" of the chimney, on the top of which is placed a board to keep out rain. The ends of the trenches are also kept open until cold weather makes it advisable or necessary to close them. To remove surface water a ditch

STORAGE OF FRUITS AND VEGETABLES 319

is dug around and away from the mound to a lower level. (Fig. 72.) Unless this precaution is taken the vegetables are almost sure to decay.

Several small mounds of this size are better than one large

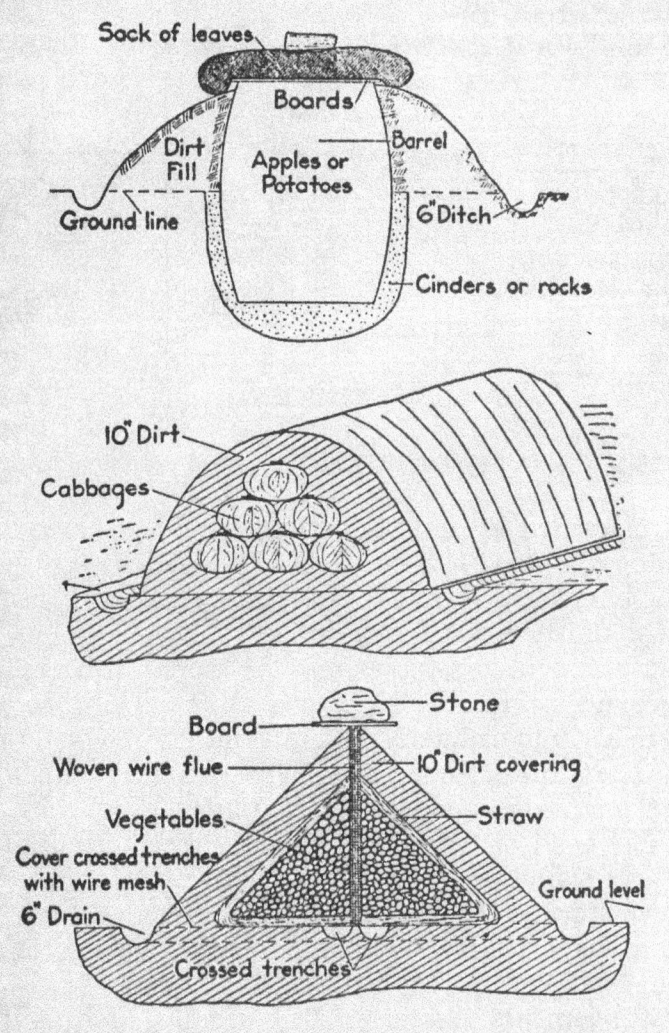

Fig. 72. Types of outdoor fruit and vegetable storage.

one because it is advisable to remove the entire contents of a mound than to try to close a large one after it has been opened during winter. Rarely can such a mound be made water-tight after once being opened; moreover, vegetables and fruits are more likely to spoil in a large mound. For home use it is advisable to have assorted vegetables in each small mound; for sale only one kind. Beets, carrots, turnips, potatoes, parsnips and salsify keep well in such storages. Apples will keep but after long storage usually have an earthy flavor when eaten raw.

Cabbage may be stored in long piles or in mounds. The

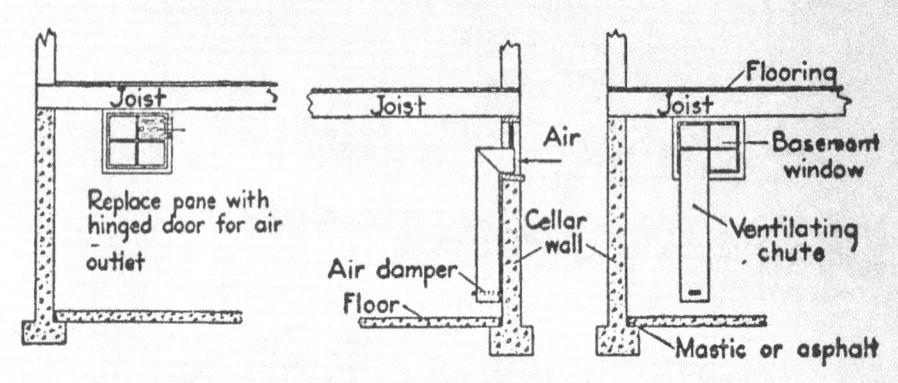

Fig. 73. Ventilating system of house storage system.

latter are formed by banking earth over the heads turned upside down and with their stems removed to prevent rot. After the burying is complete a trench should be dug on each side of the finished pile for drainage.

Apples and cool vegetables may be stored in barrels sunk in the ground, preferably so placed after packing and when cold, as in early morning. A pit is dug either in a well drained level place or on a slope, lined with gravel or cinders for drainage and, after being put in place, lined with more gravel. The head of the barrel is covered with a burlap sack full of leaves and weighted down with boards and stones.

My own storage room is built in the basement of my house beneath a bay window. Though it has given satisfaction for about ten years the style illustrated in Figs. 73, 74 offers some improvements. Failing a bay window such as mine a storage room may be built in a corner so as to take advantage of two walls already built. Preferably two windows should be available, though one may be made to serve. The partition walls should be built of two layers of sheathing with building paper between, nailed to each side of the studding and the space between filled with insulating material such as planer shavings or sawdust. The door should be double thickness with building paper between.

Never is it advisable to have anything of wood permanently fixed to the floor; the dampness will rot it, and the extra surfaces and corners will form lodging places for molds, plant disease germs and dirt. A clear floor favors cleanliness and convenience in arranging storage receptacles. Portable slat floors in sections may be used to support barrels, boxes and crates. They also favor ventilation beneath them and facilitate removal for cleaning. Wall shelves are good for carrying small boxes, crates, and canned goods. My own are of 12"-wide boards. For ventilation's sake I think cypress slats 1" thick and separated by 1" or 2" spaces might be better, especially if painted with linseed oil and dried before putting them in place. It would be advisable to have the uprights supported by brackets and from the ceiling rather than from the floor as rot at the bottom could thus be avoided and the floor would be freer of obstructions.

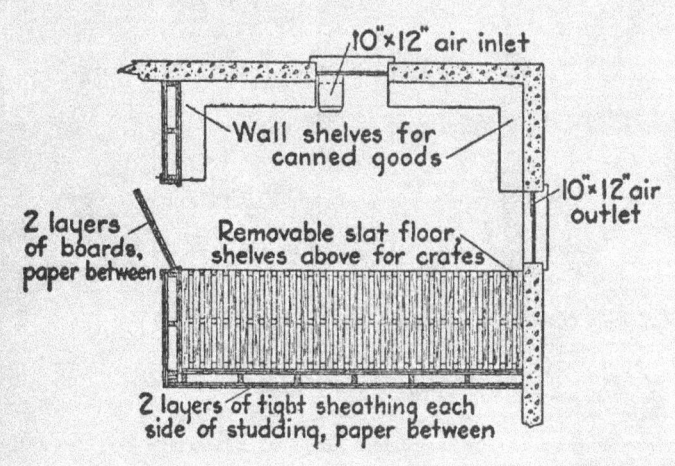

Fig. 74. Storage room in corner of house cellar.

Ventilators are essential. They may be made of wood, or preferably when possible of tile, built into the foundation wall, the intake opening near or through the floor. In a cellar already built wooden ventilators are easily placed by removing a pane in a window and building a wooden tube to near the floor from the lower half and another from the upper half across the ceiling to the farther side of the room, each provided with a damper to reduce the current in cold weather.

When larger quarters than a residence cool room are needed an underground storage may be built in a bank, between two closely adjacent rises of ground, beside the foundation of an already erected building or as a separate cel-

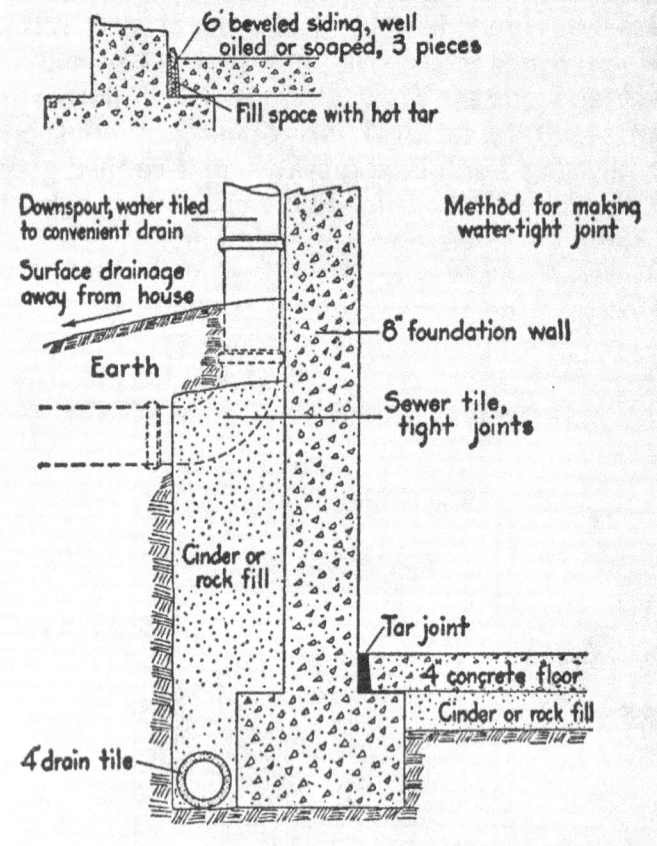

Fig. 75. Water-tight joint of cellar floor.

lar. The last is usually most costly because it demands most excavation and largest amount of wall to build. As a barn is on every farm, is conveniently located, and usually has a masonry foundation the following description by Donald Comin has been condensed from *The Bimonthly Bulletin* of the Ohio Experiment Station (Number 153).

All perishable produce storages require insulation of the walls and ceiling to prevent rapid temperature changes.

STORAGE OF FRUITS AND VEGETABLES 323

Fortunately the earth provides simple means of meeting the problem when low temperatures are not necessary. It is a poor insulation against the flow of heat, but because of its heat capacity it modifies the temperature of the surrounding air, tending to maintain storage temperatures close to that of the earth itself—40° to 55° F.

In summer and early fall the earth temperature is lower than that of the air, so it absorbs heat and thus cools the storage room. In late fall and winter it is warmer than the outside air so releases heat and thus raises the temperature

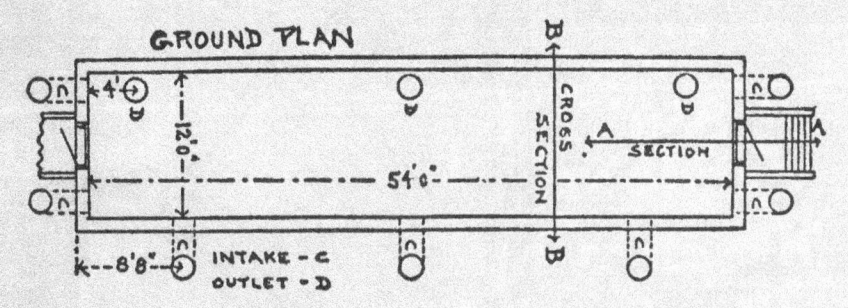

Fig. 76. Ground plan of fruit and vegetable storage.

of the room. Thus the influence of the earth on the incoming air replaces insulation necessary in above-ground storages.

The natural water movement in the soil maintains desirable humidity in such storages. Incidentally, evaporation also tends to cool the stored produce.

Under-ground storage affords ideal conditions for those perishables that require considerable ventilation. Incoming cold air removes the heat from stored produce at a rate dependent upon the volume passing through as well as the difference in temperature of the air and the produce. At the same time the earth floor and the walls raise or lower the temperature of the ventilating air all day with hardly any appreciable effect upon the soil temperature.

The result of these forces is to lower the temperature of the stored product rather rapidly while the ventilators are open, provided the outer temperature is low enough. When the ventilators are closed, the temperature inside gradually returns to that of the ground, so frost injury may be prevented. The great capacity of the earth to modify incoming air temperatures allows for maximum ventilation with minimum danger from too rapid falls of temperature in the storage.

The limitations of this type of storage should not be under-estimated. During seasons when low air temperatures

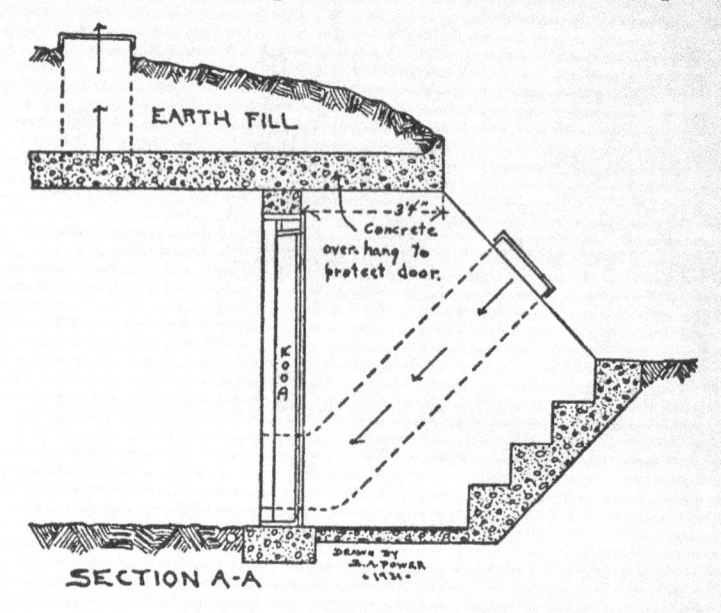

Fig. 77. Entrance to storage cellar. Arrows indicate air flow of ventilation.

arrive early, it is easily possible to lower and maintain the temperature of the produce stored below ground against the counteracting higher temperature of the earth, but difficulties arise in seasons of persistent high temperatures. It is an advantage so to place the structure as to utilize prevailing winds as aids to ventilation.

The storage illustrated in Figs. 75-78 was built at the Ohio Experiment Station for use in storage studies of vegetables. Next to the barn a 12" x 54" section of the approach was removed by team and scraper. A minimum of hand labor was necessary to straighten the earth wall in preparation for placing the forms to hold the concrete. Rough, used lumber was utilized to build the forms. Cut lengths of 6" saplings were used as supporting timbers. The earth wall on the north side and the stone wall of the barn eliminated several forms. After an over-night set of the concrete poured for the walls the forms were removed and used to support the roof slab. Thus the amount of form material was greatly reduced.

The walls and footings were poured of a lean mixture (cement 1 part; bank-run gravel, 6). The roof slab was made of a one to three mixture. Some used brick and stones were thrown in with the concrete. A thin layer of heated road tar was brushed on the finished roof slab as waterproof. The walls and footings required no reinforcement, but the roof slab was necessarily reinforced with construction steel since truck loads of 5 tons or more pass over the barn approach. Accepted plans of reinforcement, as supplied by construction steel manufacturers were followed.

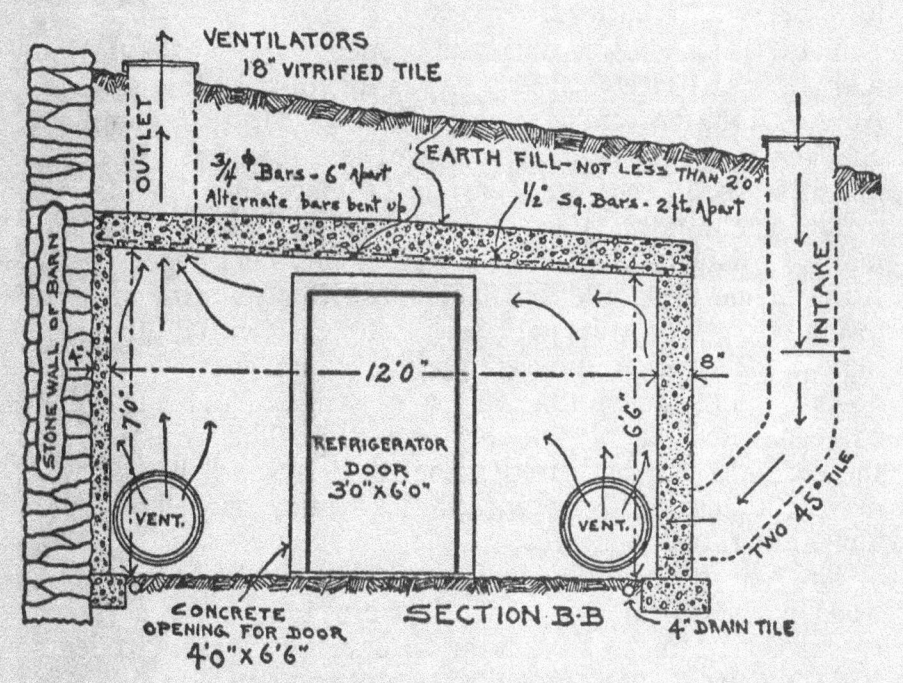

Fig. 78. Cross section of storage cellar, showing ventilating system.

One door opening was placed in each end of the storage —one square foot of door area for each 117 cubic feet of storage volume. Under normal conditions this will probably prove ample for ventilation purposes. Commercial cold storage doors were installed as they were found to be most satisfactory and as economical as an equal protection by other means.

Seven inlet and three outlet ducts were installed as supplemental and night ventilators. "Seconds" of 18" vitrified sewer tile were used for this purpose. Their location and number may be used to suit individual conditions. The number used in this storage may be excessive, though they will insure adequate ventilation under any condition and will prevent dead air pockets forming in the corners and elsewhere. The air-intake area equals one square foot for each 340 cubic feet of storage capacity; the out-takes, one to 794 cubic feet. By installing greater numbers of out-takes and lengths or an exhaust fan or other contrivance to aid air movement, the ratio of vent size to storage capacity could have been materially raised.

A 1/20 horsepower, airplane-propeller type, exhaust fan with 105-watt electrical in-put was installed in this storage. At prevailing current costs this fan will effect 25 complete changes of air an hour at a cost of less than 1¢. [In other localities such costs would rarely be more than double.]

The earth floor is expected to maintain a relative humidity satisfactory for the storage of any cool vegetable or fruit. A line of 4" tile was laid just below the ground level inside the storage at its periphery, connected with drain openings at each door and covered with cinders.

The completed storage has a minimum 2' of earth fill, covered with sod, on the roof and ends. This will prevent the entrance of frost through the roof slab and the sod will present a pleasing appearance to the exterior as well as favor low temperature beneath.

The tile vents at the surface of the ground have strong wooden and metal covers which are uninjured by horse or truck. Coarse mesh screens in all openings prevent rodents and trash from entering while the storage is being ventilated at night. Home made padded plugs fit into the tile openings from the inside and prevent air leakage during sub-zero weather.

This storage should especially appeal to the farmer who has small quantities of perishable crops and is not justified in spending much for storage. Its material cost is low and it

STORAGE OF FRUITS AND VEGETABLES 327

can be constructed with the equipment and labor found on most farms. The small investment likely would be paid from savings to be secured by such storage facilities—in many cases within five years.

Cabbage is not injured by slight frost or even a 20° field frost if not too rapid, but colder and fast freezing in field or storage are destructive. Only late planted late varieties, sound and not over-ripe store well. A good potato storage is also good for cabbage. Heads may be stored in crates, baskets or on slatted shelves not deeper than three layers and just above 32° F. If wrapped in their leaves and paper they keep crisp.

Potatoes are best stored in bushel crates, not in large piles because they heat, keep poorly or may sprout in the centers of the piles. They should be kept dark to prevent greening and spoiling their flavor. The best temperature is 37° to 40° F. but where large quantities are stored air temperature should be 3° or 4° lower to offset natural heat. As freezing destroys potatoes no frosted tubers should be stored; so all touched by frost even while in the ground should be discarded. Wet spots in sacked potatoes usually mean frosted and rotted ones.

Potatoes should not be over-ripe if to be stored long. Late varieties usually store better than early ones because they rarely become fully ripe. Moist air prevents potatoes from growing, so when temperatures cannot be held as low as recommended frequent wetting of walls and floors will prevent shrinkage and sprouting. Good storage should keep shrinkage below 5%. Summer digging and prompt storage of early potatoes is best practise. Potatoes dug before the vines are dead or at once upon maturity have brighter skins, show less bug injury, store better, make better seed and are better to eat than those left long in the ground after once ripe.

Cellars are cooler than the ground in the field. Great care is necessary to summer store in crates in good basement or room storage with perfect darkness. The chief danger in the field is sun-scald which is followed by soft rot. Potatoes should be shielded from sun heat and light which are most dangerous at the ground surface. They should therefore be picked up within a few minutes or immediately covered with vines, weeds or sacks, raised above the ground surface and promptly taken to storage.

Onions must be thoroughly ripe when harvested, always in dry weather, cured by drying a few hours in the field then under cover and in a breeze until thoroughly dry before being permanently stored. Storage must be dry and cold—just above 32° F., but the bulbs will stand a little freezing and thawing. Well cured onions are firm and not easily dented at the stem end by the thumb when held in the hand. They should show no sprouts or new roots. They are best for storage when topped at about 1½".

Root crops (beets, carrots, turnips, rutabagas, winter radish and kohlrabi—only the best—should be stored in pits, cellars or caves, tightly covered boxes, crocks or in sand. They must be kept covered and cold to prevent shriveling.

Parsnips, salsify, parsley, and horseradish may be left in the ground all winter where they grew. To prevent excessive alternate freezing and thawing, which destroys them, they should be covered lightly during the period of light fall frosts and thaws with straw, cornstalks or other material, which should be removed when severe weather arrives so they may freeze solid and then re-covered to keep them thus until spring when they may be dug. All except parsley may be stored in cellars, but to be kept crisp they should be kept frozen.

The pit method is better than the storage room for these root crops. For storage in pits they should be dug after freezing weather begins, preferably late, stored in piles, covered to prevent alternate freezing and thawing and not placed in the pits until the ground has frozen and they are also slowly frozen solid. The plan is to keep them continuously frozen or to have the least possible alternations of freezing and thawing. The boxes and barrels used to store them should be covered with burlap or carpet, then a mouse-proof cover and finally fallen leaves, straw or other light insulation. When removed from the pits the vegetables may be thawed slowly in cold water over night, after which they may be kept in ordinary storage for several weeks.

STORAGE OF FRUITS AND VEGETABLES

Squash, pumpkins and sweet potatoes must be gathered before even the slightest frost touches them, thoroughly ripened and cured before being stored; also free from bruises, else they will rot. They may be kept in crates or on shelves in a thoroughly dry place at 50° to 55°. When first put in they must be given abundant air at 80° to 90° F. for about two weeks to dry the skins. As a substitute for this they may be laid on deep straw in cold frames or on a dry knoll and covered during chilly nights with sailcloth or other heavy fabric for about three weeks. Sweet potatoes are best stored in baskets or crates. Those that tend to shrivel at the stem ends should be stored with paper coverings or in closed boxes.

Celery, head lettuce, leeks and endive may be rooted in earth in a cellar, watered occasionally and kept until Christmas or later. Root crops, cauliflower and Brussels sprouts may also be stored similarly.

Tomatoes may be kept until Thanksgiving Day if stored before frost touches them. Those that have not reached the glistening stage should not be stored as they will not ripen. They are good for pickle making. In storage the others may be placed on straw in cold frames and covered in cold weather or placed on shelves in storage at temperatures of about 55°.

Apples and pears store like potatoes except that 34° to 36° F. is the best temperature range. Apples are more easily damaged by bruising than are potatoes. Only sound fruit of late varieties free from skin breaks, scab and worm holes are safe to store. They should be picked while hard ripe, handled carefully, in padded baskets or fruit-picking bags and never dropped into containers. After picking they should be sorted for quality and only sound fruit saved for storing, packed in clean boxes, baskets or crates, stacked in storage, but never so full as to cause bruising.

Most varieties scald in storage unless preventive measures are taken. This disease kills the fruit skin which turns brown and later starts rot. It is best prevented by packing the fruit in shredded oil paper scattered through the package or by wrapping each apple in oiled paper. One pound of paper per bushel is enough. Wrapping also prevents spreading of rot.

As the red apple surfaces do not develop scald, it is important to allow the fruit to remain on the trees until well colored before picking. Most red varieties should be at least 75% red before picked.

Apples ripen much faster off than on the trees, hence prompt cooling and storing is important. After picking they should never be left in the sun, but immediately placed in the shade until they can be sorted and stored.

Early winter varieties will generally keep until mid December; late ones until February or March. Usually fall varieties cannot be stored later than mid November.

51

ESSENTIALS OF SPRAYING AND DUSTING

KNOWLEDGE of the habits of the pests of the farm, orchard and garden is essential if one is to proceed intelligently toward their control. . . . Until the organism doing the damage is somewhat understood, attempts at control are only guesswork.

A. FREEMAN MASON,
In *Spraying, Dusting and Fumigating of Plants.*

HUNDREDS, if not thousands, of amateur gardeners and even many professionals are annually disappointed because their spraying or dusting produces either no noticeable results or perhaps does actual damage to their plants. Despite the fact that tens of thousands of other gardeners obtain satisfactory results they look upon the practise as waste of money, time, work and materials and because of such experiences often belittle the practise, whereas they should condemn themselves because they either used wrong materials, or applied right ones in wrong ways, for wrong purposes, or at wrong times. For instance, it is silly to use an insecticide to kill a plant disease, or to employ a fungicide to destroy insects.

The following survey indicates the functions, extent and limitations of spraying and its dry counterpart, dusting. However, before we discuss that phase of the subject let us review precautions that may be taken to increase the resistance of plants and increase the efficacy of dusting and spraying and also reduce the time, money, work and materials necessary to use.

Cleanliness of orchard, garden, vineyard and berry patch is one of the greatest preventives of plant diseases and insect attacks. It should characterize every month of the growing season—and more too. By rights it should start before the garden is made and continue after the season closes. The more débris left from one month to the next and especially from one season to the next the more likely is the place to be infested by bugs and blights. This is because such material not only is of no use or beauty, not only a consumer of plant food and water but it is shelter, breeding and feeding material for insects, particularly those that naturally infest the specific kinds of plants of which it is composed. The sooner it is destroyed the better.

It also enables the plant diseases that attack such plants to pass the winter upon or in it in a so-called "resting stage."

During the growing season plants of too poor development to be of use for food—the runts—seem often to be preferred by insects to better developed plants and thus form sources of infestation as long as they remain. Similarly, plants poorly fed, inadequately cultivated and those allowed to grow too thickly become more or less emaciated and therefore highly susceptible to the attacks of both bugs and blights, particularly when weeds are abundant and water and plant food thereby reduced.

In view of all this it need scarcely be pointed out that the runt plants, débris and the remains of those that have passed their utility be promptly removed together with thinnings and weeds. They may be either burned or placed on the compost pile. Either method will destroy the insects and plant diseases upon them or deprive these pests of food on which to thrive and propagate. The removal will also prevent the loss of plant food and moisture these plants would take from the soil, thus benefiting the plants that remain.

During the growing season in the vegetable garden such removal should be followed by the sowing or the planting of a succession crop (Chapter 32) or of a green manure (Chapter 29). This latter will not only make the subsequently grown plants of larger size and enhanced quality but more resistant to blight and bug attack. Correct tillage and overhead irrigation (Chapter 19) when rain falls will also greatly aid in these directions.

One other highly important thing that also greatly influences plant health is sunshine. Plants well sunned are far sturdier than those poorly lighted. Because of this they resist disease better and are less harmed by bugs than are weaker ones. Hence one excellent reason for thinning the plants in vegetable gardens and planting berry plants and trees at distances which enable the specimens to develop to normal size without being crowded by their neighbors.

But preventives and precautions, though excellent, are not sufficient to preclude the attacks of either insects or plant diseases. It is necessary to be prepared with the right materials and means of application with which to fight when attacks are made, and also to know when, where, and how to apply each kind. Luckily only a few inexpensive materials are necessary and also fortunately it is easy to identify the enemies to be fought—at least so far as control is concerned. Before we discuss either materials or methods of application, however, let us understand that efficiency is dependent more upon their employment as preventives than as remedies to correct or control the trouble after it has gotten a start. Vigilance to recognize the first symptoms of attack is the watchword.

During the half century that intelligent control of plant diseases and destructive insects has been developing, hundreds of investigators and operators have contributed the results of their investigations, discoveries and observations and published their findings for the benefit of the general public. Some have discovered important "weak spots" in the "life histories" of insects and diseases; others have tested materials to determine the best and safest to use and the proper times to apply them; still others have worked to simplify apparatus and methods of application. Among the more recent developments one of the most important, because of its effectiveness, is what has been dubbed the "one spray control."

It is not intended that this term shall mean that any one spray with any one material or mixture will prevent the attack of all kinds of plant diseases, chewing and sucking insects for a whole season! That's too large an order! It is out of the question to expect any single spray or dust applied, let us say, in early spring to prevent the attacks of insects or plant diseases that do not make their appearance until perhaps late spring or summer. No single spray can accomplish such a feat! The life histories of plant pests differ too widely to expect that. Yet investigators have been working to develop spray combinations that will take care of groups of destructive pests, including both sucking and chewing insects and plant diseases that begin and carry on their active destruction at about the same time. For instance, a group that attacks apple trees in early spring includes tent caterpillar, plant lice, bud moth and case-bearer larvæ, San José and oyster-shell scales and various plant diseases that affect the foliage and get a start while the dormant buds are swelling or shortly before the leaves appear.

Though formerly economic entomologists spoke and wrote as if this or that insect must be fought thus and so as an individual species; and though the plant pathologists did the same thing concerning the control of plant diseases it is only recently that they began to coöperate and to test the efficacy of combination sprays to control these various groups of pests. Not only have they proved that one specific combination will take care of them but that when properly done and at the right time only one spraying should be necessary.

Formerly the recommendation was to spray first with a "winter strength" or "dormant strength" solution while the buds are still dormant but on mild winter days; second, with a different or more dilute solution when the leaves are beginning to unfold. Today they strike an average, so to speak, between these two periods and recommend, first, one spraying at just the time when the dormant buds are beginning to burst and, second, instead of one of the more drastic old time sprays of former years they use one of the newer, milder, yet equally effective spray materials used in accordance with the instructions of the manufacturers.

This recommendation is the result of years of infinite patience on the part of the scientists, more especially of the economic entomologists. These men and women have counted literally millions of microscopically small insect eggs after having been treated with assorted solutions and dusts and at various specified times in order to determine the percentage of those in which the insects have or have not been killed. They have also examined millions of buds sprayed at various periods of development from fully dormant to fully open and with various solutions to discover how much or how little injury was done to them. Their figures have been tabulated, published in bulletins of various experiment stations and their conclusions epitomized in only a few sentences which are still further condensed as follows:

1. As spring approaches the shells of various insect eggs—especially of those kinds which naturally hatch in early spring—gradually become porous in order to enable the developing insects to breathe.

2. Because of this porosity the right kind of spray—a caustic or a miscible oil—will penetrate the shells and kill far more insects than could be killed earlier in the winter while the shells are relatively too resistant for such penetration.

3. The buds swell and the insects emerge from the eggs at about the same time and the insects crawl to, hide and protect themselves among the bud scales and unopened leaves, upon which latter they begin to feed.

4. The young leaves and the other incipient growths are so tender when they first begin to develop that old style sprays often destroy them unless diluted perhaps too much to be effective.

5. The new spray materials, after countless tests, have not only proved themselves to be highly efficient but harmless to the new growths of plants.

6. Sprays of specified kinds may be harmlessly and efficiently combined in order to combat both diseases and insects so that only one application, when done at the proper time, need be made, thus reducing the amount of work, time and therefore cost and yet not impair the efficiency of any one of the ingredients.

7. Conversely, sprays of other specified kinds (as was previously known at least in part) must not be combined because this would make the mixture less effective or, perhaps,

ESSENTIALS OF SPRAYING AND DUSTING

actually injurious to the foliage, especially while young and succulent.

8. "Stickers" or "spreaders" added to sprays of certain kinds accomplish two main purposes; namely, they increase the adhesiveness and lengthen the period from days to weeks during which the spray will cling to the plants and thus continue effective regardless of even heavy rains.

In spite of such knowledge it is easy to waste time and money by using ineffective materials, by the incorrect use of materials, or by using these latter too early or too late. In order to avoid such mistakes and losses it is necessary to know the answers to the following questions; to answer them, however, one doesn't need scientific training but merely one's own eyes. So far as combating the various pests is concerned the following questions cover the practical points:

1. Is the trouble caused by bacteria? How are bacterial attacks recognized?

If the surface of the leaves is unbroken, smooth, without spots, holes or swellings or mildew-like growths; if the malady seems to be wholly inside the tissues which are more or less discolored (brown, as in fire-blight in pear, apple and quince) or wilted (as in cucumber, watermelon and cantaloupe) the malady is probably bacterial or physiological. In all such cases spraying is useless because the diseases are internal. However, in some cases, notably those of fire-blight, trees and bushes may be saved provided the affected parts are cut off and burned. In order to do this properly it is imperative to make the cuts several inches below the lowest noticeably affected parts and also lower than any discoloration of the tissues that are to be seen between the bark and the young wood. Furthermore, it is essential that every cut be immediately disinfected as explained in Chapter 36.

2. Is the trouble caused by a fungus?

When the surface of the leaf, especially the under side, or the other young growths have a more or less powdery, downy, or "mildewy" appearance the cause is a fungous disease. As the feeding parts of these parasitic plants are wholly beneath the surface they cannot be reached by any kind of spray or dust. Therefore nothing can be done to save affected parts; but much may be done to prevent the spread of the disease to other parts of the same plants and to other plants. Spraying or dusting with fungicides will destroy the spores (or minute, seed-like bodies) upon which the spread of the disease depends. Similarly, by having the fungicide present on the foliage before these spores alight the entrance of the disease into the plant may be prevented. This is the most important point to remember and to act upon so far as fungous diseases are concerned. (Figs. 79, 81.)

3. Do the leaves and other green parts have irregular notches or holes that look as if bitten in them?

If so the trouble is caused by chewing insects—caterpillars, beetles, grasshoppers or "slugs" (as the larvæ of various insects are called, though true slugs and snails, which are not insects, do similar damage). As a group the chewers are the easiest of the four classes of plant pests to control. After their presence is discovered all that is necessary is to spray or dust the foliage and other green parts with a "stomach poison" such as arsenate of lead, Paris green or hellebore so the insects will get it when they begin to feed.

Fig. 79. Air pressure sprayer. Galvanized iron tank. Brass is far better because resistant to spray chemicals.

4. Are the leaves more or less contorted, concave above and convex below, crumpled, the under sides thickly populated by tiny, usually green or black insects or are the tips of the growing shoots similarly covered with such insects—plant lice or aphis?

5. Are similar, but larger insects—"true bugs"—found mostly beneath the surface of the foliage (usually, however, not so thickly congregated) but the leaves without holes or notches bitten in them?

6. Are the woody twigs covered by minute, more or less round or elliptical objects that resemble tiny shells—scale-lice or scale-insects?

ESSENTIALS OF SPRAYING AND DUSTING 337

These three last classes (4, 5 and 6) cannot be killed or controlled by stomach poisons because they do not bite off, chew or swallow pieces of plant tissue but merely suck the juices of the plants. The only ways they may be checked or destroyed are by spraying or dusting with caustics, miscible

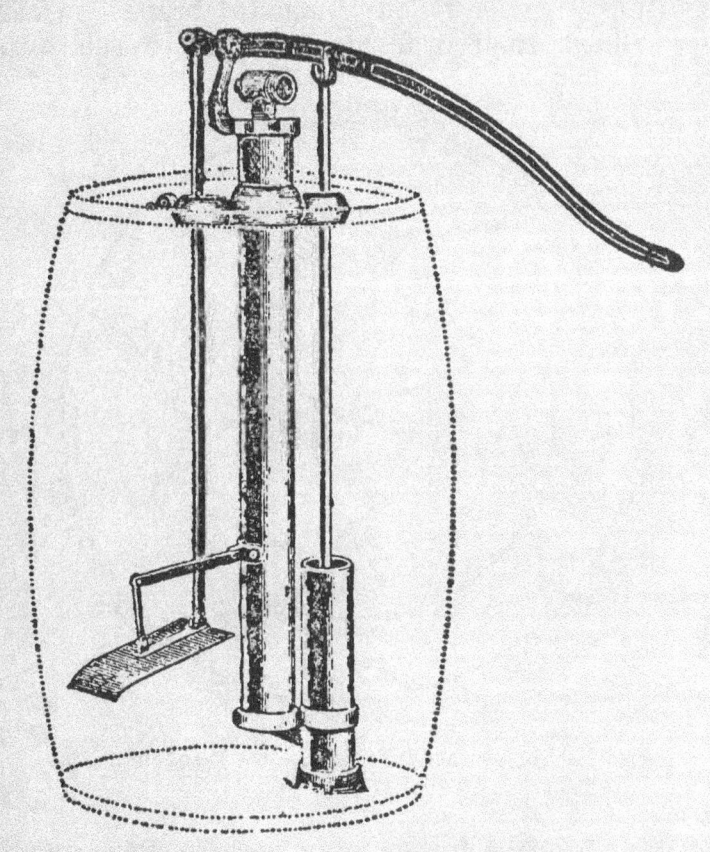

Fig. 80. Single action barrel sprayer, suitable for 100-tree or smaller fruit garden. Note agitator, pump, cylinder and air chamber.

oils or dusts that actually strike them or by a gas that poisons them through their breathing apparatus.

Once having noticed the way in which a plant pest works it is easy to choose the right material with which to fight it. Commercial plant growers are, however, not content merely to keep on the alert all the time to discover the first signs of attack; they take action *beforehand*. Experience has taught them to know when to expect such visitations. So the

338 FIVE ACRES

amateur or the beginner will do well to follow suit as soon as his experience suggests the wisdom of such action.

Many commercial growers make or mix at least some of their materials because they use large quantities and the time and cost elements are essential for them to consider. Others think it safer to buy standard brands in sufficient supply to meet their probable needs. As a rule amateurs

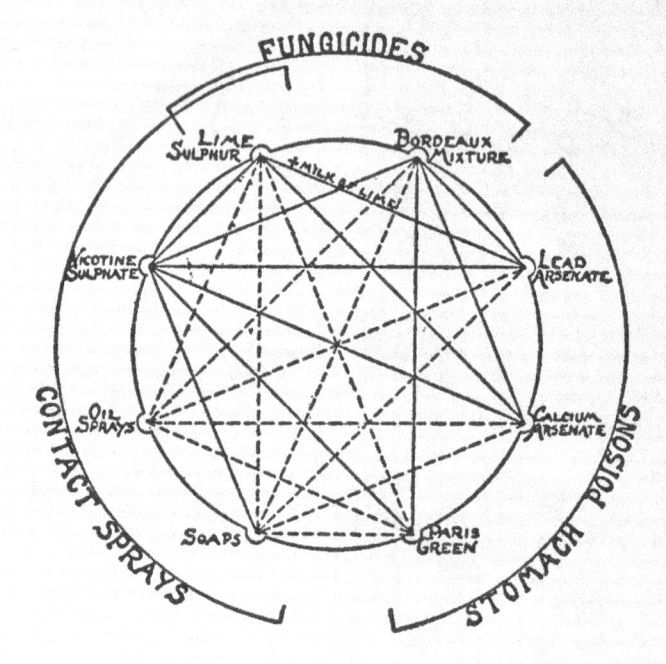

Fig. 81. Sprays that may (solid lines) and must not (dotted lines) be mixed.

should not adopt the former plan because they use so little that the cost and time elements are excessive in comparison. Furthermore, since mixing demands a knowledge of chemistry and a technic that must be acquired experimentally they had far better buy the ready mixed brands either direct from the makers or at garden, department and other stores from coast to coast. For the same reasons they should not mix spray materials of any kind without knowing exactly which chemicals each one contains because certain kinds when mixed become coagulated or ineffective or even do damage to the foliage when combined. The accompanying chart shows

ESSENTIALS OF SPRAYING AND DUSTING

at a glance what kinds can and what ones can not be combined with safety. (Fig. 80.)

For instance, never mix a miscible oil with anything that contains sulphur—lime-sulphur wash ("lime sulphate" of the chart), Bordeaux mixture, or any arsenate such as that of calcium, lead or copper (Paris green). Instances of safe mixtures are lime-sulphur and nicotine sulphate, and either or both of these with lead arsenate. Each of these latter combinations may be used as a "one spray control" just when

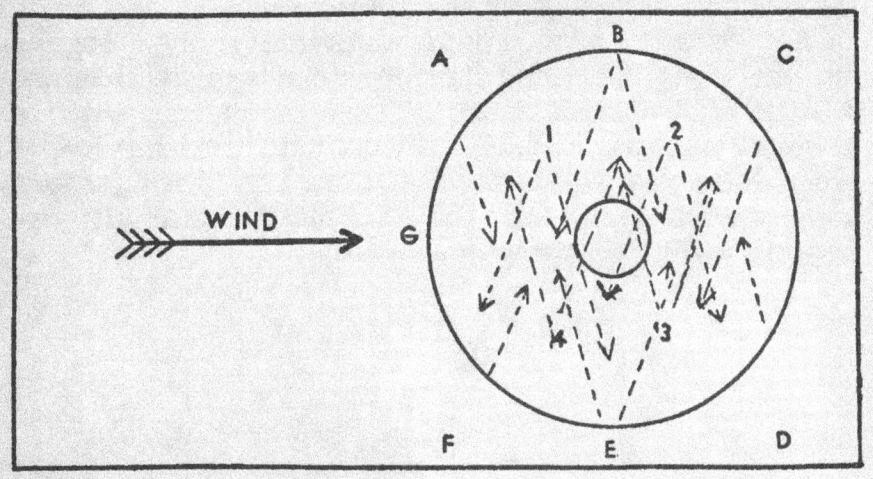

Fig. 82. Diagram showing effective way of spraying on windy days.

the buds of trees and bushes are beginning to swell. Each combination will control a varied assortment of insects and plant diseases. The nicotine kills plant lice, the arsenate kills the chewing insects and the Bordeaux (or the lime-sulphur wash) destroys the spores of plant diseases and prevents the spread of these maladies if they have already made a start.

As most amateurs and others who need only small quantities of insecticides and fungicides will use proprietary preparations it is always safe to follow the directions which manufacturers supply with their products and to avoid mixing any one with any other except as indicated by the chart for fear of injuring each and perhaps also the plants upon which sprayed or dusted.

When necessary to spray against the wind the "cross-fire" method shown in Fig. 82 will do a good job. The large circle represents the tree-spread; the small one, the trunk. On approaching the tree the spray is directed from A and B across the wind as shown by the broken arrows. If the tree is large the operator steps beneath the outer branches to 1 and 2 to spray toward the opposite side. If the tree stands alone he steps to D, E, F, 3 and 4; if in a row, he moves to the next tree and so on to the end. Here he turns back on the opposite side and sprays the D, E, F, 3 and 4 side of each tree. By this method the small area at G is the only part of the tree that need be sprayed with the wind.

In working this method the operator must spray, 1, slightly into the wind; 2, clear to the opposite side of the tree, and 3, from below upward so as to reach the undersides of the leaves as much as possible; for here most fungous infections occur. Here also poisons applied to destroy chewing insects are protected against rain. One spraying thus carefully done is worth more than many done poorly.

… APPENDICES

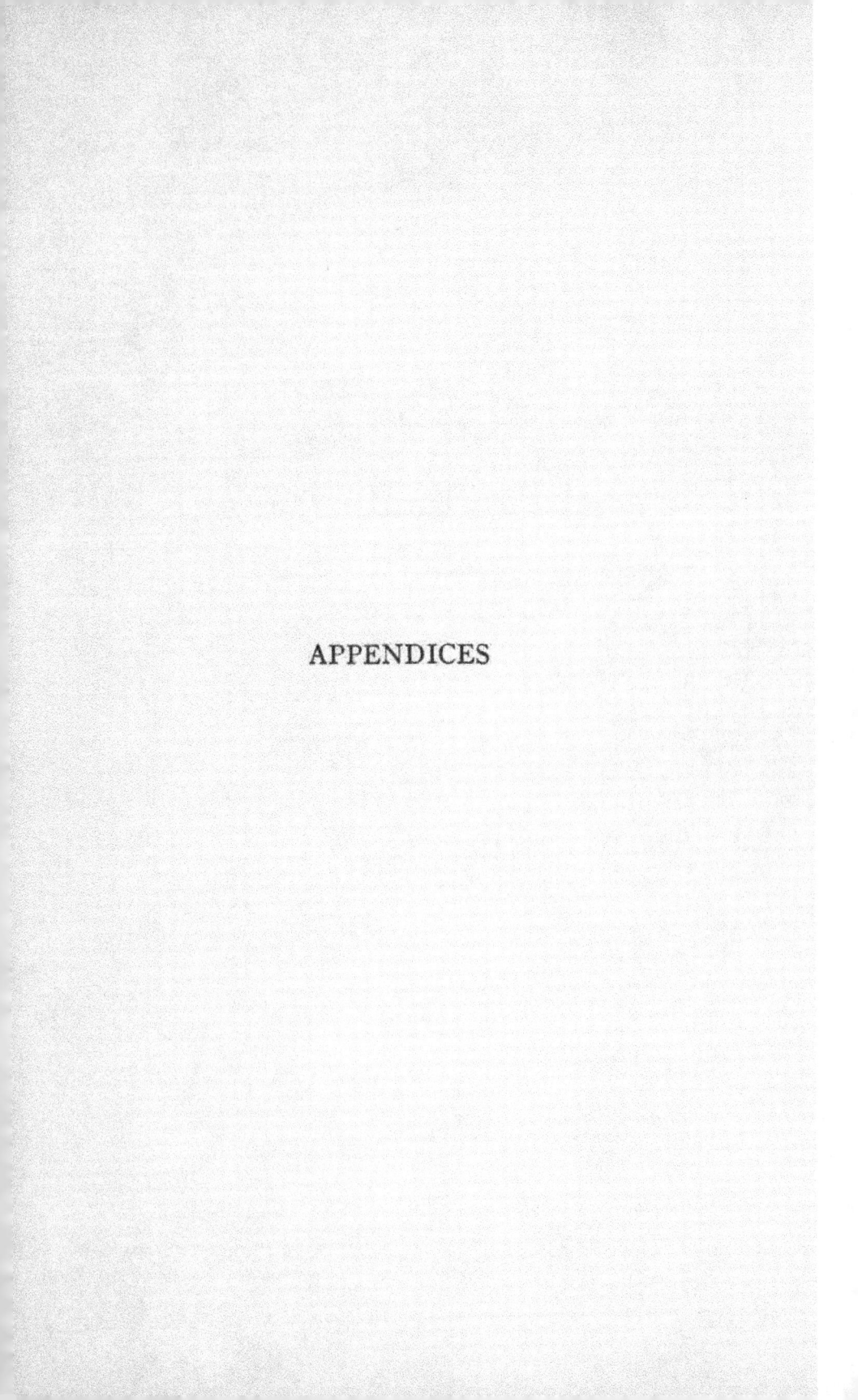

SEEDS RECOMMENDED FOR THE LONG-ROW FARM GARDEN

(120 by 200 feet. Rows 3 feet apart except where noted, Fig. 8.)

No. of row	Part of row	Vegetable	Amount
		First planting—	
Row 1	½	Asparagus, *Mary Washington* [1]	50 roots
	⅓	Rhubarb, *Victoria* or *Linnæus*	20 roots
	⅙	Onions, *Egyptian perennial*	1 lb. sets
Rs. 2–5	All	Early Ohio potatoes (use clean northern-grown seed)	1 bu.
Row 6	All	Onions, *Southport Yellow Globe*	1 oz.
		(Marked with *Scarlet Globe* radish [2]) .	1 oz.
Row 7	½	Onions, yellow (sets)	3 lbs.
	½	Spinach, *King of Denmark*	1 oz.
Row 8	⅓	Leaf lettuce, *Black Seeded Simpson* (seed for 2 plantings)	1 oz.
	⅓	Turnips, *Purple Top Strap Leaf*	½ oz.
	⅓	Kohlrabi, *Early White Vienna* (seed for 2 plantings)	1 oz.
Row 9	½	Peas, *Surprise*	1 lb.
	½	Peas, *Daisy*	1 lb.
Row 10	All	Peas, *Little Marvel*	2 lbs.
		Second planting—	
Row 11	⅓	Cabbage, *Golden Acre*	36 plants
	⅓	Cabbage, *Marion Market*	36 plants
	⅓	Head lettuce, *May King*	60 plants
Row 12	⅔	Cabbage, *Wisconsin All Season* (seed for 3 plantings)	½ oz.
	⅓	Spinach, *New Zealand*	1 oz.
Row 13	½	Beets, *Early Wonder* or *Crosby's Egyptian*	1 oz.
	⅓	Carrots, *Chantenay* (seed for 3 plantings)	1 oz.
	⅙	Parsley, *Champion Moss Curled*	1 pkt.
		(Row 13 marked with *Scarlet Globe* radish, see Row 6)	

[1] When the perennials are planted in a separate area, omit Row 1. The onions and rhubarb are easily transplanted, but it will be necessary to secure a new supply of asparagus roots if a new location is desired, as the old roots are undesirable for transplanting.

[2] The term "marked with" means that the radish seed is sown thinly with the other seed in the same row. (Chapter 32.)

No. of row	Part of row	Vegetable	Amount
Row 14	⅔	Parsnips, *Improved Guernsey*	1 oz.
		(Marked with *White Icicle* radish)	1 oz.
	⅓	Swiss chard, *Lucullus*	1 oz.

Third planting—

Rs. 15–18	⅓ each	Sweet corn, *Mammoth White Cory* [3]	½ lb.
	⅓ each	Sweet corn, *Howling Mob*	½ lb.
	⅓ each	Sweet corn, *Narrow Grain Evergreen*	½ lb.
Row 19	All	String beans, *Stringless Green-Pod* (seed for 2 plantings)	2 lbs.

Fourth planting—

Row 20	All	String beans, *Stringless Green-Pod*	See R. 19
Row 21	½	Carrots, *Chantenay*	See R. 13
		(Marked with *Scarlet Globe* radish) ...	See R. 6
	½	Beets, *Detroit Dark Red* (seed for 2 plantings)	2 oz.
Row 22	All	Cabbage, *Wisconsin All Season* (seed) ...	See R. 12
Row 23	⅓	Pepper, *Ruby King*	30 plants
	⅓	Lima beans, *Henderson's Bush*	½ lb.
	⅓	Lima beans, *Dreer's Bush* or *Pole*	½ lb.
Row 24	All	Tomato, *Marglobe* (4' from Row 23) ...	50 plants
Row 25	⅓	Tomato, *Marglobe*	16 plants
	⅔	Muskmelon, *Hale's Best* (6' from Row 24)	1 oz.
Row 26	⅙	Squash, *Giant Summer Straightneck*	½ oz.
	⅚	Cucumbers, *Chicago Pickling* (6' from Row 25)	1 oz.
Row 27	All	Watermelon, *Kleckley Sweets* (9' from Row 26)	2 oz.
Row 28	All	Squash, *Delicious* (winter) (9' from Row 27)	2 oz.
Row 29	All	Sweet potatoes, *Nancy Hall* or *Yellow Jersey* (9' from Row 28)	160 plants

[3] If yellow corn instead of white is desired, *Golden Sunshine, Whipple's Early Yellow,* and *Bantam Evergreen* may be substituted for *Mammoth White Cory, Howling Mob,* and *Narrow Grain Evergreen* respectively.

APPENDICES

DISTANCES FOR FRUIT PLANTS

Almond		20'
Apple, dwarf, on Doucin stock		15'–25'
" " on Paradise stock		10'–12'
" standard, erect growing (e.g., Wagener)		30'–35'
" " majority of varieties		35'–40'
" " spreading (e.g., R. I. Greening)		40'–50'
Apricot		20'
Blackberry	4' x 6' to 6' x 8'	
Cherry, Duke		25'–30'
" sour		15'–20'
" sweet		30'–40'
Cranberry	1' x 2' to 2' x 2"	
Currant, in rows	4' x 5'	
" in checks	5' x 5'	
Fig		20'–25'
Filbert and hazelnut		10'–12'
Gooseberry, in rows	4' x 5'	
" in checks	5' x 5'	
Grape, small growing (e.g., Delaware)		8'–10'
" moderate growing (e.g., Concord)		10'–12'
" rank growing (e.g., Agawam)		12'–15'
Grapefruit		25'–30'
Lemon		25'–30'
Loquat		15'–25'
Mulberry		25'–35'
Nectarine		20'
Orange, dwarf varieties		10'–15'
" ordinary "		25'–30'
" St. Michael		20'–25'
Peach		20'
Pear, dwarf		10'–15'
" standard		25'–30'
Pecan		40'–50'
Persimmon, Japanese		20'–25'
" American		25'–30'
Plum		20'
Quince		10'–15'
Raspberry, black, in rows	3' x 6'	
" " in checks	5' x 5'	
" purple, in rows	4' x 6'	
" " in checks	5' x 5'	
" red and yellow, in rows	3' x 5'	
Strawberries		See Chapter 48
Walnut, Black		40'–50'
" English or Persian		30'–40'

CHOICE VARIETIES OF FRUITS

The following varieties have been chosen because of their high quality and therefore suitability to amateur, home gardens.

Apples: Benoni, Blenheim, Chenango, Cortland, Delicious, Early Harvest, Early Mackintosh, Esopus Spitzenberg, Fall Pippin, Fameuse, Golden Delicious, Golden Russet, Gravenstein, Grimes Golden, Hubbardston, Jonathan, Lowland Raspberry, McIntosh, Northern Spy, Peck Pleasant, Porter, Primate, Rambo, Roxbury Russet, St. Lawrence, Shiawassee, Williams, Winesap.

Apricot: Alexander, Alexis, Blenheim, Budd, Early Golden, Harris, Montgamet, Moorpark, Peach, Royal, St. Ambroise.

Cherry, sour: Dyehouse, Early Richmond, English Morello, Large Montmorency, Ostheim, Wragg.

Cherry, sweet: Bing, Black Tartarian, Burbank, Napoleon, Lambert, Governor Wood, Schmidt, Windsor, Yellow Spanish.

Cherry, duke: Late Duke, Louis Philippe, Magnifique, May Duke, Olivet, Royal Duke, Reine Hortense.

Crab: Excelsior, General Grant, Hyslop, Martha, Transcendent, Whitney.

Nectarine: Boston, Downton, Early Newington, Early Rivers, Early Violet, Elruge, Hardwicke, Humboldt, Hunter, Newton, Pitmaston Orange, Stanwick, Victoria.

Peach: Belle, Carman, Chairs, Champion, Early Crawford, Fitzgerald, Foster, Frances, Greensboro, Hiley, Late Crawford, Morris White, Niagara, Oldmixon Cling, Oldmixon Free, Ray, Rochester, Salwey, Smock, Stump, Waddell, Yellow St. John.

Pear: The following varieties are best when grafted on pear stocks: Bartlett, Belle Lucrative, Onondaga, Seckel.

Pear: The following do equally well as standard and dwarf trees: Anjou, Beurre Superfine, Clapp, Comice, Madeleine, Josephine de Malines, Osband, Pound, Tyson, White Doyenne.

Pear: The following usually fail as dwarfs unless double worked (that is with quince roots, pear trunk and the desired variety grafted on the pear trunk): Beurre Bosc, Marie Louise, Sheldon, Winter Nelis.

Plum, American: Desoto, Downing, Hammer, Hawkeye, Miner, Stoddard, Surprise, Terry, Wayland, Weaver.

Plum, European: Bavay, Golden Drop, Imperial Gage, Italian Prune, Jefferson, Middleburg, Mirabelle, Pearl, Peters, Reine Claude, Shropshire, Tragedy, Washington, Yellow Egg.

DWARF PEARS

The following varieties do better as dwarf trees than as standards: Angouleme, Beurre Diel, Easter Beurre, Glout Morceau,

Louise Bonne de Jersey, Vicar of Winkfield. (See lists in Appendix 16.)

The following varieties are offered by the leading nurseryman who specializes in dwarf fruit trees: Anjou, Bar-Seckel, Cayuga, Clapp, Comice, Dana Hovey, Dr. Reeder, Duchess d'Angouleme, Flemish Beauty, Louise Bonne de Jersey, Seckel, Souvenir, Vermont Beauty, Wilder, Worden-Seckel and several others of less conspicuous merit.

GRAPE VARIETIES

In the following list the order of ripening is indicated by a number: "One" means earliest (in Southeastern New York about August 20) and "ten" latest (about October 20). To be on the safe side, if you reside north of New York City I suggest that you omit 10, 9 and perhaps 8 because in some years the growing season is shorter than others so these may fail to ripen; if you live north of Albany perhaps 7 and 6 should also be omitted. Names in italics are noted as long keepers.

Black varieties: Fredonia, 1; Moore's Early, 2; Worden, 3; *Barry*, 3; *Herbert*, 4; *Campbell*, 4; *Wilder*, 5; Eumelan, 6; *Mills*, 6; *Canandaigua*, 10. Red varieties: Moyer, 1; Brighton, 2; *Lindley*, 2; *Gœrtner*, 3; Delaware, 4; Brilliant, 4; *Salem*, 5; *Vergennes*, 5; *Agawam*, 5; *Caco*, 6; *Goethe*, 7; Iona, 7; *Catawba*, 7; *Urbana*, 8. "White" varieties: Ontario, 1; Portland, 1; Winchell (or Green Mountain), 1; Diamond, 3; Niagara, 4; Pocklington, 5; *Empire*, 5; *Dutchess*, 7.

BUSH BERRIES

Blackberry: Agawam, Ancient Briton, Blowers, Early Harvest, Eldorado, Kittatinny, McDonald, Mercereau, Ward.

Blueberry: Adams, Cabot, Dunfee, Greenfield, Grover, Harding, Katharine, Pioneer, Ralph, Rubel, Sam.

Currant: Boskoop Giant (black), Cherry, Diploma, Fay, Filler, Lee (black), Perfection, Pomona, Red Cross, White Grape, Wilder.

Dewberry: Gardena, Lucretia, Mayes, Premo.

Gooseberry: Chautauqua, Crown Bob, Downing, Oregon, Pearl, Poorman, Red Jacket, Smith.

Mulberry: Downing, Hicks, Merritt, New American, Stubbs, Townsend.

Raspberry, red: Cuthbert, Erskine, Herbert, June, La France, Marlboro, Ohta, Ranere (or St. Regis), Latham.

Raspberry, black: Gregg, Hilborn, Honeysweet, Kansas, Plum Farmer, Tyler.

Raspberry, purple: Cardinal, Columbian, Royal Purple, Shaffer.

Raspberry, "white": Carolina, Golden Queen.

STRAWBERRIES

Early varieties: Dorsett, Fairfax, Premier, Southland.
Midseason varieties: Aberdeen, Big Joe, Glen Mary, Marshall.
Late varieties: Aroma, Big Late, Chesapeake, William Belt.
Very late varieties: Gandy, Orem, Pearl.
Everbearing varieties: Champion (or Progressive), Super Giant, Lucky Strike, Mastodon.

VEGETABLE MATURITY TABLE

(Number of days between sowing and gathering)

Artichoke, globe (perennial), Artichoke Jerusalem	180	Leek	130 to 150
		Lettuce	50 to 70
		Mustard	20 to 30
Bean, bush	40 to 60	Okra (or gumbo)	110 to 120
Bean, pole	50 to 70	Onion	130 to 150
Bean, shell	60 to 70	Parsley	90 to 100
Bean, lima	70 to 80	Parsnip	140 to 160
Beet, round	60 to 70	Pea	60 to 100
Beet, long	150	Pepper	130 to 150
Brussels sprouts	100 to 130	Pumpkin	100 to 130
Cabbage, early	100	Radish, forcing	20 to 30
Cabbage, late	150	Radish, summer	30 to 50
Cantaloupe	110 to 130	Radish, winter	60 to 90
Celery, early	100 to 125	Rutabaga	70 to 90
Celery, late	130 to 150	Salsify	100 to 130
Chard	40 to 50	Scorzonera	100 to 130
Chicory	150	Spinach	40 to 60
Corn, sweet	70 to 100	Spinach, New Zealand	60 to 80
Cress (peppergrass)	20 to 30	Squash, summer	50 to 70
Cucumber	120 to 130	Squash, winter	120 to 130
Eggplant	140 to 160	Tomato	130 to 150
Endive	120 to 140	Turnip	50 to 70
Fetticus (corn salad)	40 to 50	Watermelon	110 to 130
Kohlrabi	90 to 100		

APPENDICES

DATA FOR GARDEN PLANNING

Vegetable	Amount of Seed (Oz.) for 50 Feet of Row	Date of First Sowing	Number of Sowings	Unit Distance Between Rows	Weeks to Maturity	Date of Last Sowing in Southeastern N.Y.
Beans, bush	4	May 15 to 25	2 to 5	1 or 2	6 to 10	8/1
Beans, pale	4	May 15 to 25	1	3 or 4	8 to 15	6/1–15
Beets	1	Apr. 1 to 15	2 to 4	1 or 2	8 to 12	8/1
Broccoli	1/8	May 1 to 15	1	2 or 3	18 to 20	5/30
Brussels Sprouts	1/8	May 1 to 15	1 or 2	2 or 3	9 to 12	6/15–20
Cabbage, early	1/8	Apr. 1	1	2 or 3	7 to 10	5/1
Cabbage, late	1/8	May 1 to 15	1	2, 3 or 4	10 to 15	6/1
Carrot	1/2	Apr. 1	1 or 2	1 or 2	8 to 12	7/15 and 8/1
Cauliflower	1/4	May 1 to 30	1 or 2	2 or 3	8 to 12	5/30
Celery, early	1/8	Apr. 1 to 15	2 to 4	3 or 4	15 to 18	5/15
Celery, late	1/8	May 1 to 15	1	3 to 5	16 to 20	5/15
Celeriac	1/8	May 1 to 15	1	3 to 5	16 to 20	5/15
Chicory	1/8	Apr. 1 to 15	1	2	20 to 25	4/30
Corn, early	2	May 15 to 20	2 to 6	2	8 to 10	7/15–8/1
Corn, late	2	May 15 to 20	2 to 4	2 or 3	10 to 12	6/1
Corn salad	1	Apr. 1 or Aug. 1	1 early, 1 late	1	8 to 10	9/1 to 10/1 or 3/15–31
Cress (peppergrass)	1/2	Apr. 1	2 or 3	1	3 or 4	5/15
Cucumber	3/4	May 15 to 30	1 or 2	3 or 4	8 to 12	5/30
Dandelion	1/8	Apr. 1 to 15	1	1 or 2	20 to 24	4/15
Eggplant	1/8	May 20 to 30	1	2 or 3	8 to 12	6/10
Endive	1/2	Apr. 1 to July 1	1 early, 1 late	1	12 to 15	8/1
Kale	1/8	May 1 to 15	1	2	12 to 15	6/1
Kohlrabi	1/8	Apr. 1	1 early, 1 late	1 or 2	18 to 20	5/1
Leek	1/4	Apr. 1 to 30	1	2	18 to 20	5/1
Lettuce	1/4	Apr. 1 to 30	2 to 6	1	6 to 12	8/1–15
Melon, Musk	1/4	May 15 to 30	1	3 or 5	12 to 15	6/10

APPENDICES

DATA FOR GARDEN PLANNING—(Continued)

Vegetable	Amount of Seed (Oz.) for 50 Feet of Row	Date of First Sowing	Number of Sowings	Unit Distance Between Rows	Weeks to Maturity	Date of Last Sowing in Southeastern N.Y.
Melon, Water	½	May 15 to 30	1	5 or 6	12 to 15	6/10
Mustard	¼	Apr. 1 to May 1	1 to 3	1 or 2	5 or 6	{ 3/15 to 4/15 8/1 to 9/30
Nasturtium	½	May 1 to 15	1 or 2	1	10 or 12	5/30
Okra	1	May 15 to 30	1	2 or 3	10 or 12	5/30
Onion	½	Apr. 1 to May 1	1	1	15 to 25	5/10
Onion Sets	16	Apr. 1 to May 1	1	1	5 to 8	5/1
Parsley	⅛	Apr. 1 to 30	1	1	12 to 15	5/1
Parsnip	¼	Apr. 1 to 15	1	1 or 2	20 to 25	5/1
Peas, dwarf	16	Apr. 1 to 30	2 or 3	1 or 2	10 to 12	8/1
Peas, tall	16	Apr. 1 to 30	2 or 3	3 or 4	12 to 14	5/1
Pepper	⅟₁₆	May 20 to 30	1	2	6 to 9	6/10
Potato	40	Apr. 15 to 30	1 or 2	2 or 3	8 to 12	6/20
Pumpkin	¼	May 15 to 30	1	8 to 12	15 to 18	6/1
Radish, forcing	½	Apr. 1 to 15	6 to 8	1	4 to 6	6/1
Radish, summer	½	Apr. 15 to May 15	2 or 3	1	5 to 7	5/15
Radish, winter	½	July 15 to Aug 15	1	2	12 to 15	8/15
Rutabaga	⅛	June 15 to 30	1 or 2	1 or 2	10 to 15	7/15
Salsify	½	Apr. 1 to 15	1	1 or 2	20 to 25	5/15
Spinach	½	Apr. 1 to 30	1 to 3	1	8 to 10	5/1
Spinach, N. Zealand	½	May 20 to 30	1	2 or 3	6 to 8	6/15
Squash, summer	¼	May 20 to 30	1	4 or 5	8 to 10	6/1
Squash, winter	¼	May 20 to 30	1	5 to 8	12 to 15	6/1
Swiss Chard	1	Apr. 1 to 30	1	1 or 2	7 or 8	4/1 or 8/1
Tomato	⅟₁₆	May 20 to 30	1	3 or 4	7 or 10	6/10
Turnip	¼	Apr. 1 to 15	1 to 3	1 or 2	8 to 10	3/15–30 or 8/1

350

APPENDICES

THE EARLY HARDY VEGETABLE GARDEN

Vegetable	Seed or Plants to 50 Feet of Row (Ounce or Number)	Depth to Sow (Inch)	Distance Between Plants in Row (Inch)	Limit Distance Between Rows
Beet	1 or 100	1	4 to 6	1
Broccoli *	25 to 35	½	18 to 24	2
Cabbage, early †	25 to 35	½	18 to 24	2
Carrot	½	¼	3 to 4	1
Celery	½	¼	2 to 3	1
Chicory	½	½	3 to 4	1
Cress (peppergrass)	½	¼	½	½
Dandelion	½	¼	6	1
Endive	½	½	12	1
Kohlrabi	25	½	18 to 24	2
Leek	½	½	3 to 4	2
Lettuce plants	25 to 50	—	12 to 24	1
Lettuce seed	¼	¼	12	1
Mustard	½	½	3 to 4	1
Onion seed	½	½	2 to 3	1
Onion seedlings	150	—	3 to 4	1
Onion sets	16	1 to 2	1 to 2	1
Parsley	½	¼ to ½	4 to 6	1
Parsnip	¼	½ to 1	3 to 4	2
Peas, dwarf	16	1 to 2	2 to 4	1 or 2
Peas, tall	16	1 to 2	2 to 4	2 or 3
Potato	40	3 to 4	12 to 15	2
Radish, forcing	½	½	1 to 2	1
Radish, summer	½	½	2 to 3	1
Salsify	½	½	2 to 3	1
Spinach	½	½	2 to 3	1
Swiss Chard	½ or 50	1	12	1 or 2
Turnip	½	¼ to ½	4 to 6	1

* Lettuce plants may alternate with these plants if set on the same day. The lettuce will have matured before the other plants need the space.

† The "unit distance" between plants is a personal choice—12 or 15 inches, as a rule, though 10 may answer in some cases.

VEGETABLES TENDER TO FROST

Beans, wax: Pencilpod, Rustproof Golden, Burpee's Kidney Wax. *Beans, green:* Masterpiece, Bountiful, Burpee's Stringless Greenpod. *Beans, dwarf lima:* Fordhook, Dreer's Bush Lima, Dreer's Wonder, Burpee's Improved. *Beans, green, pole:* Burger's Stringless Greenpod, Lazy Wife, Kentucky Wonder (or Old Homestead). *Beans, wax, pole:* Golden Cluster, Kentucky Wonder Wax. *Beans, pole lima:* Giant Podded, Sunnybrook, Leviathan, Siebert.

Cantaloupe· Honeyball, Bender's Surprise, Hearts of Gold, Fordhook, Rockyford, Emerald Gem, Osage (or Miller's Cream).

Corn, early: Golden Bantam, Metropolitan, Golden Evergreen, Aristocrat, Black Mexican.

Corn, late: Country Gentleman, Mammoth Late, Stowell's Evergreen.

Cucumber: Davis' Perfect, White Spine, Klondike.

Eggplant: Black Beauty, New York Improved.

Okra: Dreer's Little Gem, Creole (or White Velvet), Perkins's Longpod.

Pepper, hot: Long Red Cayenne, Small Chili, Tabasco. *Pepper, sweet:* Pimiento, Neapolitan, Sunnybrook, Ruby King, Ruby Giant.

Spinach: New Zealand (not a true spinach but a good hot weather substitute).

Squash, summer: Crookneck, Cocozelle, Pattypan (or Simlin), Vegetable Marrow (easiest to prepare).

Tomato: Matchless, John Bayer, Marglobe, Livingston's Globe, Avon, Bonny Best, Stone.

VEGETABLES THAT WITHSTAND FROST

Artichoke, French or *Globe:* Large Green Globe.

Artichoke, Jerusalem. Listed by few seedsmen, usually without variety name.

Beet, round: Crosby's Egyptian, Crimson Globe, Detroit Dark Red. *Beet, long:* Long Dark Blood, Henderson's Half Long, Long Smooth Dark Red.

Broccoli: White Cape, Calabrese Sprouting, Henderson's Riviera.

Brussels Sprouts: Matchless, Long Island, Half Dwarf.

Cabbage, early: Early Jersey Wakefield, Charleston (or Large Wakefield). *Cabbage, midseason:* All Seasons, Early Summer, Succession. *Cabbage, late:* Autumn King, Danish Roundhead, Flat Dutch.

Carrot, early: Earliest Shorthorn, Early Scarlet Horn, Guerandi (or Oxheart). *Carrot, late:* Chantenay, Danvers, Intermediate.

Cauliflower: Henderson's Snowball, Dreer's Earliest Snowstorm, Dry Weather.

Celery, early: Easy Blanching, Golden Plume, Rosy Plume, White Plume. *Celery, late:* Giant Pascal, Winter Queen, Golden Dwarf, New Rose.

Chard, Swiss: Giant Lucullus, Lyons.

Chicory: Large Rooted, French Endive, Witloof.
Cress (or *peppergrass*): Extra Curled.
Endive: Broad Leaved Batavian, Giant Fringed, White Curled.
Kohlrabi: Earliest Erfurt, Early White, Early Purple Vienna.
Kale: Long Standing, Dwarf Green Curled, Dwarf Scotch.
Leek: Musselburgh, Giant Carentan, Prizetaker.
Lettuce, loose leaf: Grand Rapids, Black Seeded Simpson, Early Curled Simpson. *Heading, early:* All Heart, May King, Wayahead, Tennisball. *Midseason, butterhead:* California Cream Butter, Deacon, French Unrivaled. *Crisp head:* Wonderful, Mignonette, Iceberg. *Cos* or *romaine:* Trianon, Paris White, Kingsholm.
Onion, yellow: Strasburg, Prizetaker, Southport Yellow Globe, Danvers Yellow Globe. *Onion, white:* Barletta (or White Queen), White Portugal, Mammoth Silver King. *Onion, red:* Southport Red Globe, Large Red Wethersfield.
Mustard: Fordhook Fancy, Southern Giant Curled, White London, Chinese.
Parsley: Perfection, Fern Leaved, Champion Moss Curled.
Parsnip: Guernsey (or Student), Large Sugar (or Hollow Crown), Early Round.
Pea, dwarf, early, smooth seeded: Alaska, Eureka, Electric, Eight Weeks. *Peas, earliest, dwarf, wrinkle seeded:* American Wonder, Prosperity (or Gradus), Little Marvel, Nott's Excelsior, Thomas Laxton. *Peas, midseason:* Senator, Abundance, Lincoln, Laxtonian. *Peas, late, dwarf:* Dwarf Telephone (or Carter's Daisy), Dwarf Champion, World's Record, Potlach. *Peas, late, tall:* Champion of England, Prince Edward, Telephone, Senator. *Peas, edible podded:* Melting Sugar, Giant Sugar.
Radish, forcing: Crimson Ball, Ruby, Perfection, White Box, Crimson Giant, White Olive, French Breakfast. *Radish, summer:* Chartier, Icicle, Delicacy, White Delicacy. *Radish, winter:* Scarlet China, Celestial (or White China), Black Spanish.
Rutabaga: Golden Heart, Long Island, Purple Top.
Spinach: Long Standing, Long Season, Prickly Seeded, Viroflay (or Thick Leaf), Victoria.
Salsify: Sandwich Island, Long White.
Scorzonera: Long Black.
Turnip: Aberdeen, Milan, Snowball, Golden Ball, Flat Dutch.

GARDEN TOOLS AND ACCESSORIES

The following list will help you choose items that you may need. Probably you will not want everything specified, probably also you may wish others. By all means have the equipment that will fit your individual case. In such instances as hoes and rakes have two or more sizes so as to do the work with least expenditure of time or effort.

Basket, picking
Cultivator, Norcross
Dibble, flat style
Dusting machine
Dusting materials
Edger, grass
Fertilizers
Flower pots
Fork, digging
Fork, manure
Fork, transplanting
Grafting chisel
Grafting mallet
Grafting wax
Hoe, common
Hoe, scuffle
Hoe, Warren
Hose, nozzle and reel
Hotbed mats
Knife, asparagus
Knife, pruning
Labels, garden, pot and tree
Ladder, step
Lawn mower
Lawn roller
Lawn sprinkler
Line, garden
Manures, dried
Pruner, pole

Rake, bamboo or rubber-toothed
Rake, garden
Rake, self-cleaning
Rake, steel
Rake, wooden garden
Saw, pruning (swivel style)
Scythe
Shear, grass
Shear, hedge
Shear, pruning, single hand
Shear, pruning, two hand, all steel
Shovel, D-handle
Shovel, long handle
Sickle
Spade, D-handle
Spade, long handle
Sprayer, compressed air
Spraying materials
Stakes, bamboo
Stakes, wooden
Supports, plant
Trowel, steel shank
Watering-pot
Weeder, Excelsior
Weeder, Hazeltine
Weeder, Lang
Wheelbarrow with removable sides
Wheelhoe and seed drill
Window-box, metal, self-watering

FERTILIZER APPLICATIONS FOR FRUIT TREES

Age of tree	Amounts of fertilizer to apply per tree		
	Nitrate of soda	Acid phosphate	Muriate of potash
	oz.	oz.	oz.
One year	4	12	4
Two years	8	16	8
Three years	12	20	12
Four years	16	24	16
Five to eight years	24	32	20
Nine to twelve years	40	80	32
Thirteen to thirty years	80	128	48

APPENDICES

GARDEN INSECT PESTS AND METHODS OF CONTROL

By GEORGE M. LIST, *Department of Entomology, Colorado Agricultural College.*

Crop	Insect	Stage which does damage	Type of injury	Control method given under numbers at foot of table
Asparagus	Asparagus beetle	Adult, larva	Feed upon new growth and mature stalks	1, 10
Beans	Mexican bean beetle	Adult, larva	Feed upon leaves and pods	3
Cabbage Cauliflower and Kohl-rabi	Cabbage aphis	All stages	Suck juice from leaves	4
	Cabbage worm	Larva	Eat foliage	2
	Cabbage maggot	Larva	Tunnel in base of stem and roots	5
	Cutworms	Larva	Cut off young plants	6, 7
Cucumbers	Cucumber beetle	Adult	Eat leaves and stem	8
	Aphis	Adult	Suck juices from stem and leaves	4, 12
	Squash bug	Adult, nymph	Suck juice from stem	4, 9
Eggplant	Fleabeetle	Adult	Eat young leaves	10, 11
Lettuce	Cutworms	Larva	Cut off young plants	7
	Grasshopper	Adult, nymph	Eat young leaves	7
Melon	Same as cucumber			
Onion	Maggot	Larva	Burrow in the bulb	No satisfactory method
	Thrip	Adult, nymph	Suck juice from foliage	12
Potato	Colorado potato beetle	Adult, larva	Eat foliage	3, 13
	Flea beetles	Adult, larva	Adults eat foliage, larvae feed on tubers	10, 11
	Tomato psyllid	Adult, nymph	Suck juice from plant; poisons plant	14
Peas	Aphis	Adult, nymph	Suck juice from plant	4, 12
Squash	Same as cucumber			
Tomato	Horned tobacco worm	Larva	Eat leaves and fruit	1, 9
	Tomato psyllid	Adult, nymph	Suck juice from plant	14
	Cutworms	Larva	Cut off plants	6, 7
Corn	Corn earworm	Larva	Eat into corn ear	No satisfactory method
	Corn-root worm	Larva	Tunnel into base of stalks and roots	15

APPENDICES

Methods of Control

1. Dust mature plants with a mixture of air-slaked lime (5 parts) and lead arsenate (1 part).
2. Dust with pyrethrum or derris powders.
3. Spray with magnesium arsenate or zinc arsenite (1 ounce) and water (3 gallons).
4. Dust with a mixture containing 5% nicotine sulphate.
5. Drench stems and roots of plants with one-half cup of solution containing 1 ounce corrosive sublimate (bichloride of mercury) to 8 or 10 gallons of water.
6. Wrap stem of plant, from roots to first leaves, with paper at time of transplanting.
7. Mix two-thirds ounce of white arsenic or Paris green with 1 pound bran. Dilute one-sixth pint of molasses with a small amount of water and with this moisten the bran into a crumbly mass. Scatter along the rows of plants at night. For grasshoppers scatter broadcast before 10 A. M.
8. Dust with gypsum or hydrated lime (19 parts) and calcium arsenate (1 part).
9. Hand pick.
10. Dust with hydrated lime (5 parts) and calcium arsenate (1 part).
11. Spray with Bordeaux mixture, 4-4-50 formula. This mixture can be purchased in the dry form from druggists or seedsmen.
12. Spray with a solution containing 2 tablespoonfuls of nicotine sulphate, 1 ounce soap, and 3 gallons of water, about every 10 days after insects appear.
13. Spray with Paris green (1 ounce), water (3 gallons).
14. Spray with lime-sulphur solution (1 part), water (40 parts).
15. Rotate crops. Never plant corn following corn.

WATER DEFINITIONS AND EQUIVALENTS

(See Chapter 15.)

A gallon of water weighs 8.36 pounds and equals 231 cubic inches. A cubic foot of water weighs 62.5 pounds and equals 7.48 gallons.

An "acre-inch" of water is the amount necessary to make a layer of water 1-inch deep on an acre (43,560 square feet). It requires 27,154 gallons to make an acre-inch of water.

"Pounds pressure" and "feet of head" are terms which indicate the pressure of a column of water. For instance, depth of water in a tank at sea level will create a specific amount of pressure which may be measured on a gauge at its base. Height above sea level causes variations due to pressure of the air; the greater the elevation, the less the pressure. Width or extent of tank is immaterial; the height of the water in the tank is what determines the pressure. As it increases the pressure increases; as it lessens the pressure lessens. Contrary to popular belief the supply pipe from a source lower than the tank *should enter at the bottom,* not over the top.

At sea level a column of water 2.3 feet high will produce 1 pound of pressure per square inch; a column 23 feet high will produce 10 pounds of pressure per square inch, no matter whether the tank be 1 foot or 100 feet in diameter.

To determine how large to build a tank or a cistern to supply the various needs of a farm the figures given in Table 1 will be found of great aid. See Chapter 15.

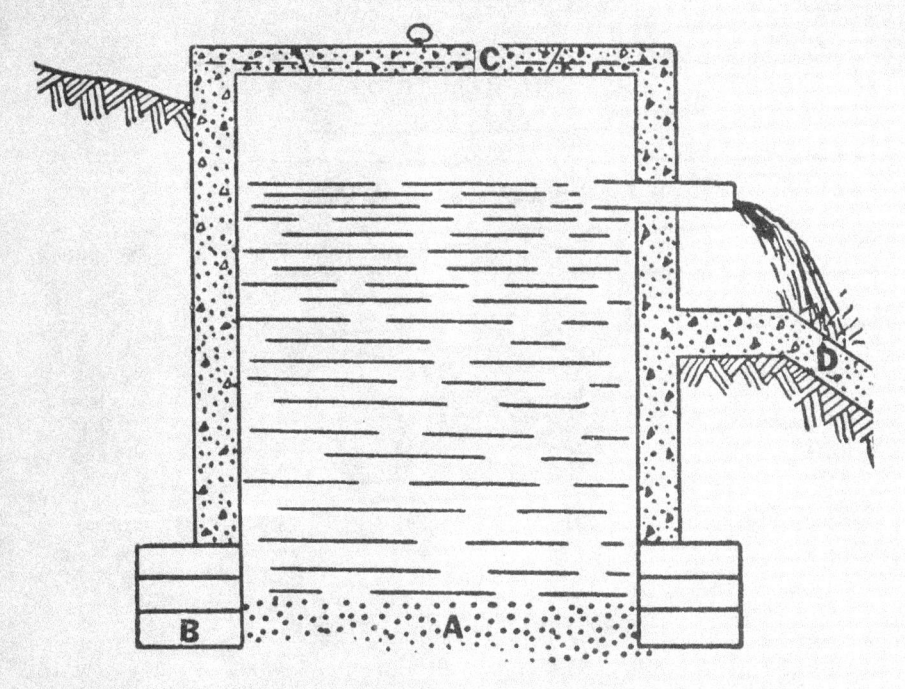

Fig. 1. Protection of a Sidehill Spring.

In order to maintain the purity and to protect the water of a spring from contamination, it is advisable to build a concrete tank. A, Sandy bottom. B, Substantial footing. C, Cover. D, Spillway. This last is one of the most important features of the construction because it prevents washing away of the soil and thus undermining the foundation. When such a tank is used for a gravity or a hydraulic ram system the outlet pipe should either enter through the footing or the side but its opening should be at least 4″ above the bottom, preferably pointing downward and certainly protected to prevent anything but water from gaining entrance. (See Chapter 15.)

TABLE 1.—WATER CONSUMPTION ON THE FARM

Water carried	8 gals. per person per day.
Pump at kitchen sink	10 " " "
Faucet at kitchen sink	12 " " "
Running hot and cold water in kitchen	18 " " "
Complete plumbing with water under pressure	30 " " "
Bathtub *	8–20 gals. each time used.
Closet *	3–5 " " " "
Lavatory (wash basin in bathroom)	1–2 " " " "
Sprinkling lawn	8 gals. per 100 square feet.
Soaking lawn	20 " " " "
Cow	15 gals. per day.
Horse	10 " "
Hog	2 " "
Sheep	½ " "

* Water for these purposes is included in the item above entitled "Complete Plumbing and Water under Pressure."

TABLE 2.—YIELDS OF VEGETABLES AS INFLUENCED BY SOIL REACTION [1]

(Acre Basis)

Calcium Limestone, to the Acre

Crop	No lime	1,000 pounds	2,000 pounds	4,000 pounds
Beans, lima	130	2,464	2,842	4,667
Beans, string	.....	1,690	3,122	3,578
Beets	1,488	25,425	32,080	30,848
Cabbage	31,840	31,840	33,008	33,728
Carrots	544	18,560	20,080	21,920
Corn, sweet	1,063	3,662	4,849	6,020
Eggplant	2,029	4,742	7,496	14,438
Pepper	2,374	6,742	7,658	7,540
Pepper (another year)	7,232	16,352	13,744	16,080
Sweet potato	6,384	11,840	14,000	14,640

Magnesian Limestone to the Acre

Crop	No lime	1,000 pounds	2,000 pounds	4,000 pounds
Beans, lima	130	2,736	3,726	4,200
" string	.....	2,491	3,843	3,022
Beets	1,488	25,072	35,600	36,544
Cabbage	18,032	30,768	24,528	21,728
Carrots	544	22,160	22,240	25,280
Corn, sweet	1,063	5,776	5,041	4,203
Eggplant	2,029	11,408	10,502	9,485
Pepper	2,374	7,269	5,607	5,456
Pepper (another year)	7,232	14,080	17,312	22,380
Sweet potato	6,384	13,808	11,696	10,864

[1] See Chapter 30.

TABLE 3.—SIZES AND WEIGHTS OF TILE

Truck loads of tile (4,500 to 6,000 pounds) run about as follows: 3"—1,200; 4"—1,000; 5"—650; 7"—400; 8"—250 to 300; 10"—165–175; 12"—120 to 130.

TABLE 4.—APPROXIMATE WEIGHT OF TILE OF VARIOUS SIZES

Size	Weight of one tile	Weight of 1,000 tile	Size	Weight of one tile	Weight of 1,000 tile
3"	4½	4,750	8"	18½	18,500
3½"	5½	5,500	10"	26	26,000
4"	6	6,000	12"	34	34,000
5"	9	9,000	14"	40	40,000
6"	11½	11,500	16"	47	47,000
7"	15	15,000			

TABLE 5.—SIZES OF TILE FOR MAIN DRAIN AND AREAS DRAINED BY EACH

Acres Drained

Fall in inches per 100 ft.	3 in. Tile	4 in. Tile	6 in. Tile	8 in. Tile	10 in. Tile	12 in. Tile
60 inches	18.6	26.8	74.4	150.0	270.0	426.0
40 "	15.1	21.8	60.4	128.0	220.8	346.0
30 "	12.9	18.6	51.6	108.8	189.6	298.4
24 "	11.9	17.0	47.7	98.0	170.4	269.0
20 "	10.9	15.6	43.4	90.0	156.0	246.0
17 1-7 "	10.0	14.5	39.9	83.0	144.4	228.1
15 "	9.3	13.4	37.2	77.0	135.0	213.0
13 1-3 "	8.1	12.6	35.0	72.5	127.0	200.5
12 "	7.3	11.9	33.1	69.2	120.6	190.5
8 "	6.7	9.5	26.6	56.0	97.3	154.4
6 "	5.7	8.2	22.8	48.0	83.9	132.5
4 4-5 "	5.1	7.5	20.4	42.4	74.4	117.0
4 "	4.6	6.9	18.4	38.2	65.5	107.0
3 "	4.1	5.9	16.5	32.6	60.3	90.7
2 2-5 "	3.7	5.2	14.8	30.1	54.0	81.6
2 "	3.3	4.7	13.3	28.0	48.6	74.0
1 1-2 "	2.9	4.1	11.4	24.0	41.9	65.0
1 1-5 "	2.6	3.7	10.2	21.2	37.2	56.0
4-5 "	2.1	3.0	8.5	16.8	30.8	47.0
3-5 "	1.9	2.8	7.4	15.0	25.0	40.8

TABLE 6.—NUMBER OF FEET OF TILE NEEDED TO DRAIN AN ACRE

6 rods	or	100 feet	apart	equals	436 feet	to acre
5.4 "	"	90 "	"	"	484 "	" "
4.7 "	"	80 "	"	"	544 "	" "
4.2 "	"	70 "	"	"	622 "	" "
3.6 "	"	60 "	"	"	726 "	" "
3.0 "	"	50 "	"	"	871 "	" "
2.4 "	"	40 "	"	"	1089 "	" "
1.8 "	"	30 "	"	"	1450 "	" "

TABLE 7.—SIZES AND WEIGHTS OF PIPE FOR DISTRIBUTION MAINS IN SMALL SPRAY IRRIGATION SYSTEMS

Size of Pipe (Inches)	Amt. of Water (Gallons per Minute)	Approximate Wt. per 100 Ft. (Pounds)	Size of Pipe (Inches)	Amt. of Water (Gallons per Minute)	Approximate Wt. per 100 Ft. (Pounds)
¾"	3 to 4	113	2"	23 to 28	375
1"	5 to 7	175	2½"	39 to 65	600
1¼"	8 to 14	225	3"	66 to 110	775
1½"	15 to 22	275	3½"	111 to 175	995

FACTS ABOUT ICE

Ice forms from water at a temperature of 0 centigrade or 32° F.

A cubic foot of solid ice weighs 58 pounds; a cubic foot of water, 62½ pounds; therefore ice is 9/10 the weight of water.

Ice occupies 1/10 more space than the volume of water that forms it. One ton of solid ice occupies 36 cubic feet, but as stored in icehouses it requires about 45 cubic feet of space.

Ten pounds of ice has about the same cooling effect as 100 pounds of cold well water. Every pound of ice in melting abstracts 142 units of heat or enough to cool 10 pounds of water 14.2 degrees F.

Cooling by ice is due to melting. The faster ice melts the greater becomes its value as a cooling medium. This is why salt is mixed with ice; its great power to absorb water hastens the rate of melting. Salt can be used profitably at the rate of 1 part to 3 of crushed ice.

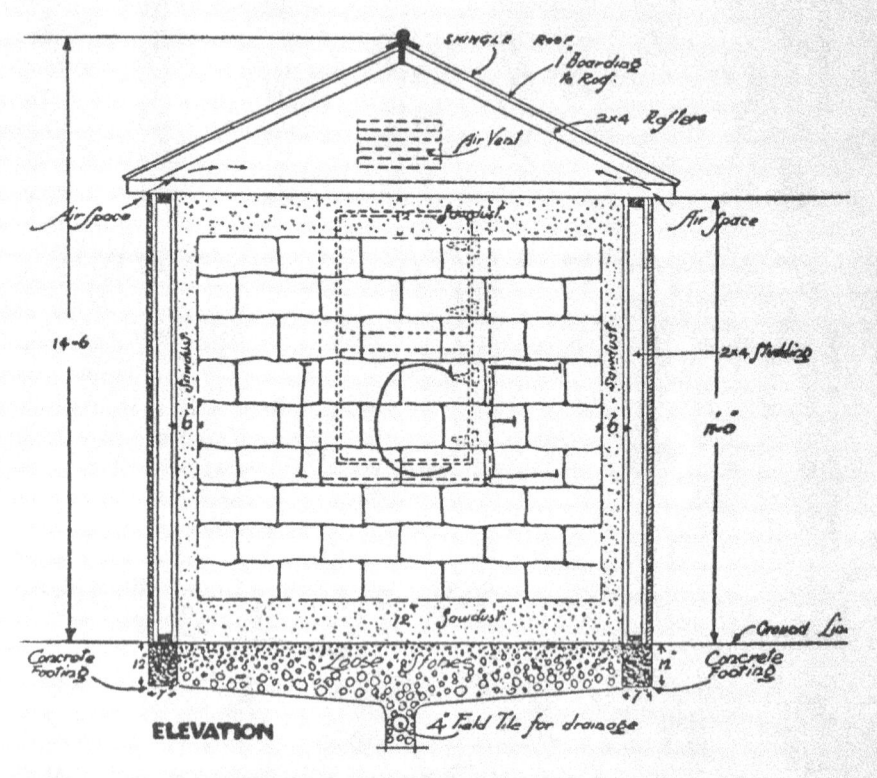

Fig. 2. Standard Type of Icehouse.

In sustaining power, ice in a sheet 2″ thick will usually bear a man; one 5″ thick, a team of horses and a load weighing 2 tons.

The best quality of ice is obtained from pure, clean, quiet water. Snow and slush are injurious to quality and keeping power.

Ice has a specific heat of 0.5, or exactly half that of water; i. e., the temperature of ice changes twice as fast as that of water.

The quantity of ice required on the farm will depend on location, number of cows milked, methods of handling the product, kind and site of ice-house, size of family and whether or not a household refrigerator is used.

To supply average family needs about 5 tons of ice should be stored; i. e., about 50 cakes 22" square and 12" thick, allowing for considerable waste. For dairy purposes 3,000 pounds per cow will be needed where whole milk is cooled, and 1,000 pounds where only cream is cooled.

Cost of storing varies with localities and distance ice must be hauled—about $1.50 a ton when near-by.

TABLE 8.—MATERIALS FOR BUILDING AN ICE HOUSE

Sills and wall plates	8 pieces, 2" x 4" x 11' 10"
Studding	28 " 2" x 4" x 10' 8"
Corner posts	4 " 4" x 4" x 10' 8"
Outside matched boards, 1"	536 sq. ft.
Rough boarding, 1"	536 "
Waterproof insulating paper	536 "
G. & G. boarding, prepared with waterproofing material	524 "
Rafters	18 pieces, 2" x 4" x 7½'
Rafters	9 " 2" x 4" x 13' 2"
Rough boarding, 1"	195 sq. ft.
Shingles	195 "
Ridge boards	3 pieces, 1" x 4" x 1¼'
Ridge roll	1 " 13'
Doors	2 " 3' x 3'
Door framing	2 pieces, 1½ x 3" x 14'
Door framing	3 " 1½ x 3" x 3' 2"

Planer shavings for packing between studding.
  Sawdust for filling around and above ice.
  Hardware including nails, hinges, screws, bolts, etc.

VALUE OF ELECTRIC MOTORS

The electric motor is the most dependable and economical power for operating farm water pumps, provided the cost of current is not excessive. Pumps are of two classes, shallow and deep-well. The first are designed to lift water from wells or cisterns the bottoms of which are not more than 22' below the level of the pumps themselves; the second, where the vertical lift from the water to the pump is greater.

Shallow-well pumps may be placed above or near the wells or cisterns from which they draw water; e. g., a basement where they are protected from freezing. Deep-well pumps must be placed above the well, though the pressure tank, if any, may be located elsewhere. A pump jack, if used, with either a shallow- or a deep-well pump, must be placed over the well or cistern when used with a deep-well pump.

Changes from engine to electric motor to operate pumps rarely involve much change of equipment. In most cases the drive pulley

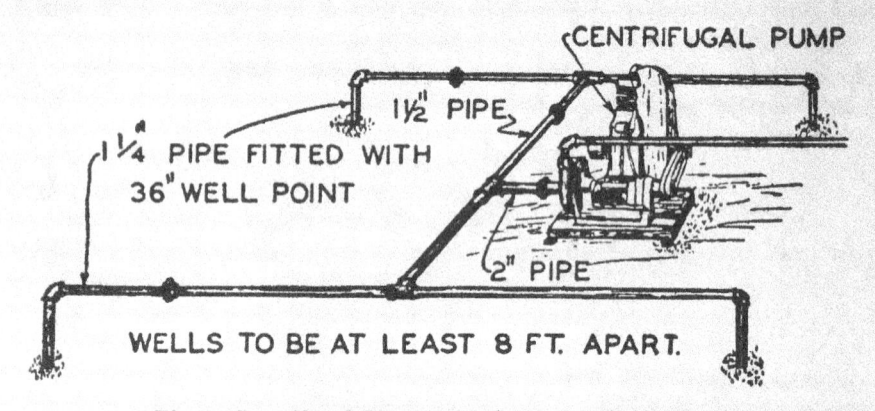

Fig. 3. Centrifugal Pump with Several Driven Wells.

When considerable water must be pumped the arrangement shown here will often be found less costly than to sink one tube of total, equal capacity because, when soil conditions are equal, it is easier to drive the smaller pipes than a large one and the flow is likely to be surer. Shallow wells, 15' to 20' deep, may be connected together and connected with one pump.

must be replaced by a larger one to secure pump speed. Where automatic operation of the motor is desired, a switch to control the motor must be installed. When belt drives are used, some style of belt tightener will be needed.

Always the wiring for motor installation should be done by a competent electrician and fuse plugs of proper capacity used to prevent damage to motors.

The amount of water pumped by electric current or a gallon of gasoline per kilowatt-hour depends upon various factors such as size and class of pump, depth of well, and pressure against which the water is pumped.

In tests of 5 pumps at the Indiana experiment station (reported in Circular 184) the average consumption with a 15' lift at 30 pounds gage pressure was 1.25 kilowatt-hours per 1,000 gallons of water pumped. Similar results, secured on 5 farm-installed, shallow-well pumps where the water pumped was measured by meters, showed a consumption of 1.44 kilowatt-hours per 1,000 gallons of water pumped, with pressures varying from 20 to 40

APPENDICES

pounds. Energy required for pumping with deep-well pumps is greater than that with shallow-well pumps and increases with the increase in lift of the water.

The average of 8 field survey records taken on shallow-well pumps driven by gasoline engines showed a consumption of 1.05

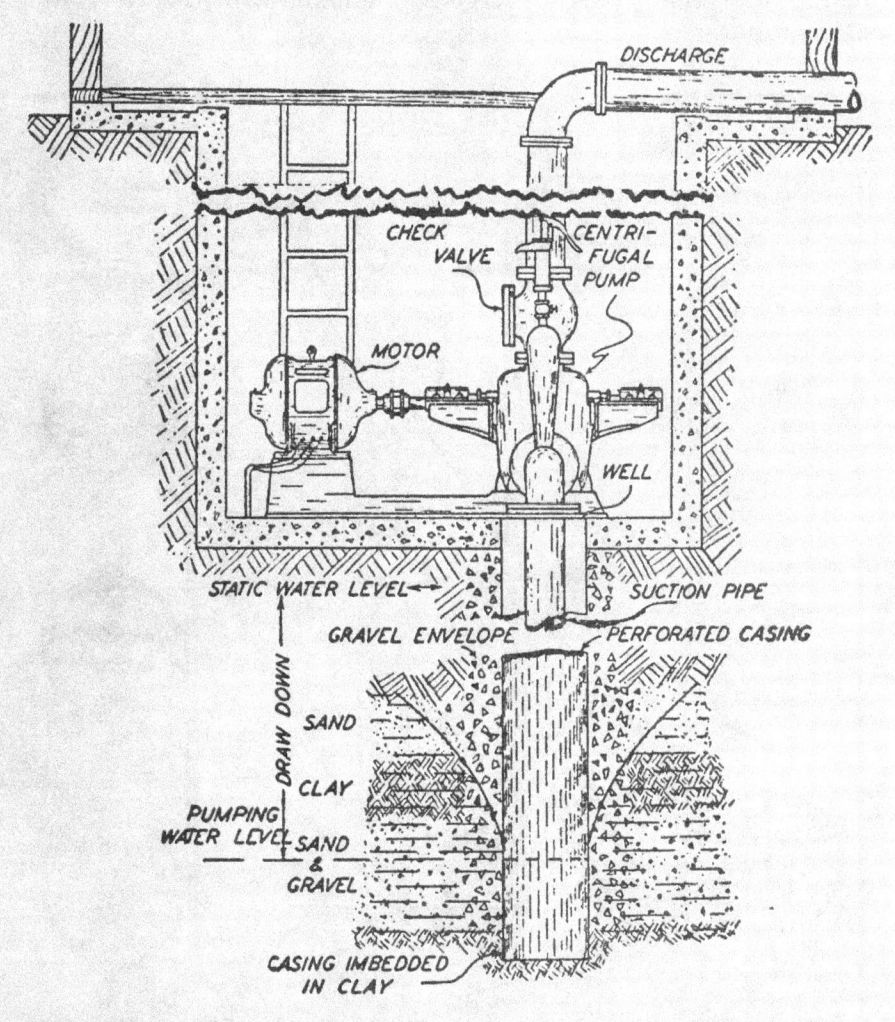

Fig. 4. Correct Mounting of Centrifugal Electric Pump. (See Chapter 15.)

gallons of gasoline per 1,000 gallons of water pumped *against no pressure*, and similar records taken on 5 shallow-well pumps driven by electric motors showed a use of 1.56 kilowatt-hours per 1,000 gallons *against pressures of 20 to 40 pounds*.

364 APPENDICES

With gasoline costing 12¢ a gallon and electric energy costing 5¢ a kilowatt-hour, the cost of pumping 1,000 gallons of water under the above conditions were $0.78 for electric power and $0.126 for gasoline power. The results of the tests reported would indicate that under identical pumping conditions (both pumping against the same pressure) a greater difference in cost of power would result.

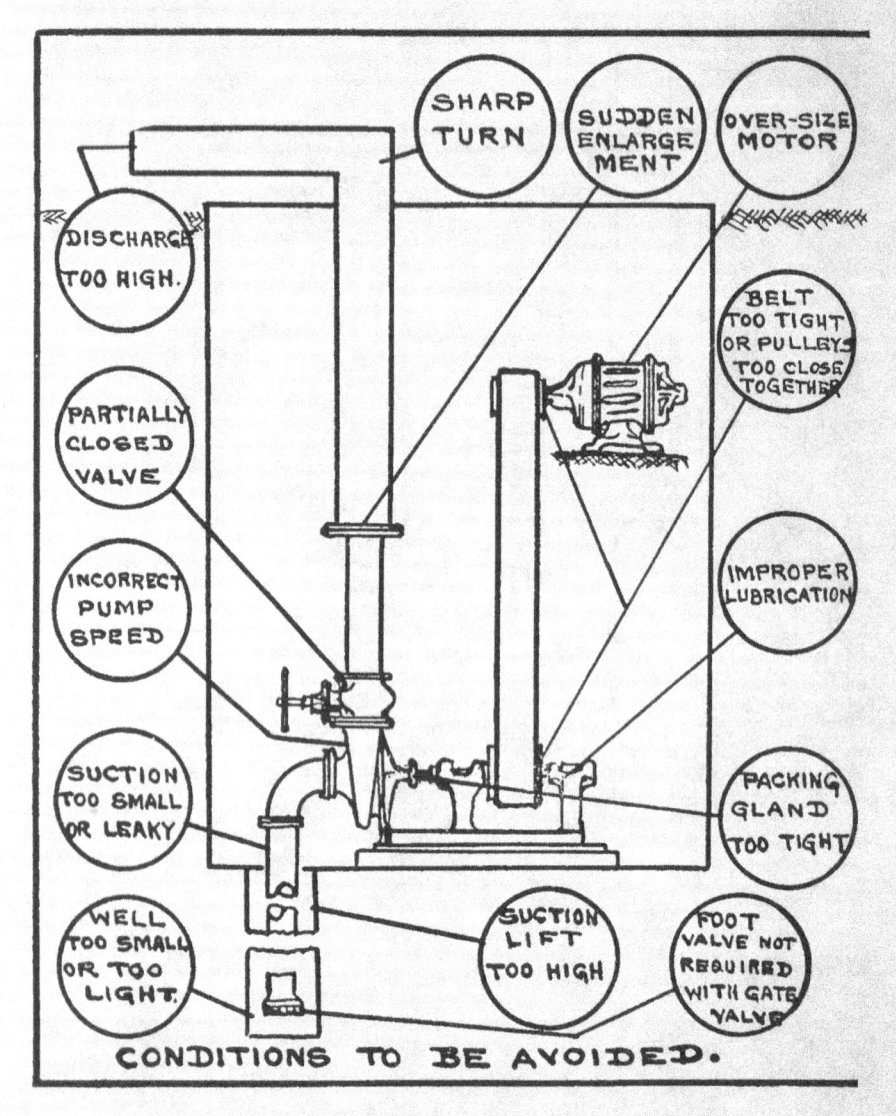

Fig. 5. Errors in Mounting Centrifugal Electric Pump.

APPENDICES

FERTILIZER MIXTURES [1]

It is important to understand the distinction between a carrier (without which no fertilizer can be mixed) and a filler or makeweight which may or may not be necessary. In every case the carrier IS a necessary part of the mixture since all three of the plant food elements must be combined with some other material before they can be handled or stored. For example both nitrogen and ammonia in their pure form are gases which must be absorbed and combined with some other substance before they can be used in a fertilizer mixture.

Every farmer is familiar with the odor of ammonia which he smells on entering a horse stable after it has been closed over night. This is from the ammonia gas being given into the air from the manure. When land plaster (gypsum) is sprinkled over the floors a part of the ammonia is absorbed and may be handled with the manure, in which case the gypsum is the carrier. So likewise of other plant food elements, potash and phosphorus.

The amount of inactive material in a fertilizer depends both upon the analysis or percentage composition and upon the character of the material from which the mixture is made. A high analysis fertilizer (Table 9) might be made up as shown under formula A.

TABLE 9.—FERTILIZER FORMULAS

A	Pounds	B	Pounds
Nitrate of soda	500	Nitrate of soda	500
Superphosphate	1,000	Superphosphate	1,000
Muriate of potash	100	Muriate of potash	200
Kainit	400	Filler (sand or muck)	300
This makes a 4-8-5 mixture	2,000	This also makes a 4-8-5 mixture	2,000

If stored, especially in a damp place, fertilizer A would cake, so the lumps would have to be broken and screened before using. To avoid this the same formula may be made of slightly different materials and a few pounds of filler added to preserve a good physical condition, as shown in formula B.

The manufacturer might also use sulphate of ammonia, tankage or fish scrap or some combination of these materials as the

[1] The diagram, tables and paragraphs have been condensed from Maryland Experiment Station Bulletin, written by A. G. McCall. (See Chapter 28.)

sources of nitrogen. Instead of deriving all the phosphorus from superphosphate he might use some bone or tankage to supplement the superphosphate, and wood ashes as the source of part of the potash. In high analysis fertilizers, therefore, it is necessary to use such filler to insure a good physical condition.

TABLE 10.—PERCENTAGE AMOUNTS OF BASIC MATERIALS TO USE IN MIXING FERTILIZER

Basic Material	1	2	3	4	5	6	7	8	9	10
Nitrogen Carrier										
Nitrate of soda (18%)	112	224	336	448	560	672	784	896	1008	1120
Sulfate of ammonia (25%)	80	160	240	320	500	480	560	640	720	800
Dried blood (10%)	200	400	600	800	1000	1200	1400			
*Tankage (5%)	400	800	1200	1600						
*Fish scrap (8%)	250	500	750	1000	1250	1500				
Phosphoric Acid Carrier										
Acid phosphate (16%)	125	250	375	500	625	750	875	1000	1125	1250
Acid phosphate (14%)	143	286	429	572	715	858	1001	1144	1287	1430
*Ground bone (23%)	87	174	261	348	435	522	609	696	783	869
Potash Carrier										
Potassium sulfate (50%)	40	80	120	160	200	240	280	320	360	400
Potassium chloride (muriate 50%)	40	80	120	160	200	240	280	320	360	400
Kainit (12%)	168	336	504	672	840	1008	1176	1344	1512	1680
Wood ashes (4%)	500	1000	1500							

* Tankage, fish scrap, and bone contain both nitrogen and phosphorus.

TABLE 11.—COMPOSITION OF MATERIALS USED IN MIXING FERTILIZERS

Fertilizing Material	Ammonia	Phosphoric acid	Potash
	Per cent	Per cent	Per cent
Nitrogen carriers			
Nitrate of soda	18 to 19		
Sulfate of ammonia	24 to 25		
Dried blood	10 to 16		
Tankage	5 to 12	3 to 18	
Dried fish scrap	8 to 10	6 to 8	
Phosphoric acid carriers			
Acid phosphate		12 to 16	
Raw ground bone	3 to 5	20 to 25	
Potash carriers			
Potassium sulphate			48 to 52
Potassium muriate			48 to 52
Kainit			12 to 12.5
Wood ashes		1 to 2	2 to 8

APPENDICES

A low analysis fertilizer (Table 12) may be made as follows:

TABLE 12

	Pounds
Nitrate of soda	125
Superphosphate	1,000
Muriate of potash	40
	1,165
Filler (muck or sand)	835
This mixture is a 1-8-1 formula.	2,000

The fertilizer *ingredients* in this low grade formula are all high grade, but in compounding the low analysis brand it is necessary to use an excessively large amount of make-weight filler. On this make-weight stuff the buyer must pay freight, cartage and the cost of application to his crop. It does neither him nor his crop any good, so it does not pay him to handle it.

The amount of filler might be reduced in this case by using tankage or fish scrap for the nitrogen carrier and kainit instead of muriate as the potash carrier but this would not increase the amount of plant food in the fertilizer. In this case the additional weight of inactive material would be in the form of a carrier instead of a filler.

TABLE 13.—OUTSIDE OF THE FERTILIZER BAG

High Analysis / Low Analysis

High Analysis			Low Analysis
167 pounds	Net weight		167 pounds
Potato Special	Brand name (unimportant)		Grass and Grain
4—8—5	Formula (important)		1—8—1
Ammonia 4.0%	Guaranteed analysis		Ammonia 1.0%
Available Phos. acid 8.0%			Available Phos. acid 8.0%
Potash 5.0%			Potash 5.0%
Fertilizer company	Name and address of fertilizer company		Fertilizer company

INSIDE OF THE FERTILIZER BAG

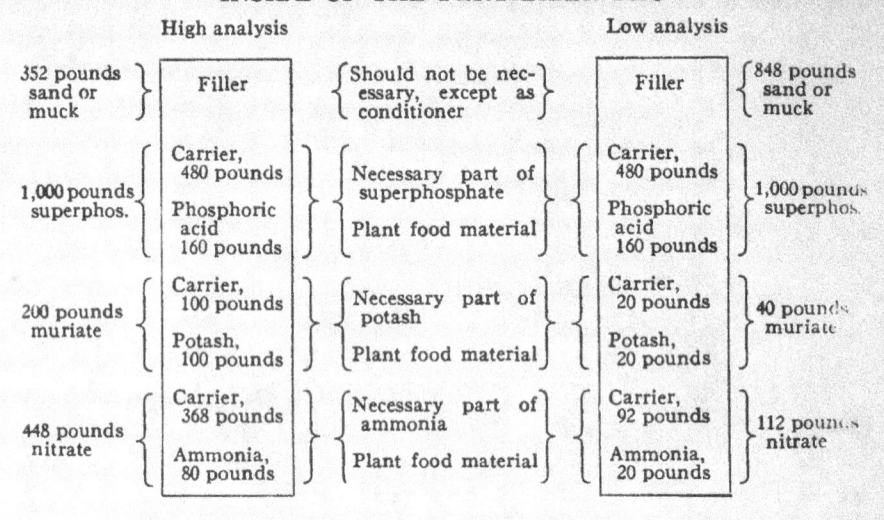

High analysis   Low analysis

	High analysis			Low analysis	
352 pounds sand or muck	Filler	Should not be necessary, except as conditioner	Filler	848 pounds sand or muck
1,000 pounds superphos.	Carrier, 480 pounds / Phosphoric acid 160 pounds	Necessary part of superphosphate / Plant food material	Carrier, 480 pounds / Phosphoric acid 160 pounds	1,000 pounds superphos.
200 pounds muriate	Carrier, 100 pounds / Potash, 100 pounds	Necessary part of potash / Plant food material	Carrier, 20 pounds / Potash, 20 pounds	40 pounds muriate
448 pounds nitrate	Carrier, 368 pounds / Ammonia, 80 pounds	Necessary part of ammonia / Plant food material	Carrier, 92 pounds / Ammonia, 20 pounds	112 pounds nitrate

CAPACITIES OF ROUND AND RECTANGULAR CISTERNS

(*See Chapter 15*)

An easy way to determine the cistern capacity needed during any period is to multiply the square-foot area of roof by inches of rainfall and divide by 1.6. For instance, a 30' x 40' roof and 1½" rainfall. $\frac{1200 \times 1.5}{1.6} = 1,125$ gallons. A family of 5 which uses 5 gallons daily per person will require 9,125 gallons annually, or 4,562 gallons each half year. If a minimum of 1,125 gallons is collected during the half year of low storage, 3,437 gallons will have to be in storage when this period starts; i. e., the cistern would have to be that large or larger. If, however, it is desired to have a 6-months' reserve, the cistern should hold 4,562 gallons. In this latter case the year's supply would be collected and stored during half of the year and used during the other.

When the rainfall during a year is only 28", it would be possible to collect 12" of water during 6 months of storage period, or 9,000 gallons, as found by the formula above. A circular cistern 8' deep and 10' in diameter with a capacity of 4,700 gallons or a rectangular one 8' square and 10' deep (Table 14) with a capacity of 4,888 gallons (Table 15) should be ample when the water is collected.

APPENDICES

TABLE 14.—CAPACITY OF PLAIN CYLINDRICAL CISTERNS AND TANKS

Capacity of cistern with diameter indicated

Depth of cistern ft.	6 feet gals.	6 feet bbls.	7 feet gals.	7 feet bbls.	8 feet gals.	8 feet bbls.[1]	9 feet gals.	9 feet bbls.	10 feet gals.	10 feet bbls.	11 feet gals.	11 feet bbls.	12 feet gals.	12 feet bbls.
6	....	..	....	..	2,256	72	2,855	91	3,525	112	4,265	135	5,076	161
7	....	..	2,015	64	2,632	84	3,331	106	4,113	131	4,976	158	5,922	188
8	....	..	2,303	73	3,008	96	3,807	121	4,700	149	5,687	181	6,768	215
9	1,904	60	2,591	82	3,384	107	4,283	136	5,288	168	6,398	203	7,614	242
10	2,115	67	2,879	91	3,760	119	4,759	151	5,875	187	7,109	226	8,460	269
11	2,327	74	3,167	101	4,132	131	5,235	166	6,463	205	7,820	248	9,306	295
12	2,537	81	3,455	110	4,512	143	5,711	181	7,050	224	8,531	271	10,152	322

[1] One barrel equals 31½ gallons.

TABLE 15.—CAPACITY OF RECTANGULAR CISTERNS

Capacity of cistern with cross-section indicated

Depth of cistern ft.	6 x 6 feet gals.	6 x 6 feet bbls.	8 x 8 feet gals.	8 x 8 feet bbls.[1]	10 x 10 feet gals.	10 x 10 feet bbls.
6	....	....	2,880	91	4,500	143
7	1,890	60	3,360	106	5,250	166
8	2,160	68	3,840	122	6,000	190
9	2,430	77	4,320	137	6,750	214
10	2,700	85	4,800	152	7,500	238
11	2,970	94	5,280	167	8,250	262
12	3,240	102	5,760	183	9,000	286

PRECAUTIONS AGAINST A WET CELLAR

Methods of preventing a wet cellar are shown in the accompanying diagram. In excavating for a basement, a trench should be cut far enough back into the earth walls to permit laying a line of 4″ tile all around the foundation footing. If the basement floor is concrete it should be laid to slope toward a floor drain which may be connected with the tile line around the footing.

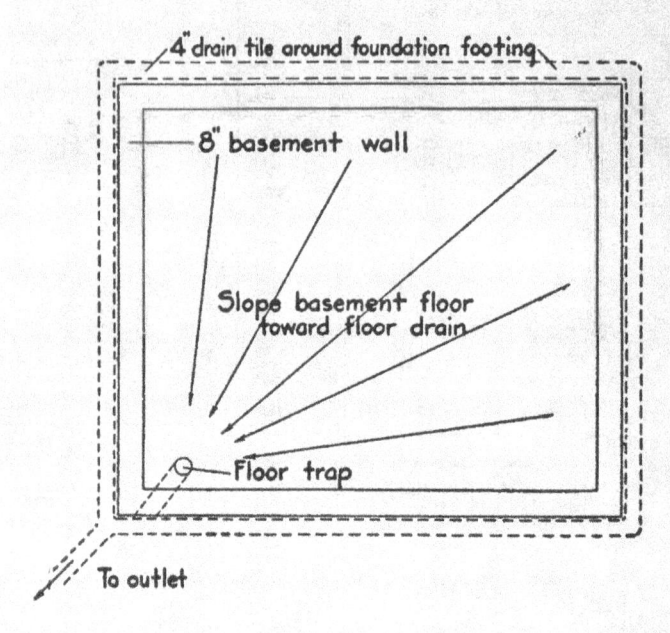

Fig. 6. Precaution Against Wet Cellar.

In order to keep water away from the foundation wall it is important that the rain water from the roof be conducted away from the house. The downspout should be connected to a sewer-tile drain, with cemented joints, leading to a convenient outlet. Earth should also be piled up slightly around the house so surface drainage will be away from the building. If the foundation wall is made of stone, concrete block, tile or other masonry these should be laid in 1:3 Portland cement mortar, and all joints thoroughly filled. For a concrete foundation wall, not more than 5 gallons of mixing water should be added for each sack of cement used, and the sand and stone should be proportioned to give the best workability.

TABLE 16.—VEGETABLE PLANTING CHART

When growing vegetables for sale it is important to know the amounts of seed to sow to the 100' of row and to the acre; also the average yields. These are shown herewith. (See Chapter 40.)

Vegetable	Seed 100 Ft. Row	Seed for Acre	Plants Per Oz. Seed	Space Between Rows (inches)	Space in Rows (inches)	Days from Seeding to Maturity	Expected Crop 100 Ft.
Asparagus seedings	1 oz.	3 lbs.	500	14-24	4-6	(4 yrs.)	300 lbs.
Asparagus plants				48	36	(3 yrs.)	300 lbs.
Beans, bush	2 lbs.	60 lbs.		30-36	2-3	50-75	5 bu.
Beans, pole	1 lb.	30 lbs.		40-48	6-8	60-75	5 bu.
Brussels Sprouts		5 oz.	4000	24-36	16-22	100	
Beet	2 oz.	8 lbs.		14-24	1-3	60-70	2½ bu.
Cabbage		4 oz.	5000	24-36	16-22	100-140	90-150 lbs.
Chinese cabbage	1 oz.		2000			60-70	85 heads
Carrot	1 oz.	3 lbs.		16-24	1-3	70-90	2 bu.
Cauliflower		5 oz.	4000	24-30	20-24	75-90	75 heads
Celery		5 oz.	8000	24-40	4-6	120-130	150 stalks
Chard	1 oz.			30-36	10-12	50-60	10 bu.
Sweet corn	1 lb.	12 lbs.		34-42	5-6	75-95	12 doz.
Cucumber	1 oz.	3 lbs.		48-60	36-48	60-70	2 bu.
Eggplant		8 oz.	2000	24-30	18-24	80-90	200 fruits
Kale	1 oz.	5 lbs.		24-32	18-24	55-65	3 bu.
Lettuce	½ oz.	4 lbs.		12-18	4-8	65-85	4 bu.
Lettuce (Cos)	½ oz.	4 lbs.		12-18	4-8	60-70	100 heads
Cantaloupe	1 oz.	3 lbs.		70-80	48-60	80-95	60 fruits
Watermelon	2 oz.	4 lbs.		90-100	72	80-90	40 fruits
Mustard	1 oz.	4 lbs.		14-24	4-6	40-50	3 bu.
Okra	2 oz.	10 lbs.		24-40	12-18	50-60	
Onion	1 oz.	5 lbs.		18-24	3-4	100-140	2 bu.
Onion (sets)	2 lbs.	8 bu.		18-24	3-4	80-100	2 bu.
Parsnips	1 oz.	4 lbs.		18-24	3-4	200	2 bu.
Parsley	1 oz.	4 lbs.		12-20	4-6	50	200 plants
Peas	1½ lbs.	150 lbs.		24-36	1-2	60-80	1 bu.
Pepper		2 lbs.	1000	20-30	18-20	115-130	2 bu.
Pumpkin	5 oz.	4 lbs.		95-110	60-80	120	50 fruits
Radish	1 oz.	4 lbs.	12-18	12-18	1-2	25-30	100 bunches
Spinach	1 oz.	8 lbs.		18-24	3-5	40-50	2 bu.
Spinach (New Zealand)	1 oz.	8 lbs.		36-48	18-24	50-60	2 bu.
Squash (bush)	2 oz.	5 lbs.		42-48	42-48	55-65	200 fruits
Squash (vining)	4 oz.	3 lbs.		70-90	60-90	90-110	80 fruits
Tomato		4 oz.	2000	40-60	36-40	110-130	10 bu.
Turnip	1 oz.	2 lbs.		12-20	2-4	45-60	2 bu.

POWER FROM A STREAM

Many a small stream, if harnessed, would furnish power enough to generate electricity for lights and small household appliances. Factors which determine the power a stream can develop, writes W. H. Sheldon,* are height of dam, and quantity of water flow.

* Quarterly Bulletin of the Michigan Experiment Station (Vol. XIV, No. 1)

APPENDICES

The height is limited by the height of the banks. It may be measured by using a carpenter's level and a rule. The quantity of water flowing may be determined by means of a weir dam (Fig. 7).

Fig. 7. Weir for Measuring Flow of Water.

The weir consists of a rectangular notch cut in a temporary dam across the stream. The notch should be centered and its width not more than ⅔ that of the dam, but sufficiently wide, so the water flowing over the lip will be not more than 10" nor less than 1" deep. This depth of water should not be more than ½ the drop outside. The edges of the notch should be beveled on the sides and the bottom (Fig. 8) to afford as little resistance as possible to the flow. It is essential that the dam be high enough so the water will be quiet before reaching the weir and that the weir be wide enough so the flow will not be too rapid.

As the water surface above the dam slopes toward the notch the depth of the water flowing over the weir must be measured up-stream. To do this set a stake (A) 4' back of the weir with its top exactly level with the lip of the weir. An ordinary rule (B) set on this stake will measure the depth of water flowing over the lip. The depth of the water being known the volume of the flow is determined from the formula $Q = 3 \times w \times d\sqrt{d}$. Q equals

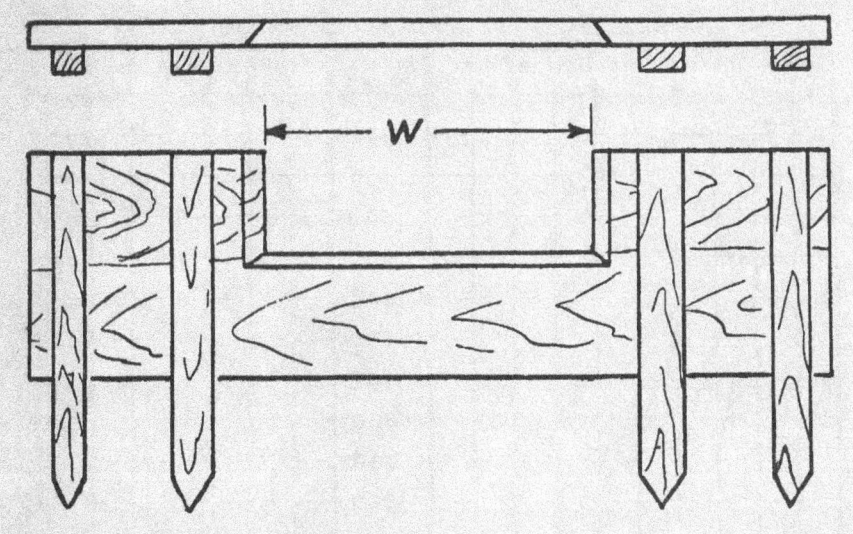

Fig. 8. Details of Water-measuring Weir.

discharge in gallons a minute; w, width of notch in inches; d, depth in inches flowing over weir.

Having determined the gallons a minute and the available head in feet, the power that may be developed is: $H.P. = \dfrac{Q \times H}{4,000}$. Q equals gallons a minute and H, available head in feet.

The width of the weir, depth of water flowing over it and the available head being known the horse-power may be read directly from the graph (Fig. 9). Example: If a weir notch is 25" wide and 4" of water flows over it what power will the stream develop with a 7' dam? By the formula:

$Q = 3 \times 25 \times 4 \sqrt{4} = 600$ gallons a minute $H.P. = \dfrac{600 \text{ gal.} \times 7 \text{ ft.}}{4,000} = 105$ H.P.

Turning to the graph (Fig. 9) follow the curve near the left side marked 25" weir to intersection of vertical line under 4" Water Depth. On a horizontal to the right of this intersection, find the quantity 600 gallons a minute in the center column. Follow the same horizontal line to the right until it intersects the vertical line under 7' of head which is just to the right of the curve marked 1 H.P. This indicates that the stream will develop slightly more than 1 H.P.

1. With a given depth the amount of water flowing over the weir is proportional to the length of the weir; 2. With a given fall the power a stream will develop is proportional to the quantity

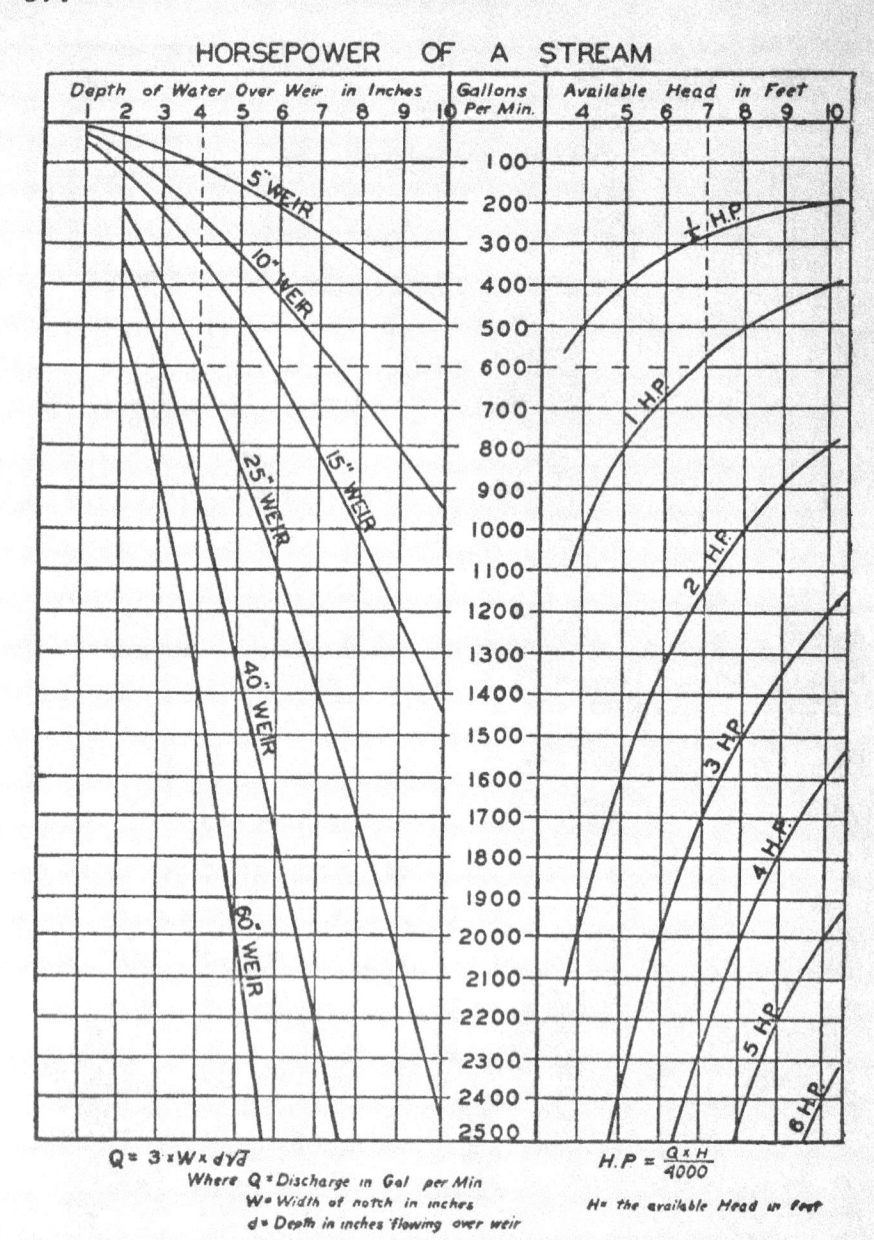

Fig. 9. Measuring Horsepower of a Stream.

of water available. 3. With a given quantity of water, the power a stream will develop is proportional to the fall used.

Streams that develop four horse-power or more will operate a lighting plant without a storage battery by using a turbine with a generator which will deliver a constant voltage. Smaller streams which develop at least one horse-power will operate battery-charging light plants satisfactorily, but those which develop less than one horse-power are better for operating hydraulic rams for pumping water. It is usually not advisable to attempt to harness a stream unless the dam can be built at least 4' high because a special, large capacity turbine or undershot wheel would be required to develop even a small amount of power.

After determining that a stream will develop ample power, the expense of harnessing it should be carefully considered; for instance, the location of the dam site should be reasonably near the buildings where the electricity will be used in order to avoid the necessity of a long transmission line. If the banks are widely separated, a long dam will be required and a large area will be inundated, or if the banks are very porous a dam that will hold back the water will be costly to build.

FARM EVALUATION

Farms cannot be judged by score cards which give percentages or "weights" to desirable and undesirable features; because, for instance, a farm that might score perfect on every item except water supply would be worthless if there were no water or if the water were unfit for human and animal consumption. Hence, assigning weights or percentages might lead to erroneous conclusions. With the subject matter of Chapters 6, 7, 8, and 9, especially, in mind, a better method will be to examine all the features of two or more farms and then view the matter in a broad, common sense way, making sure to cover the main points summarized as follows:

Location of farm. Owner. Address. Distance to shipping station, trading center. Condition of highways, in spring, in winter. Distance to schools, churches, nearest neighbors. Is telephone available? R. F. D.? Electric current for lighting and power. Total area of farm. Acres in crops; that can be used for crops; in pasture; in woods; in waste land; in roads, buildings, lots, swamps, lakes, etc.; in stump or brush land. Kinds of timber. Ease or difficulty of getting out timber or wood. Topography as regards economy of cultivation, irrigation, danger from erosion, flooding.

TABLE 17.—NUMBER OF PLANTS REQUIRED TO SET AN ACRE OF GROUND WHEN PLANTED AT GIVEN DISTANCES.

To determine the number of plants to set at distances other than shown, multiply the two distances apart and divide 43,560 sq. ft. (to the acre) by the answer. For instance, to set a peach orchard with trees 20' apart each way multiply 20 x 20 and get 400 sq. ft. Divide this into 43,560 and get the answer 108.9. (See various chapters on planting.)

Inches	Plants	Inches	Plants
2 × 8	392,040	15 × 42	9,953
2 × 10	313,632	15 × 48	8,712
2 × 12	261,360	18 × 30	11,616
3 × 8	261,360	18 × 36	9,680
3 × 10	209,088	18 × 42	8,297
3 × 12	174,240	18 × 48	7,260
4 × 8	196,020	20 × 36	8,712
4 × 10	156,816	20 × 42	7,467
4 × 12	130,680	20 × 48	7,534
5 × 12	104,544	24 × 36	7,260
6 × 6	174,240	24 × 48	5,445
6 × 8	130,680	24 × 60	4,356
6 × 10	104,544	24 × 72	3,630
6 × 12	87,120	24 × 96	2,722
8 × 10	78,408	Feet	
8 × 12	65,340	3 × 3	4,840
10 × 10	62,726	3 × 4	3,630
10 × 12	52,272	3 × 5	2,905
10 × 15	41,817	3 × 6	2,420
10 × 18	34,848	3 × 8	1,815
10 × 24	26,132	4 × 4	2,722
10 × 30	20,908	4 × 5	2,178
10 × 36	17,424	4 × 6	1,815
12 × 24	21,780	4 × 7	1,556
12 × 30	17,424	4 × 8	1,361
12 × 36	14,520	5 × 5	1,742
12 × 42	12,446	5 × 6	1,457
12 × 48	10,890	5 × 8	1,089
15 × 30	13,939	6 × 6	1,210

TABLE 18.—VEGETABLE STORAGE CHART

In ordinary storage of vegetables it is important to know what degrees of temperature and humidity are most favorable, also the length of time each species can be held in good condition. This table specifies these items for the most important crops and makes cogent comments on methods of handling. (See Chapter 50.)

Vegetable	Temperature (Degrees Fahr.)	Humidity	Storage Period	Remarks
Beets	34-38	Medium	Nov. to Apr.	Dark place. Keep best in sand.
Brussels Sprouts	34-38	Medium	Nov to Dec	Dig. Plant close together in sand, sprouts unpicked.
Cabbage	34-38	Medium	Nov to Apr	Sound, hard heads only. Demands good ventilation. Ball head stores best
Carrots	34-38	Medium	Nov to Apr	Dark place Keep best in sand
Celery	34-38	Medium	Nov to Feb	Difficult; dig with roots, stand upright in moist sand. Darken 3 weeks before use.
Onions	34-38	Dry	Sept. to Apr	Shallow crates or on slat shelves. Free ventilation. Dry before storing
Potatoes	34-38	Medium	Nov to Apr	Dark Pile in bulk Good ventilation.
Turnips	34-38	Medium	Nov to Apr	Pile in bulk. Avoid bruising
Pumpkins and Squash	50-55	Dry	Nov to Apr.	Avoid bruising.
Sweetpotatoes	50-55	Dry	Nov to Apr	Cure at 70° for 3 weeks before storing Discard all bruised.

Natural fertility as evidenced by kind and character of forest and vegetative (even weed) growth. Present fertility as evidenced by growth of crops. Physical condition of the soil, adaptability to legumes, other kinds of crops. Natural drainage. Artificial drainage. Depth of soil. Kind of surface soil. Kind of subsoil. Water supply, source, quantity in dry summer months; winter months; cost of upkeep; supply in pastures. Buildings as suited to kind of farming to be followed. Adaptability to some other type; cost of upkeep; arrangement for economy of work. Desirability of dwelling as a home. Condition of fences, kind as regards cost of upkeep. Farm roads. Shape of fields. Nearness to farmstead. Kind of orchards, condition. Adequacy of trees for home use. Climate, as to growing season, days available for farm work, healthfulness. Neighborhood, character of people, available labor supply. Possibility of increase or decrease of land value. Possibility of selling. Possibility of renting. Desirability of farm as a strictly business investment. Desirability of farm as a home place to retire to. Adaptability of farm to changing economic conditions necessitating change of type; to enlargement of business; to diversification or improved organization of the business; to high yields of crops and desirability of livestock production. Sureness of market for major crops grown. History of farm as regards management of land with respect to keeping up fertility. History of region as to development and speculation in lands as affecting present price. Number of other well developed, successful farms in immediate vicinity. How long have they been farmed? What are some of the operators' difficulties? How soon can the farm be made a going concern? Are taxes on this farm reasonable? Can the present owner transfer a sound title?

VALUES OF FARM MACHINES AND TOOLS *

The value of an implement, tool, or machine may be found by using the original cost, estimating the life of the machine and the probable residual junk value. Each machine, tool or implement depreciates at a different rate. Depreciation also depends upon the care given and the repairs made. Hence it is impossible to state what rate should be applied in any specific case. However, Tables 19 and 20 and the graph shown in Figure 10 may help in estimating depreciation for various types of equipment. For general inventory uses a uniform rate of 10% will answer perhaps as well as anything. It is an easy rate to work with. Where greater accuracy is necessary, tables that apply to each machine should be used. These may be obtained through the manufacturers or other organizations interested in each kind.

Most machines ("assets") depreciate or become obsolete. Obsolescence arrives when a machine is out of date or a more effective new type is introduced. The average life of most farm implements is shown in Table 18. Depreciation applies to decrease

* Quarterly Bulletin, Michigan Experiment Station, Vol. XIII, No. 4.

TABLE 19.—AVERAGE LENGTH OF LIFE OF FARM IMPLEMENTS *

Implement	Average Years of Life	Average Acres Covered Per Year
Walking plow	11.7	32.9
Sulky plow	8.1	30.9
Springtooth harrow	11.0	71.1
Spiketooth harrow	14.0	46.3
Disc harrow	13.0	35.2
Land roller	16.0	65.9
Grain drill	16.4	46.3
Corn planter:		
1-row	11.7	4.1
2-row	14.0	16.9
Cultivator:		
1-row	14.0	16.9
2-row	12.5	39.3
Cabbage transplanter	12.8	12.
Mower	14.8	28.0
Hay rake	14.5	43.0
Hay tedder	14.0	21.6
Bean harvester	12.9	16.9
Grainbinder	15.4	35.2
Corn binder	10.8	21.1

Farmers' Bulletin 1182, U. S. D. A.

in value due to wear and tear. It is usually a more important factor than obsolescence. Hence attention is centered on it, but assets should also be reduced in value proportionately as they become obsolete.

Of various methods of calculating, the "Straight Line" (Table 20) is probably the simplest. By it the rate of depreciation is de-

TABLE 20.—STRAIGHT LINE METHOD OF CALCUATING DEPRECIATION

Years in Use	Rate of Depreciation	Total Depreciation Cumulatived	Present Value
1	9%	90	910
2	9%	180	820
3	9%	270	730
4	9%	360	640
5	9%	450	550
6	9%	540	460
7	9%	630	370
8	9%	720	280
9	9%	810	190
10	9%	900	100

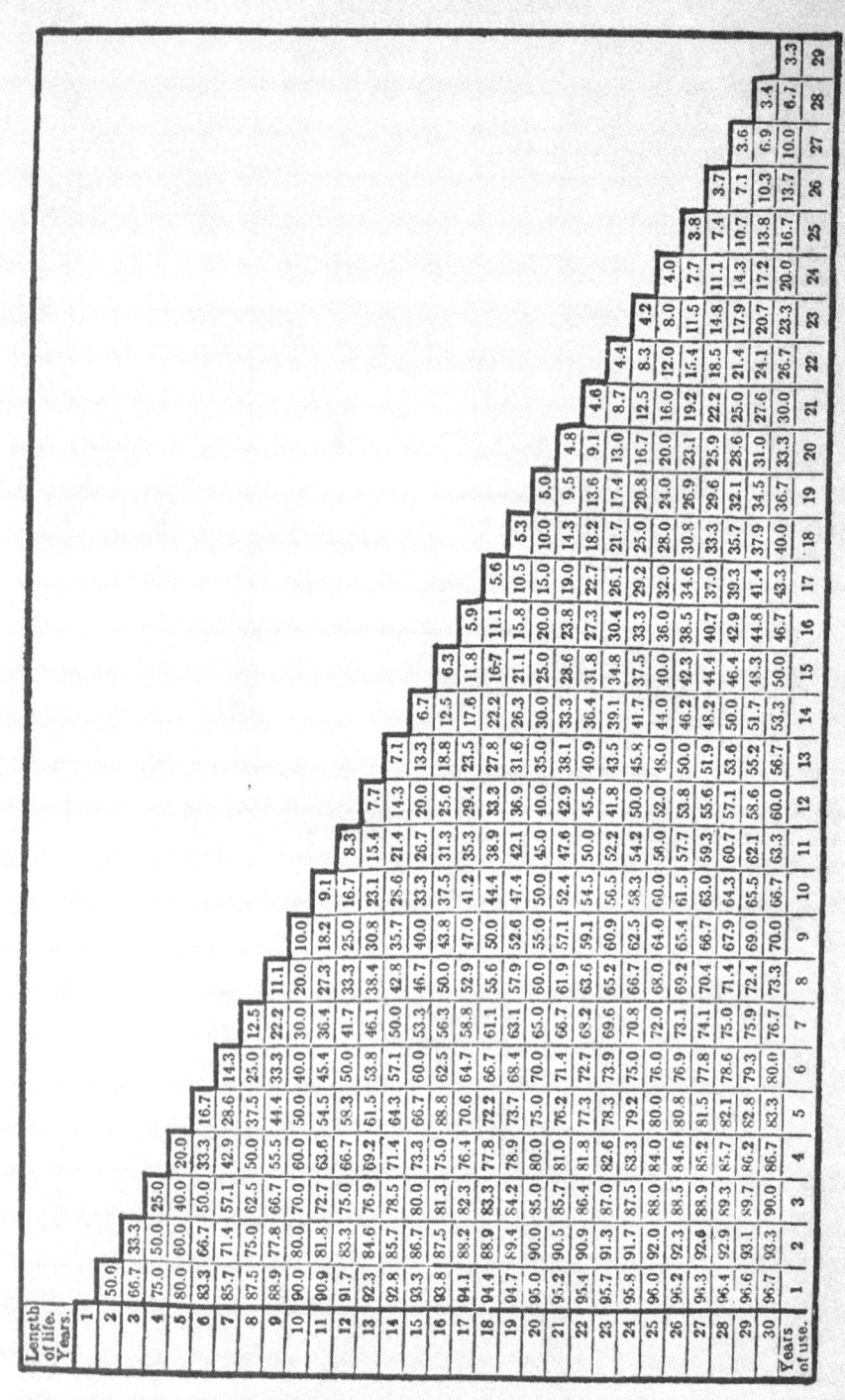

Fig. 10. Longevity and Values of Tools. (See text.)

termined by dividing the "total depreciation" by the estimated life of the asset. Total depreciation is obtained by subtracting the junk value from the original cost. If the original cost is $1,000, the estimated life 10 years and the junk value $100., the total depreciation would be $900. and the rate of depreciation 9% per annum. Hence the value of the asset at the end of the first year would be $910., at the end of the second, $820., and so on.

In general, the useful life of farm implements is governed by such factors as design, quality, construction, climate, soil, topography, use, and character of maintenance. In every case it is necessary to consider each factor in appraising. Exact rules are impossible.

Figure 10 will be useful in estimating implement values. To use it find the probable length of life figure in the left hand column, then on the bottom line, the number of years used. Where the line of figures meets the column above the "years used" figure will be the per cent of original cost. For instance, to find the present value of a $10. machine whose probable life is 15 years but which has been used 7 years, find 15 in the left-hand column and 7 in the bottom row, follow the line and the column of figures to where they meet at 53.3. This signifies that the value of the implement at the time is 5.33% of its original cost; namely, $5.33.

When accurate cost figures on repairs are known, the value of machines or tools can be indicated as follows: A machine that costs $1,000. and which has an estimated life of 10 years represents a yearly overhead cost of $193. This figure is found in the graph (Fig. 11) by locating in the left margin the value $1.000. initial investment, tracing the dotted line horizontally from this point until it reaches the oblique "life line," then following the vertical dotted line to the base line where the yearly overhead cost will be found; namely, $193. This cost includes the amount of estimated repairs (4% of the initial investment). Therefore, $193. minus $40. (4% of $1,000. initial cost of machine) equals $153. yearly overhead cost, less the estimated repairs. Hence, $153. plus the *actual* repairs for any one year equals the total yearly overhead cost.

Most farm machinery costs less than $200: tools only a few dollars—often only cents. To interpret overhead costs accurately from the graph on machines costing less than $200. multiply the initial cost by 10 and follow the chart with a machine costing, say, $130. whose estimated life is 15 years thus: $130. x 10 = $1,300. Trace the $1,300. investment line on the graph to the 15-year oblique "life line" and drop a line from this point to the base line, where the amount $206.25 will be found. But this sum

APPENDICES

is 10 times too much, so it must be divided by 10 to show the annual estimated overhead cost; namely, $20.62.

The overhead cost of tools that cost less than $20. may be made by multiplying by 100 and, after making the calculation as already explained for the $130. machine, dividing by 100.

Having determined the yearly overhead costs from the chart, follow the line upward to the "hours a year" line. Then follow a line from the intersection of these lines horizontally to the right hand margin and there read the value of *hourly* overhead costs.

The same answer can be obtained by dividing the yearly overhead cost by the estimated number of hours the machine is to be used during the year.

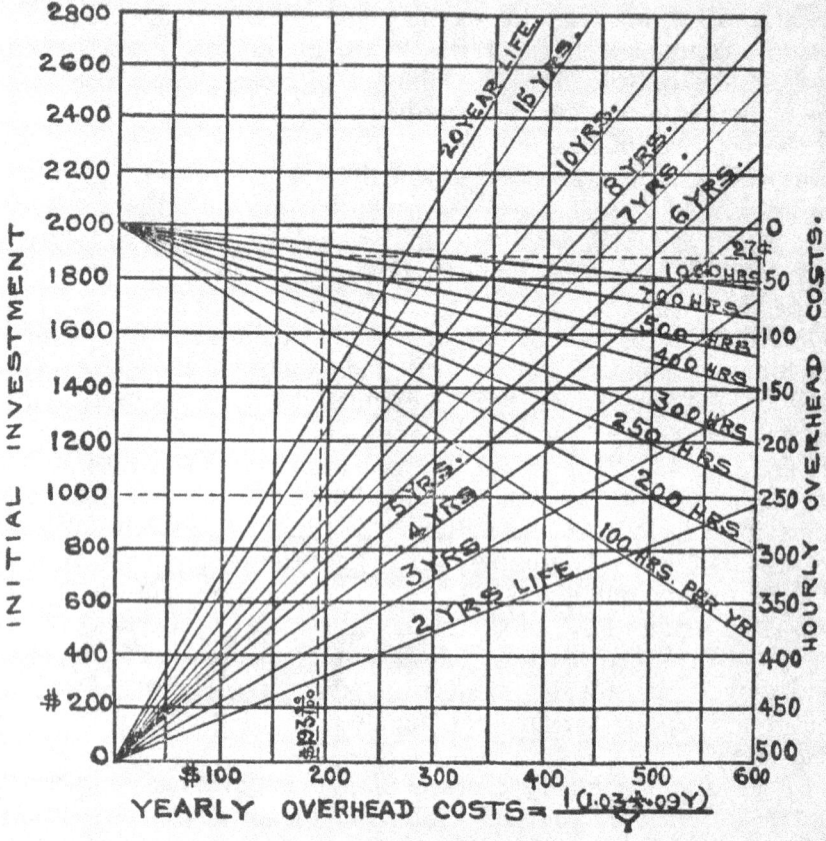

Fig. 11. Over-head Tool Costs.

An idea of the importance of keeping the soil covered by some crop will be apparent from study of the figures in Table 20 and will emphasize the necessity of taking proper care of the land, especially where there are slopes. See Chapter 26.

APPENDICES

TABLE 20.—PER CENT OF RAINFALL IN RUN-OFF, AND RATE SOIL IS LOST BY SHEET EROSION FROM LAND UNDER DIFFERENT SYSTEMS OF MANAGEMENT [1]

Treatment	Per cent of rainfall in run-off	Tons of soil eroded per A. in six years	Years necessary to erode 7 inches of soil
Not cultivated—no crop	48.9	207	29
Plowed 4 ins. deep. Fallowed. No crop	31.3	247	24
Plowed 8 ins. deep. Fallowed. No crop	28.4	214	28
Corn annually	27.4	106	56
Wheat annually	25.2	39	150
Rotation—corn, wheat, clover	14.1	13	437
Sod—blue grass	11.5	1.7	2,547

[1] Duley, F. L., and Miller, M. F. Erosion and surface run-off under different soil conditions. Mo. Agr. Expt. Sta. Res. Bul. 63:1–50. 1923.

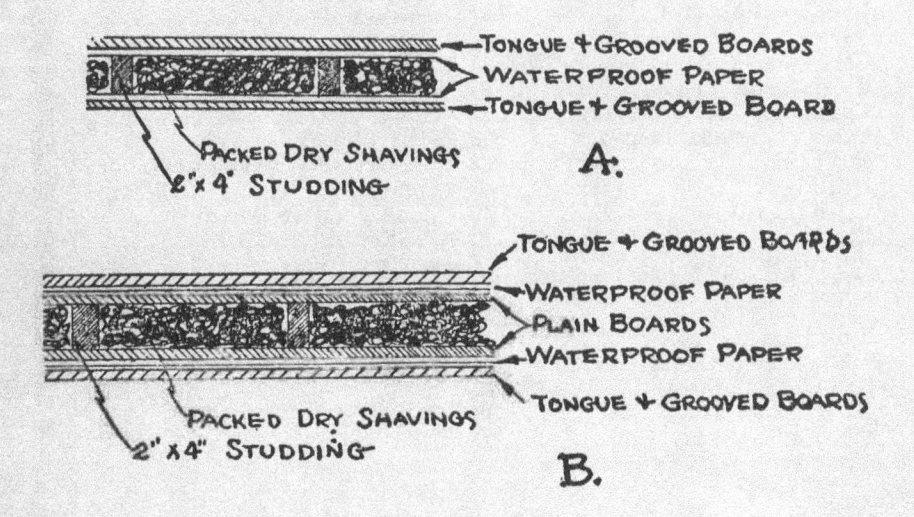

Fig. 12. Details of Basement Walls.

A. Where only a small amount of protection is necessary this type of wall will be satisfactory. B. For greater protection against both heat and cold this form is advisable, either below or above ground.

COLD STORAGE AND FREEZING TEMPERATURES

The best temperatures for the storage of any product depend somewhat upon its condition when stored, length of time to be stored, condition of air circulation, and humidity in the storage room, etc., but the temperature given in the following table for the various products and articles named are bound to be about the average of the best present practice in cold storage work:

TABLE 21.—COLD STORAGE AND FREEZING TEMPERATURES FOR VARIOUS PRODUCTS

Products	Deg. F.	Products	Deg. F.
Apples	30	Grapes	36
Asparagus	33	Hams (not brined)	20
Berries, fresh (few days only)	40	Hogs	30
Bulbs	34	Ice cream (few days only)	15
Butter	14	Maple sugar	45
Cabbage	31	Maple syrup	45
Canned fruits	40	Meats, fresh (10 to 30 days)	30
Canned meats	40	Meats, fresh (few days only)	35
Carrots	33	Meats, salt (after curing)	43
Celery	32	Milk (short carry)	35
Cheese (long carry)	35	Nursery stock	30
Cider	32	Onions	32
Corn (dried)	45	Parsnips	32
Cream (short carry)	33	Peaches (short carry)	50
Cucumbers	38	Pears	33
Currants (few days only)	32	Peas (dried)	45
Cut roses	36	Plums (one to two months)	32
Dried beef	40	Potatoes	34
Dried fruits	40	Poultry (after frozen)	10
Eggs	30	Poultry, dressed (iced)	30
Fish, fresh water (after frozen)	18	Poultry (short carry)	28
Fish, not frozen (short carry)	28	Poultry (to freeze)	0
Fish (to freeze)	55	Sauerkraut	38
Fruit trees	30	Strained honey	45
Fur and fabric room	28	Sugar	45
Game (after frozen)	10	Syrup	45
Game (short carry)	28	Tomatoes (ripe)	42
Game to freeze	0		

When a product, particularly a perishable one, is stored or frozen under artificial refrigeration much depends upon its condition when stored, the humidity of the storage room, the air circulation and the length of time to be stored, but most of all upon the temperature. This naturally varies widely with the product. The temperatures given in Table 21 have been found to be about the average of the best practice in cold storages throughout the country.

APPENDICES 385

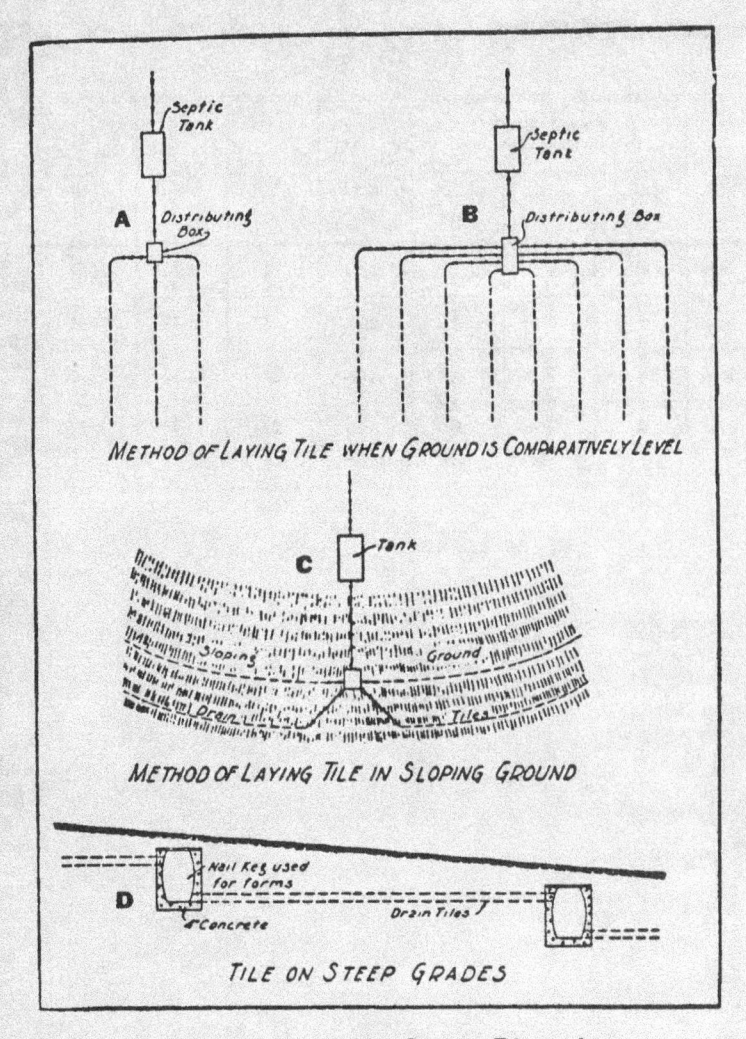

Fig. 13. Lay-outs for Sewage Disposal.

A and B, arrangements on fairly level ground. C, On somewhat sloping land. D, On a steep slope. (See Chapter 16.)

TABLE 22.—QUANTITIES OF MATERIAL NEEDED TO CONSTRUCT SEPTIC TANKS

No. of Persons	Size of Tank (Inside measurements)			Cu. Yd. of Concrete	Sacks of Cement	Cu. Yd. of Sand	Cu. Yd. of Stone
	Length Feet	Width Feet	Depth Feet				
6 or less	6	3	5	2.49	18	1.24	1.86
8 or less	7	3	5	2.77	20	1.42	2.13
10 or less	8	3	5	3.06	22	1.57	2.35
12 or less	8	3½	5	3.25	23	1.70	2.55

INDEX

Note: Books quoted in italic type; authors in Roman

Absentee farming, 11
Accounts. Farm, 63
" Interpretation of, 64
Acidity, Tests for, 172, 173
Acme harrow, 185
Acre costs, 55
" How much money per, 54
Adaptability of farm, 48
Advertising plants for sale, 261
Advice, Disinterested, 48
Agreement to buy farm, 48
Agriculture, Chemistry of, 146
Agriculture, 82
Agriculture, Soil Fertility and Permanent, 154
Alabama corn growing, 16
Alderney cow, 113
Alkali, 171
Allen, W. Lee, 267
Alterations, 20
Alum for de-coloring water, 69, 70
Amendments, 171
American Fruits, 223
American Peach Orchard, 105
Ammonium salts acidify soils, 156
Amortization of note, 60
Analysis of Leaves, 149
Anderson, M. S., 151
Aphis, Recognition of, 336
Appendices:
   Choice varieties of fruit, 345-348
   Data for garden planting, 349-350
   Distances for fruit plants, 345
   Early hardy vegetable garden, 351
   Fertilizer applications for fruit trees, 354
   Garden insect pests and controls, 355-356
   Garden tools and accessories, 353-354
   Long row garden varieties, 343
   Vegetable-maturity table, 348
   Vegetables that withstand frost, 352
   " tender to frost, 351
Apple costs, 56
" fruit buds, 217
" harvest, 330
" neglected trees, 209
" storage, 320, 329
Apple, The, 166
" trees, 207
Apples, Faulty varieties of, 311
" leading, 308
" number of varieties of, 209
Appraisal, 49
Apricot, Fruit buds of, 218
Arbors for grapes, 284
Arid Regions, Fruit Growing in, 232
Arizona Experiment Station quoted, 124

Ashes, Wood, handled like lime, 174
" " source of potash, 161
Asparagus, When to expect, 53
Aspect defined, 27
Assets, 64
Attachments, Wheelhoe, 192
Augur, Soil, 19
Author, Experience of, 2
Averages of production, 37
Ayrshire cow, 113

Back-filling, 94
Bacteria, Identifying work of, 335
Bailey, L. H., 243
Baldwin apple growing, 17
Bankruptcy, 57
Banks favored by borrowers, 61
Barometer readings, Significance of, 109
Barrel sprayer (Fig.), 337
Barry, Patrick, 88
Bearing age trees, 234
Beds, Outdoor nursery, 251
Bee-Keeping, Productive, 125
Bees (Chap.), 125
" How many, to keep, 127
Beeswax, 125
Beet storage, 328
Bennett, Ida D., 253
Berries and vegetables together, 294
" in orchards, 298
Berry boxes for transplanting, 258
" bush planting, 294
" canes, Pinching, 295
" patch cleanliness, 331
" patches developed without cost, 54
" plant diseases, 297
" " insects, 297
" " pruning, 295, 296
" plants needed for family, 298
" row distances, 293
" yields, 298
Bills, Discounted, 62
" of lading, 62
Bi-Monthly Bulletin, 322
Blackberries, 292
" When to expect, 53
Blackberry suckers, 294
Blasting, Costly, 26
Bleaching to remove water stains, 67
Blight fighting, Fire, 207
Blights, Control of, 331
Blue-prints of septic tanks, 78
Body-blight, 227
Bonds, 62
Bones for grapes, 282
Borrowing, 57
" for production, 57
" money safely, 58

387

INDEX

Boswell and White, 157
Braces, Living, in trees, 238
Break-down of trees, 218
Breeding, Corn, 38
" for eggs, 122
" Potato, 39
" Strawberry, 40
Bridge-grafting, 228; (Figs.), 228, 230
Broadcasting seed, 250
Brooding, Artificial, 119
Brussels sprouts, 329
Buckwheat, hulls and straw as mulches, 84
Budget, 61
Buds, fruit, Where to find, 217
Bud sport, defined, 40
Bugs, Control of, 331
Buildings poorly arranged, 19
Burlap, Wet, over plants, 260
Bush and cane fruits (Chap.) 291
Bush-Fruits, 291
Business, Volume of, 21, 65
Buy, When to, 50
Buying, Hasty, 48
" Legal features of, 48
" vs. renting (Chap.), 47
" smaller place than needed, 50

Cabbage storage, 320, 327
Cables, Thermal, 143
Caking of fertilizers prevented, 165
Cane fruits, Bush and (Chap.), 291
Canning, Advantages of, 266
" Vegetable, 242
Canopy grape training system, 288
Cantaloupe, a gamble crop, 240
Canvas hose irrigation, 103
Capillary water defined, 83
Capital (Chap.), 52
" deciding factor, 50
" How much, 52
" must be worked, 21
" needed, 55
" needs investment, 56
" on acre basis, 54
" origin of, 57
" reserve needed, 53
" small places need high proportion, 54
" working
" " and investment, 7
Card, F. W., 291
Carrot, Wild, indicates mismanaged soil, 26
Cash crops, 7
Cashier as confidant, 61
" Consult bank, 50
Cash *vs.* inventory, 64
Cat, 115
Catch basin (Fig.), 95
Cauliflower, a gamble crop, 240
" storage, 329
Caustics for insects, 337
Celery storage, 329
Cellar cooler than ground, 327
" floor, Water-tight, 322
" storage, 322; (Figs.), 320, 321
Cement Assoc., Portland, 78
" mixing, 79
Census, averages of milk production, 37
Chemistry of Agriculture, 146
Cherry fruit buds, 218

Cherry trees, neglected, 208
Cherries, When to expect, 53
Chewing insect work, 336
Chick, Clean, campaign, 123, 124
Chicks, Day-old, 119
" pay first season, 56
" Tenant's, 47
Choosing a farm, 7
Cigar boxes for seedage, 249
Cistern cleaning, 70
" Concrete (Fig.), 51
" Filter, 66; (Fig.), 59
" Measuring water in, 70
" treatment, 69
Cisterns, 68
Citrus fruits, When to expect, 53
Clay, Characteristics of, 147
Clay soils conserve food, 155
Cleanliness of orchard, 331
Climatic region, Familiar, 22
Clipping leaves for transplanting, 259
Clover sod for strawberries, 272
" Sweet, as green manure, 127, 170
Cluster buds, 217
Coldframes and hotbeds (Chap.), 136
Collar-rot, 227
Colloids defined, 151
Colony houses, 119
Color as index of soil value, 149
Color, Removing, from water, 69
Combination tools, 255
Comfort from wind-break, 32, 33
Comin, Donald, 322
Companion cropping, 181; (Fig.), 181
Compatible sprays, 338
Competing regions, 63
Complete fertilizer defined, 164
Compost (Chap.), 175
Comstock, J. H., 126
Conifers for wind-break, 35
Connecticut egg-laying contests, 15
Consultant advice, 48
Construction of tools, 201
Contracts, 62
Contract to buy farm, 48
Contests, egg-laying, 15
Coolidge, Calvin, 265
Cool-season vegetables, 42
Corbett, L. C., 136, 182
Corn breeding, 38
" growing record, 16
Cornell Experiment Station, 254
Corrosive sublimate antidotes, 209
Cost, Calculation of, 55
Cost of production, 21
Cottonseed meal, 159
Counselor, Advice of, 48
Cover crops, Green Manures and (Chap.), 166
Cow, grade *vs.* purebred, 113
" Requirements, 112
Crab apple trees, Neglected, 208
Credit, Powerful, 56
" rules, 58
" only advantage of, 57
Creosote for tree wounds, 212
Cross-fire, spraying, 339
Crop Production, Fertilizers and, 171
" Rotation, 178
Crop use of lime (Fig.), 73
Cropping, Companion, 181
" systems affect land, 153

INDEX

Cropping systems (Chap.), 177
" Marker, 180
Cropping, Partnership, 180
" Succession, 181
Crops as soil tell-tales, 25
" Special, 267
" tenants should plant, 47
" to avoid in orchards, 314
Crotch treatment, 216; (Fig.), 237
Cucumber, a gamble crop, 240
Cultivation, Deep, undesirable, 44
" Function of, 44
" Shallow, desirable, 44
" Time-saving (Fig.), 24
" When to start, 193
Cultivator, Norcross, 195
Curd of water, 67
Curing cistern, 69
Currants, 292, 293
Currants, When to expect, 53
" with grapes, 288
Currence, T. M., 143
Cutaway harrow, 184
Cutworms in strawberry bed, 272
Cyanamid, 162

Daisy indicates mismanaged soil, 26
Damping-off, cause and cure, 251
Day-old chicks, 119
" " " tenant's, 47
Debris, destruction of, 331
Debts may bankrupt borrower, 57
Decay follows stubs, 217
" in soil, Rate of, 156
" of heartwood (Fig.), 211
Decomposition of manure, 155
" " green manures, 168
Deed, 49, 62
Delaware County (N. Y.), farmer, 10
" (state) grape grower, 22
Deliquescent fertilizers, 165
Denver, Colorado, 247
Deodorizing water, 70
Deposit box, 62
Dewberries, 292
" When to expect, 53
Deuteronomy, 66
Dew defined, 105
Dew-point, 106
Dibble for pricking out, 259
Digging, Right and wrong, 187; (Fig.), 186
Digging *vs.* plowing, 189
Disbudding in tree building (Fig.), 220
Discounts, 62
Disease precautions, 331
Diseases of berry plants, 297
Disk harrow, 184
Distances between trees, 304
Distances for berry rows, 293
Ditch digging tools (Fig.), 89
Dog, 115
Dormant spray, 333
Dowell and Jesness, 63
Downing, Charles, 219
Drafts, Bank, 62
Drain digging (Figs.), 90, 91
" outlet, (Figs.), 86, 95
Drainage, 109
" (Chap.), 88
" effects (Fig.), 93
" essential, 19

Drainage, Loss of lime by, 173
" needed, 23
" systems (Fig.), 92
Drains, Depths of, 94
" Distances between, 94
Drawbacks, capitalized, 11
" more important than advantages, 26
Drawbacks of farm life, 1, 2
Dressings, Top, 164
Dried blood, 159
" fish, 159
Drill for seed sowing, 252
" marking, with rake, 252
" seed, 190
Drouth affects plant growth, 100
Duck raising undesirable, 117
Duggar, B. M., 37
Duley, F. L., 153
Dumb-waiter and refrigerator (Figs.), 13, 16
Dusting, Essentials of Spraying and (Chap.), 330
Dusting and Fumigating of Plants, Spraying, 330
Dusts to suffocate insects, 337

Earliness due to drainage, 89
Earnings, Net, 64
" to pay principal, 58
Ecclesiastes, 6
Edgewood, My Farm of, 11
Education, liberal, 6
Egg grading, 120
Egg-laying breeds, 119
" " contests, 15
" prices, when high, 120
" production enhanced, 120
" receipt in large markets, 124, 125
" yield, 15
Eggs, 116
" for profit, 54
Electricity, experiments in hotbeds, 143
" for hotbeds, 141
Electric hotbed heating (Fig.), 145
" pumping (Fig.), 8
Endive storage, 329
Erosion graphs (Fig.), 153
" Sheet, 152
Erwin and Haber, 99
Evaporation, checked by glass, 250
" " " mulch, 84
" " " wind-break, 32
" of water from soil (Fig.), 194
Excelsior weeder, 195
Expense control, 65
Experience dearly bought, 7
" of author, 2
Export Trade, The Farmer and The, 63

Factors for building locations, 20
" Income, 21
Fail, Ways to (Chap.), 7
Failure factors, 10
Fan grape training (Fig.), 289
Farm accounts, 63
Farm finance (Chap.), 56
Farm flocks of fowls, 120
Farm "on edge," 10

INDEX

Farm safe place to live, 4
Farm Sewage, 75
" size, Factor of, 21
" to choose, 18
Farmer's care of tools, 203
Farmer and the Export Trade, The, 63
Farming, Garden, 136, 182
Fat of the Land, 308
Feed saved by windbreak, 33
Feeding for eggs, 121
Feeds and Feeding, 112
Fence rows, Evils of, 29
Ferguson, F. L., 88
Fertility experiments, Orchard, 315
Fertility, Soil, and Permanent Agriculture, 154
Fertilizer distributors, 165
" efficiency, 165
" formula, 165
" kills fruit trees, 235
" laws, 166
" tests, 162; (Fig.), 163
Fertilizers, 177
Fertilizers and Crop Production, 171
Fertilizers, commercial, 159
" for strawberries, 273
" proprietary, 166
" synthetic, 166
" where to place, 165
Fertilizing by irrigation, 101
" for corn, 16
" for fruit, 17
Field mice poison station (Fig.), 312
Fields, Irregular and Little, 29
Figures Don't Lie (Chap.), 15
Filler trees, 304, 314
Filter cistern 68; (Fig.), 59
Finance, Farm (Chap.), 56
Findlay, Hugh, 267
Fire, Adjustment after, 64
Fires prevent frost, 110
Fire-blight, 209, 327
Firewood from neglected trees, 210
Firming soil around plants, 260
Fiscal year, 61
Fish, 116
Fisher and Fisk, 200
Flat, Sowing and planting in (Fig.), 251
Flat, taking plants from, 259
Flats, Standard, 249
Florists' Exchange, 262
Flowers for sale, 263
Foliage color index of plant needs, 213
Food for Plants, 159
Forecasts, Weather, 248
Foreclosure, 60, 61
Forks, Digging, 186
Foundations of Grape Training, 281
Fowls on farms, 120
Framework, Tree, building (Fig.), 220
Fraser, Samuel, 223
Freezing heaves soil, 111
" checked by wind-break, 33
Frost, Signs of approaching, 109
" Damage Prevention (Chap.), 105
" defined, 105
" influenced by water bodies, 108
" prediction, 108
Frosted plants, To save, 107, 111
Fruit aided by wind-break, 32
" and Vegetable Storage (Chap.), 317
Fruit Garden, Barry's, 88

Fruit Gardens, Small Farm (Chap.), 299
Fruit Growing in Arid Regions, 232
Fruit Tree Pruning (Chap.), 213
Fruit Trees, Grafting (Chap.), 223
" When to expect, 52, 301, 313
Fruits, culinary, 303
" for roadside stands, 309
" " the home, 311
" " sequence of, 303
" Succession of, 311
" Tree, Selection of (Chap.), 308
Fuel saved by wind-break, 33
Fullerton, E. L., 96
Fumigation of Plants, Spraying, Dusting and, 330
Fungicides wrongly used, 330
Fungus, Recognition of, 335

Gamble crops, 339
Garden cleansing, 331
" clubs, Growing plants for, 264
Garden Farming, 136, 182
Garden, How to Make a Vegetable, 96
Garden, Long row (Fig.), 43
Garden Pay, How to Make the, 239
Garden, The Vegetable, 253
Gardening for Profit, 7, 52
Gardening, Modern Guide to, 82, 97, 247, 258, 267
Gardening, Subtropical Vegetable, 196
Gardening, Vegetable, 175
Gases for insects, 337
Gate valve (Fig.), 67
Germination test, 245
Gilbert, E. H., 22
Girdled trees, Saving, 227, 229
Goats, 114
Goose raising undesirable, 117
Gooseberries, 292
" late planted, 293
" When to expect, 53
" with grapes, 288
Grafting, 225
Grafting Fruit Trees (Chap.), 223
" Successful, 226
" wax for tree wounds, 213
" " heater, 231; (Fig.), 230
" " making, 231
Granary of the mind, 12
Granulating fertilizers, 165
Grape Culture, Foundations of, 281
" grower in Delaware (state), 22
" pruning, 283, 289
" shoots, Worthless, 284
" training (Fig.), 285, 287, 289
Grape vine, Two trunk (Fig.), 282
Grapes (Chap.), 281
" borne on green wood, 283
" first fruits, 290
" in orchard, 303
" near trees, 303
" on walls, 284
Grass in orchards, 301
" " strawberry bed, 272
Gravel for cement, 79
Gravity water supply, 65 (Fig.),
" " system, 70
Green, S. B., 175
Greenhouse Construction, 128
" display (Figs.), 127, 132
" of hotbed sash, 135
Greenhouses (Chap.), 128

INDEX 391

Green manures for strawberries, 273
Greiner, T., 239
Ground bone, 59
Grubs, White, in strawberry bed, 272
Guernsey cow, 113
Guests *vs.* wife, 9
Guide, Kelsey's Rural, 7
Guide to Successful Gardening, Modern, 82, 97, 247, 258, 267
Guinea fowls, 118
Gypsum, 171, 174

Haber, Erwin and, 99
Handicap, Unsuspected, 20
Hardpan, 19, 25
Hardening off, 107, 247, 256
Hardiness, Developing, 107
Hares, 114
Harrows, Kinds of, 184
Hatching by hens, 119
Hauling, Horse or truck, 112
Hay costs, 56
" to improve soil, 178
Hazeltine weeder, 195
Head wall and drain outlet (Fig.), 95
Heads, Low tree, 216
" Tree, poor, 220
Health and sunshine, 332
Heartwood, Decay of (Fig.), 211
Heat, Latent, defined, 106
Heater for grafting wax, 231; (Fig.), 230
Heating hotbeds, Electric (Fig.), 245
Heaving, Soil, due to frost, 111
Heeling-in, Strawberry, 273
Henderson, Peter, 7, 52
Henry and Morrison, 112
Hens, 116
Hexagonal orchard lay-out 303; (Fig.), 302
Hill *vs.* level culture (Fig.), 94
Hired help needed, 55
Hoar frost defined, 106
Hoe, Styles of, 194
" Hand, necessary, 193
Hoffman, J. C., 246
Hog, 114
Holes for trees, 307
Holstein cow, 113
Home, Making of a, 116
" Self-supporting, 4
Honey for profit, 54
" prices, 127
Hopkins, C. G., 154
Horse cultivation of berries, 293
" Pleasure, 112
" *vs.* tractor, 186
" work, 112
Horses, Team of, 112
Horseradish storage, 328
Hose, 87
" irrigation, Canvas, 103
Hotbeds and Coldframes (Chap.), 136
Hotbed construction (Fig.), 140
" Electric heating of, 145
" frame (Figs.), 137, 138
Hour costs, 55
How to Live, 200
How to Make the Garden Pay, 239
Hudson, View of, 49
Humidity prevents frost damage, 110
Humus defined, 148
Hutchins, A. E., 85

Huxley, Thomas, 6
Hydraulic ram (Fig.), 72, 73
Hygroscopic water defined, 83

Illinois Experiment Station, 41, 219, 22-.
Imperfect strawberry flowers, 269
Improvements belong to landlord, 47
Inaccessibility, 49
Income, Gross, 65
" *vs.* outgo, 5
Incompatible sprays, 338
Increase of bee colonies, 126
Incubators, 119
Indirect fertilizers, 171
Injuries due to transplanting, 256
Injurious Insects, 206
Injury prevented, Winter, 108
Inorganic fertilizers, 159
Insect attack preventives, 331
Insects work, How, 336
Insecticides wrongly used, 330
Insects bred in wind-breaks, 31
Insects on berry plants, 297
Insects, Strawberry, 272
Inspection, 12, 49
Installments Repayment, 60
Insurance, 49, 62
Interest, 58, 60
" less important than principal, 58
Inventory, 61, 62; 64
Investment, Safe, 58
Investor of earnings, 57
Iowa irrigation experiments, 103
Iowa Experiment Station, 99
Irrigation, Amount of water to apply, 44
Irrigation, Night, advantages, 103
" bulletin, 103
" (Chap.), 96
" Costs, 102
" enhances vegetable quality, 42
" infeasible, 87
" in Iowa, 99
" overhead, 87
" " effects, 101
" " on Long Island, 97
" does not always pay, 101
" system (Figs.), 97, 98, 100, 102, 104
" thorough, 102
" to prevent frost damage, 110
" water evenly spread, 44
" when sun shines, 103
Iron in water, 67
Isolation of farm life, 2, 49
Isotherm defined, 248

Jelly, Royal, 125
Jersey cow, 113
Jesness, Dowell and, 63
Journal of Agricultural Research, 246
Junebugs or May beetles in strawberries, 272

Kainit, 161
Kansas Experiment Station, 153
Kelley, Ruth and, 221
Kelsey, D. S., 7
Kniffin grape training (Figs.), 285, 286
Koopman, Karl, 213

INDEX

Labelling, 250
Labor costs, 55
" Productive, 65
" saving important, 20
Laborers vs. white collar men, 14
Lanes, Field, 30
Lang weeder, 195
Land, Hilly, 26
Land, how handled, 25
Land, Lay and Lay-out (Chap.), 27
" need not be idle, 55
" unused is an expense, 21
Landlord, Investment of, 51
Latent heat prevents frost injury, 110
Lawns, too large, 10
Laws concerning sewage, 81
Lay and Lay-out of Land (Chap.), 27
Leader, Never cut, 216
Leaders, 219
Leasehold vs. ownership, 47
Leaves, Value of, 149
Leek, Storage of, 329
Leeks, a long season crop, 214
Legal features of buying, 48
Leghorn, Strains of, 122
Legumes as cover crops, 168
Lemons, When to expect, 53
Lender, Unscrupulous, 60
Lettuce storage, 329
Level vs. hill culture, 94
Liabilities, 64
Lice, Plant, recognition of, 336
Lie, Figures Don't (Chap.), 15
Life histories, Bug and disease, 332
Lime, amount to apply, 175
" (Chap.), 171
" chloride to deodorize water, 70
" for strawberries, 273
" in water, 67
" losses from soils (Fig.), 173
" never mix with manure, 173
" where to apply, 174
Lignification defined, 255
Litter, Forest, value of, 149
Live, How to, 200
Live stock, 112
Living, High, 57
Loan, Long Time, 60
" maturity, 59
" renewal, 60
Loans, Merits of, 148
Loan sharks, 61
Loans, Advantages of, 147
Locate, Where to, (Chap.), 22
Loganberry, 298
Long Island small places, 21
" row garden, 41; (Fig.), 43
" rows, 315
Loomis, W. E., 254
Loss, Selling farm at a, 48
Losses due to weeds, 197
" How to Avoid Nursery Stock (Chap.), 232
Lure of the Land, The, 56

Magnesia in water, 67
Manure, amount to apply, 155, 156
" experiments in Maryland, 157
" fresh or rotted, 155
" kills fruit trees, 235
" must never contact lime, 173

Manure protects trees against frost, 111
" When to apply, 155
Manures (Chap.), 154
" Green, and Cover Crops (Chap.), 166
Map for drainage, 89
" Soil survey, 22
Maple trees, 19
Marker cropping, 180
Market, roadside, Fruits for, 309
Marketing, Roadside, 317
Maryland farmer's care of tools, 203
Mason, A. F., 330
Massachusetts egg costs, 121
Mats for hotbeds, 141
Mattoon, W. R., 149
Maturity of loan, 59
Maryland Experiment Station, 157
May beetles, 272
Meadows for steep banks, 153
Meeker harrow, 185
Melilot as green manure, 170
Metabolism defined, 255
Mice injure trees, 227
" poison station (Fig.), 312
Michigan Experiment Station, 104
Michigan fruit varieties, 308
" poultry costs, 123
Milk production averages, 37
Milking must be done, 112
Milkweed, 198
Miller, C. C., 126
Minced fields, 29
Minnesota Experiment Station, 85, 143
Miscible oils for insects, 337
Missouri Experiment Station, 299
Mitchell, D. G., 11
Mixed goods (fertilizers), defined, 164
Moisture in air prevents frost damage, 110
Money crops, Reliable, 54
" makes money, 61
" making on borrowed money, 62
" safely borrowed, 58
Morrison, Henry and, 112
Mortgage, 49
Mortgages, 62
Mound storage, 318
Mulch, Paper, 84
" prevents soil heaving, 111
Mulches for strawberries, 277
Mulching experiments, 85
" explained, 84
" materials, 84
" potatoes, 45
" practical, 45
" when desirable, 45
" wind-break, 36
Munson grape training, 288; (Fig.), 287
" T. V., 281
Muriate of potash, 161
Mushrooms, 240
Myers, W. S., 159
My Farm of Edgewood, 11

Nature, Alphabet of, 14
Nebraska Experiment Station, 42
Nectarine, a smooth-skinned peach, 40
Needles, Value of pine, 149
Negotiable paper, 62
Neutral soils, 171
New Hampshire Experiment station, 76
" Jersey Experiment Station, 171

INDEX

New man made by farm life, 14
" York Experiment Station, 122
" York fruit varieties, 308
Nitrate formation in soil, 168
" of soda, neutral in effect, 161
Nitrates lost in drainage, 152
Nitrogen, costly, easily wasted, 160
" growth maker, 159
" not fixed, 161
Norcross cultivator, 195
Note, promissory, 60, 62
Novelties, Test some, 242, 244
Nursery beds, Outdoor, 251
" Stock Losses, How to Avoid (Chap.), 232
Nursery Stock, When to buy, 236

Oats, Cost of growing, 56
Ohio egg receipts, 121
" Experiment Station, 322, 324
O'Kane, W. C., 206
One spray control, 332
Onion storage, 328
Ontario Department of Agricultural Engineering, 88
Ontario Department of Agriculture, 317
Open-center trees, 219
Operation, expenses of, 65
Option to buy, 48
Orange rust of blackberries, 297
Oranges, When to expect, 53
Orchard, American Peach, 105
" arrangements (Fig.), 302
" cleanliness, 331
" Grapes in, 303
" lay-out, 303; (Fig.), 305
" Laying out an, 306
" Mixed, 302, 303
" Remaking a Neglected (Chap.), 206
" Soil preparation for, 305
" Vacancies in, 210
Orcharding, Productive, 30
Orchards, Crops to avoid in, 314
" developed without cost, 54
Organic fertilizers, 159
Oregon Experiment Station, 124
Ornamental planting, Too much, 10
Outdoor storage, 318
Outgo *vs.* income, 5
Outlet, Protected drain, 96
Ownership, Advantages of, 50
" Avoid, at first, 48
" *vs.* leasehold, 47

Paddock and Whipple, 232
Paint for tree wounds, 212
Paper mulch, 84
" mulches for vegetables, 42
Parsley storage, 328
Parsnip, long season crop, 241
" storage, 328
Partnership agreements, 62
" cropping, 180
Pastures for steep land, 153
Payment, Inexorable, 57
Payments, Periodical, 60
Peach fillers, 304
" fruit buds, 218
Peach Orchard, American, 105
" trees, Aged, 52

Peach trees, bear young, 313
" " Neglected, 208
" " protected against frost, 110
Peaches, When to expect, 52
Pear fruit buds, 217
" storage, 327
" trees, 209
Pears, When to expect, 52
Peat moss as mulch, 84
" " for seed beds, 252
Pellett, F. C., 125
Pennsylvania Experiment Station, 315
Perennial weeds, 197
Perfect strawberry flowers, 269
Pets, 10
Philadelphia, Pa., 247
Phosphorus, the ripener, 161
Physiology, Plant, 37
Physiology of Plants, 299
Pickling vegetables, 242
Pickles, 266
Pigeons, 118
Pigs, 38, 114
Pike *vs.* ducklings, 117
Pinches of seed for sowing, 249
Pinching berry canes, 295
Pioneer conditions, 2
Pistillate strawberry plants, 269
Pit storage, 318, 328
Plan for drainage, 89
Planning activities, 63
" for convenience, 20
Plant lice, Recognition of, 336
Plant Physiology, 37
" setter, 260
" shifting, Pot (Fig.), 260
Planting berry bushes, 294
" board, 307
" tables, 247
Plants, Food for, 159
" for sale (Chap.), 261
" injured by transplanting, 256
" not to be transplanted, 258
" Potted, transplanted, 258
" saved, Frosted, 111
" Tender and hardy, 106
Plowing, Hired, 112
" *vs.* digging, 189
Plum fruit buds, 218
Plums, When to expect, 53
Plumping up nursery stock, 233
Pneumatic water system, 72
Pocket, Air, defined, 27
Pockets, Frost, 109
Poison station, Mouse, 312
Poisons for insects, 337
Popularity of farmer, 9
Pork, 114
" for profit, 54
Portland Cement Association, 78
Potash, fiber maker, 161
" fixed in soil, 161
" sources of, 161
Potato breeding, 39
" storage, 327
Pot plant shifting (Fig.), 260
Potted plants, Transplanting, 258
Poultry breeds, 119
" (Chap.), 116
Practical Gardening, 267
Practical Real Estate Methods, 22,
Prediction of frost, 108

INDEX

Preserving, Advantages of, 266
Pressure tank (Fig.), 71
" water system, 72
Price trends, 63
Pricking out seedlings, 259
Principal more important than interest, 58
Principles and Practise of Pruning, 213
Principles of Vegetable Gardening, 243
Production, Essential Factors of (Chap.), 37
Production, Methods to increase, 41
Production, Tomato, 261
Productive Orcharding, 30
Proficient, Becoming, 11
Profit, Factors for, 54
" first year, 56
Profit, Gardening for, 7, 52
Profits, 21
" Tenants, 51
" When to expect, 52
Propolis, 125
Protection against frost, 110
" by wind-breaks, 31
" from frost, 107
" of wind-break, 36
Pruning after tree planting, 308
" berry bushes, 294, 295, 296
" Fruit tree (Chap.), 213
" large branch, 211
" newly set trees, 215
Pruning, Principles and Practise of, 213
" Winter, produces wood, 217
Puddling seedlings, 259
Pumping, Electric (Fig.), 8
Pumpkin storage, 329

Quack-grass, 198
" " in strawberry bed, 272
Qualifications for success, 10
Quince fruit buds, 218
Quicksand, Drains in, 94

Rabbits, 114
Rabbits injure trees, 227
Racks for tools, 203
Rail fences, 29
Rake, How to use, 190
Raking, Skill in, 189
Ram, Hydraulic, 72; (Fig.), 72, 73
Raspberries, 292
" When to expect, 53
" with grapes, 288
Raspberry plants, 295
" suckers, 294
Real Estate Methods, 22, 27
Reclaiming neglected orchards, 207
Records, Effective, 64
" of departments, 65
Reconnoitering for sales, 263
Rectangular orchards (Fig.), 302
Refrigerator, Basement, 317; (Fig.), 13, 16
Region, Familiar climatic, 22
Renting, Advantages, of, 47, 48
" on shares, 50
" safer than buying, 50
Renting vs. buying (Chap.), 47
Repair grafting, 228
Repayment of principal, 58
" in installments, 60
Research, Journal of Agricultural, 246
Resources, 21
Responsibility of home ownership, 5

Retainer fee, 48
Returns, When to expect, 7
Rexford, E. E., 116
Rhode Island Red fowls, 122
Right angle, To make a, 306
Roads, Field, 30
Robey, O. E., 103
Roadside fruit varieties, 311
Roadside Marketing, 317
Roadside stand, Fruits for, 309
Roberts, I. P., 18
Rock, near surface, 25
Rogue defined, 40
Rolphs, P. H., 196
Root crop storage, 328
Roots clog drains, 95
Root system, Sweet potato (Fig.), 3
Rotation, 178
" of vegetable crops (Fig.), 179
" to control weeds, 199
Rows, Broad vs. narrow, 252
" Direction of, 30
" Long, 315
Rules for credit, 58
Runt plants, 331
Rural Guide, Kelsey's, 7
Rust of bramble fruits, 297
Ruth and Kelley, 221
Rutland, Irving, 27

Safety deposit box, 62
Sale, Plants for, 261
Sales, Daily, 265
" Fostering, 263
Salsify, a long season crop, 241
" storage, 328
Salt defined, 171
Sand, Characteristics of, 147
" Drifting, checked by wind-break, 33
" for cement, Testing, 80
Santee, Dr. E. M., 75
Sash, Double glass, 140
" Greenhouse made of (Fig.), 135
Satisfied mortgage, 49
Saving frosted plants, 111
Saw, Double edged, 205
Scale insects, recognition of, 336
" in water, 67
" kills trees, 210
Scarified seed, 170
Scion wood, 41
Scrapple, 114
Screen gate for drain (Fig.), 95
Scrub stock, 116
Sears, F. C., 30
Security, 60
Seed analysis, 197
" Cheap, 243
" Depth to cover, 252
" drill, 190
" growing, 246
" Home grown, 246
" Impure, 197
" in broad or narrow rows, 252
" Large vs. small, 245
" Scarified, 170
Seed sowing, 249
" " thick or thin, 250; (Fig.), 252
" " with drill, 252
" strains, 243
" test, 245
" Viable, defined, 247

INDEX

Seedage decided by natural signs, 248
Seedlings, Pricking out, 259
Seeds and Seeding (Chap.), 243
Selecting a farm, 7
Selecting a Farm, 18
Seller, Contract with, 48
Selling a farm at a loss, 48
Septic tank (Figs.), 77-81
Sewage disposal, 75
" " laws, 81
Shading transplanted plants, 260
Shares, Renting on, 50
Sharks, Loan, 61
Sheds for tools, 203
Sheep, 114
Shifting plants in pots (Fig.), 260
Shovels, 186
Silt basin (Fig.), 95
Size of farm, 21
Slag, 161
Sludge in water, 69, 70, 97
Smothering weeds, 198
Smudge fires prevent frost damage, 110
Snow conserved by wind-break, 33
" trap and wind-break (Fig.), 33
Social conditions, 20
Sod in orchards, 301
Soda for curing cistern, 69
Sods for transplanting, 258
Softener, Water, 67
Soil, Character of, 23
" Darkening, to favor germination, 248
" depth, importance of, 19
" fertilizer tests on, 162
Soil Fertility and Permanent Agriculture, 154
Soil heaved by freezing, 111
" heaving prevented by mulch, 111
" nature told by trees, 25
" sample taking (Fig.), 174
" shallow, liability, 19
" Sick, 177
" Surface Management (Chap.), 182
" Suitable, 21
" texture, 23
" thawing, 111
Soils and Their Care (Chap.), 146
" classified, 147
Solomon, 196
Something to Sell Every Day (Chap.), 265
Sorauer, Paul, 299
Sorrel indicates acid soil, 26
Sowing at wrong times, 240
" seed, Thick or thin, 250
" times, 247
Spades, 186
Special crops, 267
Sphagnum moss for seed beds, 252
Spit defined, 186
Spotting board (Fig.), 259
Sprayer, Barrel (Fig.), 337
Spraying, Circular system of, 101
" Dormant, 211
Spraying, Dusting and Fumigating of Plants, 330
Sprays that may and must not be mixed (Fig.), 338
Spraying on a windy day (Fig.), 339
Sprays, Commercial, 338
" Home made, 337
Spraying and Dusting, Essentials of (Chap.), 330

Spraying against the wind, 330
Sprayer, Air pressure (Fig.), 336
Spreaders, 335
Spring-tooth harrow, 185
Sprinkling, Folly of, 87
Spurs, Fruit, 217
Square orchards (Fig.), 302
Squash storage, 329
Staminate strawberry flowers, 269
Starving weeds, 199
Stew, 10
Stickers, 335
Stock, Scrub, 116
Stocks, 62
Stoddart, C. W., 146
Stone walls, 29
Store plants, 261
Storage cellar, 322
" essentials, 317
" in mounds, 318
" of Fruits and Vegetables (Chap.), 317
Storage room, Cellar, 320; (Fig.), 320
Storer, F. H., 82
Stover, Shredded, as mulch, 84
Strains of seed, 38, 244
Straw as a mulch, 84
" in vegetable garden, 42
" mulches, pro and con, 45
Strawberries (Chap.), 267
" dislike lime, 273
" pay the second season, 56
" When to expect, 53
Strawberry bed renewal, 279 (Fig.), 279
" breeding, 40
" fertilizing, 277
" flowers removed, 276
" harvest, 278
" hoeing, 276
" insects, 272, 280
" irrigation, 278
" marketing, 278
" a money crop, 268
" mulches, 277
" nurseries, 270, 271
" picking rules, 278
" plant and runner (Fig.), 276
" " setting (Fig.), 275
" planting, 274
" plants, Home grown, 271
" " perfect and imperfect, 269
" preparation for bed, 273
" production costs, 268
" runner removal, 276
" succession crop, 278
" Systems of growing, 274
" varieties, 269, 270
" When to plant, 273
" yields, 268
Streeter, J. W., 308
Stump fences, 29
Suberization, 255
Subtropical Vegetable Gardening, 196
Succeed, Who is Likely to (Chap.), 11
Succession crop after strawberries, 279
Succession cropping, 181
Successful Gardening, 82
Suckers of berry plants, 294
Sucking insect work, 336
Sulking plants, 208
Summer-houses for grapes, 284

INDEX

Sunlight distribution (Fig.), 28
Sunshine aids plant health, 332
" and irrigation, 103
Superphosphate, 101
Supplemental land, 50
Surface management of the soil, 182
Survey for drainage, 89
" Soil, map, 22
Swamp, 49
Sweet potato root system (Fig.), 3
" " storage, 329
Synthetic fertilizers, 166
Systems of grape training, 286, 288

Taft, L. R., 128
Tank, Pressure (Fig.), 71
Tankage, 159
Tar for tree wounds, 212
Taxes, 49
Team, rented, 112
Temperature raised by black soil, 248
Ten Acres Enough, 18
Tenant owns movables, 47
" profits, 51
" supplies, What, 51
Tests, seed, 245
Texture of soil, 147
Thawing of soil, 111
Thermostat, 142
Thinning to improve fruit, 17
Thistle a bad weed, 196
Thomas slag, 161
Throckmorton, R. I., 152
Tile laying, 93
Tillage, Deep, undesirable, 44
" Function of, 44, 152
" Importance of, 182
" When to start, 193
Timber, 10
Time costs, 55
" saved in turning, 30
Time-saving cultivation (Fig.), 24
Times to sow and plant, 247
Title, Clear, 49
Tomatoes planted late, 241
Tomato Production, 261
Tomatoes saved from frost, 242
Tomato storage, 329
Tomatoes till Thanksgiving, 239
Tool construction, 201
" costs, 200
" painting, 204
" sharpening, 204
" storage, 203
Tools (Chap.), 200
" buy early, 205
" Care of, 203
" Combination, 205
" Ditch digging (Fig.), 89, 91
" for grafting, 224
" Newly invented, 205
Top dressings, 164
Tractor, 293
" rented, 112
" vs. horse, 186
Transplant, Difficult plants to, 258
" When to, 258
Transplanter, 260
Transplanting (Chap.), 253
" effects, 255
" injuries to large plants, 256
" under irrigation, 101

Transplanting machine cared for, 203
" Plants for, 261
" Preparation for, 259
Trap-nesting, 122, 123
Tree fruits, Selection of (Chap.), 308
" planting, 215
" pruning, 213
" roots clog drains, 95
" setting (Fig.), 306
" setting methods (Fig.), 305
" surgeon, 207
" training, 235
" training by disbudding (Fig.), 220
Trees, Grafting fruit (Chap.), 223
" Grapes near, 303
" indicate soil nature, 25
" Low headed, 216
" Objectionable, 19
" Preventing lop-sided, 22
" Saving girdled, 227
" sizes to buy, 234
" to work over, 210
" Training, 214
" " young, 221
Trellises for grapes, 286
Trenching, Bastard, 188; (Fig.), 188
" True, 187
Trends, Price, 63
Trowel, Long lasting, 201
True bugs, Work of, 336
Truly Rural, 15
Truck for hauling, 112
Tumefaction of the cranium, 9
Turkey raising undesirable, 118
Turning time saved, 30, 41
Turnip storage, 328
Turtles vs. ducklings, 117

Usury, 57

Van Slyke, Dr. L. L., 171
Varieties, Fruit, to choose, 303
" High quality vegetable, 242
Vase-form trees, 219
Vegetable and Fruit Storage (Chap.), 317
Vegetable Crops to Choose (Chap.), 239
Vegetable Garden, The, 253
" quality enhanced by irrigation 101
Vegetable Garden, How to Make a, 96
Vegetable Gardening, 175
Vegetable Gardening, Principles of, 243
Vegetable Gardening, Subtropical, 196
" storage (Figs.), 319, 323, 324
" cool season, 42
" and cool storage, 320
" pay the first season, 56
" to pay expenses, 53
" usable in several ways, 242
" with berry plants, 294
Velvet of peach crop, 53
Verandas, Tools beneath, 203
Vetch among berry plants, 293
Viable seed, 247
Vine weeds destroyed, 198
Vineyard cleanliness, 331
Vineyards developed without cost, 54
Visitors from the city, 9
Volume of business, 21
Voorhees, E. B., 177

INDEX

Walls, Stone, 29
Walks, Field, 30
Warranty deed, 49
Warren hoe, 194
Washington, Fruit varieties of, 308
Washington, George, 47
Water by gravity, 70
" filtering, 67
" Forms of, in the soil, 83
" Functions of, 82
" from springs, 70
" increased in the soil, 87
" measuring in cistern, 70
" needed, Calculating, 69
" Objections to ground, 67
" Purifying rain, 67
" Rain, superiority of, 67
" Removing color from, 69
" Screening, 67
" supply, Gravity system, 66; (Fig.), 65
" supply, Inadequate, 49
" system, Pneumatic, 72
" table, 19
" to prevent frost damage, 110
Watering flats, 250
" pot, 87
Waterfowl raising undesirable, 117
Watermelon, a gamble crop, 240
Watermelons, seventy million, 15
Watersprouts are safety valves, 212
" become branches, 212
" for bridge grafting, 228
" Pulling off, 212
Water-tight joint for cellar, 322
Watts, G. S., 317
Waugh, F. A., 105
Wax, Grafting, for tree wounds, 213
Weak stems need potash, 164
Weather maps, 249
Weed control practises, 199
" dissemination, 196
Weeders, Styles of, 194
Weeds as soil tell-tales, 25

Weeds (Chap.), 196
" in strawberry bed, 272
" smothered, 198
Well, Location of, 30
Werner, H. O., 42
West Virginia Experiment Station, 271
Wheelhoe, advantages, 41
" styles, 191, 192; (Fig.), 191
Whipple, Paddock and, 232
Whips—young trees, 236
White and Boswell, 157
" collar man, 14
Who's Who in America, 6
Wiley, Dr. H. W., 1, 56
Wilkinson, A. E., 166
Wilt of melons and cucumbers, 240
Wilting, Temporary and permanent, 103
Windy day spraying, 339
Wing, H. H., 37
Winter strength spray, 333
Wires attached to trees, 238
Wireworms in strawberry bed, 272
Wisconsin Experiment Station, 120
Wind-break and snow trap (Fig.), 33
" distance from buildings, 34
" extent, 34
" mulching, 36
" objectionable trees in, 31
" pro and con, 30
" protection, 36
" wasteful of land, 31
" way to plant, 34
Wood-ashes, 174
Woman who developed a career, 267
Women sell goods, 266
Work, Paul, 261
Wounds, Painting, 212
Wright, Richardson, 15
Wyoming bee-keepers, 128

Y-crotches, 237
Y-crotch treatment, 216; (Fig.), 237
Yield increase due to mulching, 45

CPSIA information can be obtained
at www.ICGtesting.com
Printed in the USA
BVHW081628290721
613187BV00007B/255